YOUR HEALTH, OUR WORLD

YOUR HEALTH, OUR WORLD

*The Impact of Environmental
Degradation upon Human Wellbeing*

DIANE WIESNER PhD, MA, B.Pharm. (Hons)

PRISM · UNITY

Published in Great Britain 1992 by

PRISM PRESS
2 South Street
Bridport
Dorset DT6 3NQ

Distributed in the USA by
AVERY PUBLISHING GROUP
120 Old Broadway
Garden City Park
NY 11040

Published in Australia by
UNITY PRESS
6a Ortona Road
Lindfield
NSW 2070

ISBN 1 85327 071 7

Typeset by Prism Press, Bridport, Dorset.
Printed by The Guernsey Press Ltd, The Channel Islands.

Contents

Acknowledgements   vii

List of Tables   viii

List of Figures   ix

1.0 Introduction   1

2.0 Causes   13

3.0 Changes   34
     3.1 Climate Essentials   34
     3.2 The Greenhouse Phenomena   43

4.0 Effects   55
     4.1 General Effects   55
     4.2 Cardiovascular Disease   81
     4.3 Cancers   92
     4.4 The Skin   119
     4.5 The Respiratory System   136
     4.6 Sensory and Nervous System   154
     4.7 The Unexplained and Unexplainable   161

5.0 Threats and Hazards   180
     5.1 The Population Problem   189
     5.2 Populations: The Human and
         Natural Environment   201
     5.3 Atmospheric Pollution   218
     5.4 Water Quality   233

6.0 Interventions   253
     6.1 Interventions for Environmental
         and Human Health   253
     6.2 Chemical Hazards to Health   268
     6.3 Population Control   286

7.0 Concluding Section   292
     7.1 Economics and Politics   292

8.0 Restoring the Balance   318

Glossary   320

Index   327

Acknowledgements

This book would not have been possible without assistance. Inspiration came from a search for knowledge. Initially, my own special knowledge lay in toxic chemical principles contained in plant materials, pharmacologically active substances whose levels fluctuated with the season, soil, rainfall, location and presence or absence of mutually compatible species. Yet I also possessed training in health sciences and humanities so it was not surprising that I sought a broader approach to human health beyond the conventional medical model.

I was continually frustrated by the lack of data embracing effects on the individual due to environmental changes such as the weather, resulting from conflicts at home and in the workplace. Elderly people are noted for complaining about stiffness, aches and pains; heart attacks and traffic accidents seem to show an increase during dry westerly winds. Then there were the effects of pollens, smog, industrial chemicals and harsh sunlight itself on the skin, the lungs, the eyes and people's moods and behaviour.

I soon found that the range of material and the emphasis I sought could not be gained from any available publication or journal in the field. At the same time I was approached by Nevill Drury, a noted Australian publisher of quality arts, religion and science. The result has been this book, something I have found a mammoth undertaking, bearing in mind the scope and comprehensive nature I wished for it. I only hope that the reader is satisfied and informed as a result of my efforts.

In acknowledging assistance, I would like to thank the following individuals and organisations for formal permission to use or extract material from reports and research first published by them: *Scientific America, Choice* magazine, *New Internationalist*, Edward Arnold Ltd, and publishers of numerous professional journals. In all instances, I have attempted to make specific reference to these and other authors in the appropriate chapter and/or place in the text. In Great Britain, I would like to thank Caroline Morrison for her conscientious and interested type-setting, and Dr Julian Kenyon for his useful comments. To Nevill Drury, my tolerant and helpful editor, many thanks. In Australia, I have discussed many of the ideas and research contained in the book with professional colleagues, and I thank them all for their contributions. Thanks also to Hugh and Sophie who, unknowingly, stimulated my desire to put my findings on record. Finally, the finished product is dedicated to my parents, to whom I owe brains, an inquiring mind and the perseverance necessary to complete the task.

List of Tables

1.1 Effects on Humans — 11
2.1 Radiation — 28
2.2 Effects of Heat and Humidity — 29
2.3 Acid Rain — 29
2.4 Noxious Gases — 29-30
2.5 Diseases Transmitted by Mosquitoes — 30
2.6 Heavy Metals and Radiation Poisoning — 31
2.7 Reduction in Mortality since 1700 (England, Wales) — 32
3.1 Health Implications of Climate Change Attributable to Greenhouse Effects — 53
3.2 Annual Emissions of Carbon Dioxide (millions of tonnes) — 54
4.1 Pathogens and Health Effects (vector-borne) — 77
4.2 Wastes — 78
4.3 Pesticides/Herbicides — 78
4.4 Organochlorines — 79
4.5 Nutrition — 79
4.6 Characteristics of Population Screened for Cardiovascular Risk Factors — 90
4.7 Principal Occupational Hazards to Human Health — 115
4.8 Frequency of Occurrence of Common Cancers in Men and Women (international comparisons) — 116
4.9 Response of Normal Skin to Sunlight — 131
4.10 Classification of Photosensitivity Diseases — 132
4.11 Intensity of Radiation — 132
4.12 Cutaneous Signs of Chronic Sun Damage — 133-5
4.13 Allergic Lung Reactions Associated with Micro-organisms — 152
5.1 Populations (in millions) — 197
5.2 Average Annual Rates of Growth: Years 1950-80 — 197
5.3 Birth, Death and Infant Mortality Rates (UK) — 198
5.4 Life Expectancy Years (UK) — 198

5.5 Demographic Summary (USA)   198
5.6 Factors Contributing to Decline
       in Death Rates (UK)   199
5.7 Immigration and Emigration
       (numbers per year)   199
5.8 Housing and Population (1986 data)   212
5.9 Dwelling Standards (1986 data)   213
5.10 Urbanisation, Effects and
        Implications   213
5.11 High-Rise Settlement   214
5.12 Population & Food Output 1952/6-81   214
5.13 Nutrition in Under-developed and Developed
        Regions   215
5.14 Consumption of Animal Protein   215
5.15 Health Problems Associated with Large
        Populations   216-17
5.16 Composition of Earth's Atmosphere   231
5.17 CFC Alternatives   231
5.18 Common Pollutants of Water   251
6.1 Mean Age of Death and Infant Death Rates
       (England, 1842)   267
6.2 Deaths, Age-Standardised against the Australian
       Population and Federal Government employees   284
7.1 Preventive Policies for Australia   316

List of Figures

2.1 Relative contributions of greenhouse gases   21
2.2 Greenhouse gas projections   23
4.1 Contributions of major nutrients to total dietary
       energy intake   83
4.2 Premature Mortality: proportion of 15-year-olds
       dying before age 65 (Australia, 1950-81)   85
4.3 Age-standardised death rates from major causes
       of death (Australia, 1970-83)   85
4.4 Rates of cancer mortality for 28 industrialised
       countries   94
4.5 Percentage changes in cigarette smoking
       consumption per person   107
4.6 Sun-induced disorders of the skin   124

1.0 INTRODUCTION

Health — what is it?

Throughout the world, people understand and explain the way they feel about themselves and their immediate environment by talking about their 'health' and describing how they 'feel'. The result is a series of subjective and personal words which cover the situation or 'state' in which they find themselves. These tend to differ throughout the world, as different societies and peoples interpret and explain the various phenomena in terms they understand and which cover the range of human experiences encountered by them.

A traditional society will have few words to describe symptoms of high blood pressure, angina and diabetes. The range of diseases commonly encountered within each community will vary with level of development and sophistication, the people's lifestyle, means of production, culture and size. The extent of contact with, and isolation from, other communities, particularly those which have been exposed to more sophisticated western cultures, is also important.

Hunter-gatherers are not familiar with hospitals, CAT scans, cardiac monitors, isolation wards and specialist clinics. They do not understand modern treatment methods and have different concepts of disease. Death normally follows infection, accident or simple old age and it may be linked to some infringement of natural or spiritual laws. The degenerative and lifestyle diseases which afflict western industrialised societies are rarely encountered.

However, these societies and their communities often have extremely effective ways of coping with mental illness, anxiety and depression. They make use of the important social bonding processes possible through participation in rituals and shamanistic practices which serve to unite the members of the group.

A Definition of Health

With these differences in mind, and aware of the need to embrace the diverse views of many cultures and peoples throughout the world, the World Health Organisation has adopted a general and idealistic approach to its understanding of health and its approach to disease. It defines health as a state of complete physical, mental and social well-being, not merely the absence of disease or infirmity.

This definition has its limitations. It still does not provide a ready, objective way of measuring 'positive health'. For example, instead of measuring the health of a community in terms of its level of well-being, tables indicate the extent of illness or disability, the mortality rate (the number of deaths) or the most common causes of disease. The problems of definition are further compounded by differences in health priorities and perceptions among various socio-economic, cultural and ethnic groups within communities.

Thus, health is a resource for everyday life but not the object of living. It is a positive concept which emphasizes the social and personal resources available to individuals as much as the individual's physical characteristics and capabilities. People do not live to be healthy but they need to be healthy in order to live.

Perceptions of Health
Perceptions of health vary. To some people, health is an end in itself. Some measure their muscle bulk and power each day in a gymnasium and take fright if their weight deviates by more than a few grams from its expected level. This is not health but 'healthism': it provides no answers about social purpose; rather, it indulges the individual's whims.

To other people, including some doctors, health is a mechanical state. The body is perceived as a machine and illness as a mechanical failure which can be corrected if the appropriate technique is applied. In this view, health care is a commodity to be purchased when things go wrong. Medical practitioners who are trained primarily to diagnose and treat disease, dispense 'health' (in reality, disease treatment) on request. Politicians, sensitive to community expectations and the priorities of the medical profession, have generally accepted this mechanistic view of health and therefore underwrite increased public access to medical care and the growth of high technology, hospital-based medicine.

Yet other people see health behaviour and health status as the outcomes of social and economic forces. This 'social model' of health proposes that the essential remedy is to change the social environment, thereby increasing the knowledge, motivation, resources and opportunities for individuals and groups to improve their health status. These proposed social changes extend well beyond the sphere of influence of current health policy and services.

The Basic Requirements
The basic requirements for health are a suitable diet, decent shelter, companionship, an adequate income and a healthy environment. In addition, people require information, opportunities and social and behavioural skills. With these benefits, most individuals and groups knowingly and unknowingly play a significant role in determining the level of health they attain. In developed societies and communities enjoying basic political freedoms, people can make healthy choices about the foods, services and facilities which are available and

to a large extent determine their own health and the behaviours which influence it. These 'rights', benefits or privileges are denied to those in underdeveloped and developing societies where freedoms and individual choices are not universal and to a large extent depend on socio-economic status and education in the community.

To the individual, good health means improved quality of life, less sickness and disability, a happier personal, family and social existence and the opportunity to make choices in work and recreation. To the community, good health means a higher standard of living, greater participation in the making and implementation of community health policies, and reduced health-care costs.

Being Born Healthy, Remaining Healthy

Most babies are born healthy and can be labelled as 'free of disease'. Sooner or later, however, many are exposed to factors in their environment which lead to a change in their health status and place them at increased risk of developing disease. Where the causes of their ill-health become known and recognizable, it may be possible to remove or alter the environmental hazard, thus preventing a deterioration in their health. This is normally referred to as primary prevention. By virtue of their exposure or their own unique biological makeup, these 'high risk' individuals show changes in the normal functioning of their bodies. In some instances these changes, usually biochemical, can be detected, but often the significance of such subtle changes remains elusive.

For example, there is much controversy concerning the significance of children's exposure to wind-borne dust from crops sprayed with aerial pesticides. Some workers claim to detect effects on intellectual development, poor health and a range of behavioural and nervous system problems. Others interpret as inadequate and inconclusive the results of findings on the presence of these chemicals in the atmosphere and in the bloodstream of affected families. Just as with many issues where a universal or local environmental phenomena is implicated, scientific workers are criticised for faulty design and testing procedures.

People who usually appear well and lack any symptoms, may be experiencing influences on their health due to their environment. These influences may be biological, physical, social or emotional. A proportion of these people will, in time, become recognizably ill. Detectable changes of a biochemical or physical nature will have occurred and the disease process commenced. At this stage, it becomes possible to identify and possibly treat them.

The picture is complicated because not all children are born as normal healthy infants, even though the circumstances of their life in the womb could not be expected to have exposed them to any known danger. In the normal course of events, they might be expected to be born well and healthy. No genetic predisposition would identify them as individuals likely to be burdened by an inherited disease which will affect their life chances and the quality of that life.

During pregnancy, the foetus obtains its nutrition and eliminates wastes via the placenta which intimately links it into the mother's bloodstream. Any potentially toxic chemical or environmental factor present in her blood during that time, is likely to pass into the delicate developing bloodstream of the foetus through the placenta. The baby will be exposed to any increased levels of chemicals — such as lead, drugs like nicotine (from smoking), marijuana and cocaine or food additives — present in the mother's body. Particularly in the important first twelve weeks of the pregnancy, these chemicals will affect subsequent development of vital body organs such as the heart, bones, liver, kidneys, the nervous system and the brain. They can thereby indirectly affect the chances of the baby being born healthy.

Human Ill-Health

Good health implies the achievement of a dynamic balance between individuals or groups and their environment. Interactions occur between physical and social factors and it is important to recognise that environmental factors rarely act in isolation. This means that a social factor such as prosperity affects housing and sanitation, which are associated with exposure to micro-organisms. Education, one determinant of the level of socio-economic development of a society, impacts on hygiene practice and exposure to micro-organisms. It also affects diet and nutrition, which in turn influence the health status of the community.

Thus it is not always possible to establish a simple cause-and-effect relationship between a given factor and some specific effect on health. A number of environmental influences are involved. There is a 'web of causation' or a network of interrelated factors which work synergistically towards the outcome. When discussing environmental influences on health, these complex multifactoral relationships must be borne in mind, even though, for clarity and convenience, individual factors are usually considered separately.

Ill-Health, Illness, Disease

When people in western nations talk about illness, they think about disease, refer to disease-causing organisms and agents, and to degenerative or ageing processes. Scientists and doctors identify and seek to eliminate the cause but their skills are limited.

Many people living in non-western nations and traditional societies lack the sophisticated scientific knowledge and tools available to the developed world. However, their understanding of illness and its causes is much broader and more holistic. It embraces the total environment of the person who is ill: body, mind and emotions.

The word 'disease' describes a state of being: literally, a state in which there is dis-ease, a lack of, or a disturbance in, natural equilibrium. Disease is viewed as a process, a cycle of events through which the individual passes and whose passage depends on the severity and nature of the causal agent. That agent may be biological, chemical or physical. Attention to the cause restores the natural balance or ensures that a new, different equilibrium can be established so that health will return.

History and Disease

The fossil record indicates that prehistoric human life was subject to many of the same diseases as afflict modern communities, but it also shows that there are significant differences. Perhaps five million years ago, human-like beings began to live in small, nomadic groups which subsisted by hunting and gathering. They did not live long by modern standards: 80 per cent died before the age of 30, and only 2 per cent lived more than 50 years. The probable causes of death were a combination of high infant mortality, accidents, food shortages, predators, violence and disease. The last resulted only from those pathogens which could survive in small nomadic groups, and therefore excluded many viruses such as measles, smallpox, polio and mumps, which depend for transmission on large sedentary populations. It also excluded those diseases which depend on complex cycles with animal or insect vectors, such as malaria or bilharzia. During their short life-spans, these prehistoric ancestors were not free of sickness: dental decay and arthritis were common, as were infected fractures (Wood, 1979).

Some hunter-gatherers, such as the Australian Aborigines, preserved their way of life into modern times so that something of their traditional health status has been documented. So too have the changes brought through contact with, and assimilation into, the predominant white Caucasian population. The aborigines' life-span appears to have been short, mainly because of fairly high infant mortality. The principal damaging factors in their environment were accidents, burns and violence, problems of childbirth affecting mother or child, mild trachoma, minor parasitic infestations and perhaps gastrointestinal infections. Otherwise the people were slim, strong and well nourished.

About 10,000-14,000 years ago, the way of life of some human populations changed dramatically. The Neolithic Revolution involved development of plant cultivation and the breeding of domestic animals. People began to spend their lives in one small area, with a more or less assured food supply and in intimate contact with large numbers of others. This new form of social organisation brought with it new patterns of disease and new modes of resistance.

Food shortages, wheat and corn crop diseases, and failure of the potato crop left many hungry. New methods of crop rotation were tried, but many people abandoned their land and gravitated towards the cities in search of food and work. With the advent of factories, the towns grew even quicker. Here, new industries were beginning with the first signs of the coming Industrial Revolution.

However, their rural habits of living caused a problem when villagers moved to the growing new cities. In the changed conditions, massive accumulations of refuse and excreta began to rot and contaminate the streets because they could not be recycled naturally. Vermin thrived in the new conditions, disease-causing organisms and animal droppings soon contaminated the water and food. As a result, the late Middle Ages saw the first large-scale epidemics of cholera, plague and typhoid. For the first time, those viruses and bacteria responsible for syphilis, plague, tuberculosis and pneumonia found a fertile breeding ground in a population already debilitated by lax living conditions and poor hygiene.

The crowding together of large numbers increased competition and destroyed opportunities for rest. Reduced living space, overcrowding and the crush of bodies placed a strain on food supplies which became difficult to obtain. Overwork combined with undernourishment decreased the standard of health and further undermined the quality of life. However, it took time for public concern to identify the causes and mobilize resources to implement new measures and laws, to ensure health was maintained under the new conditions.

The Industrial Revolution itself brought smog, dirt, grime and the ravages wrought by overwork in poor conditions of lighting and ventilation. The average life-span did not extend beyond the thirties. Infant mortality was high and women frequently died in childbirth.

Disease, Post-Industrial Revolution

In the post-Industrial Revolution, pollution, degenerative and lifestyle diseases, ageing and stress have replaced infectious diseases as the principal threats to human health. Industrial pollutants such as hard plastics, chemicals like freons, polyvinylchlorides and organofluorides and chlorides used in agriculture, are now the source of enquiry as scientists seek to understand their effects on the aetiology of disease.

For example, it appears that the synthetic chlorinated hydrocarbons, like the organochlorines, have untoward long-term effects on biological systems which have not adapted to them through the normal slow processes of evolution. Massive doses may have obvious and immediate toxic effects, but there is also evidence that small quantities of some chemicals, which have no immediate discernible effects on well-being, may nevertheless be harmful if exposure occurs over a prolonged period. At the same time, detecting the influences of known and unknown contaminants, measuring sometimes minute amounts which enter and affect biological systems, makes prediction of future outcomes difficult and complex. It may become essential to develop processes which ensure complete segregation of such materials from all living species.

Clean Water, Pure Air

Concern about clean water and pure air first emerged in the mid-nineteenth century as the effects of rapid industrialisation became apparent. Cities had become dirty, polluted and unpleasant places to live. Poverty, disease and violence lurked in the streets as dangers to all who ventured out alone.

Those with a social conscience were appalled by the working conditions in the factories, by the close and unhygienic living conditions, poorly constructed buildings, rubbish littering the streets, and overcrowding. Humanitarians and compassionate people who saw the poverty and the stench were dismayed also at the long hours worked for low wages, the dirt, dank and damp of factories, child labour and conditions in the mines supplying raw materials to the factories. They agitated on the streets and in parliament and eventually forced reluctant governments to act. Regulations were imposed on factory owners, child protection measures were implemented, education and basic social support mechanisms were introduced for the first time.

Yet it was to be more than a hundred years before the interests and political energies of the descendants of this same social group were again mobilized to achieve substantial social and environmental changes in their living conditions. The true cause of the problem — the environment — has only recently been recognised because it continues today.

Prior to the twentieth century, however, pollution — smoke, smog and factory wastes — was simply taken for granted as the inevitable accompaniment to development and increased income. Very little thought was given to what happened to the dirt and rubbish puffed out by factory chimney stacks. These problems were regarded as local and not the responsibility of health authorities. Consequently, early protective legislation tended to impose district or state regulations which lacked uniformity or a national impact. Measures for water supply, drainage and sewerage were stipulated. Health standards and disease, the initiating agents for these measures, were of secondary concern. Even less thought was given to the fact that these effects might have been caused by human activities.

It was the mid-twentieth century before there was recognition that local social behaviour and the nature of the individual's social environment were important to their health. Attention is now being directed to the pollution of the environment and the changes wrought by human exploitation.

Disease, A Feature of a Settled Lifestyle
Settled living, however, also offered compensations. Storage of food meant an improved and steady supply. Permanent shelter was guaranteed and the world experienced its first 'baby boom'. In fact both birth and death rates were high, but people died young. And they still suffered from conditions such as arthritis and dental problems, violence and accidents.

Before the development of agriculture there was no evidence of deficiency diseases such as scurvy or beri-beri, since the hunter-gatherer generally enjoyed a well-balanced diet. Agriculture, however, often involved reliance on a single staple crop, resulting in a diet which was far less nutritious.

The health status of a present-day group of subsistence farmers in Papua New Guinea Highlands has been documented by Sinett and Whyte (1978). They show that the major cause of sickness and death is infectious disease. The five leading causes of death are pneumonia (45%), gastro-enteritis (12%), dysentery (10%), meningitis (8%) and malnutrition (5%). The country has a high birth rate and a high death rate. Thirty per cent of all babies born do not survive to adulthood: life expectancy for males is 58 years.

Further historical and anthropological data on health and disease indicate that health patterns differ not just between populations with different ways of life but also within them. One of the most consistent patterns is that men and women have different life expectancies. In modern industrial societies, women live several years longer than men. In prehistoric hunter-gatherer and agricultural societies, the situation appears to have been reversed with men consistently outliving women. There are other sources of health difference within societies, the most striking being those between the rich and the poor, between those living where there are plentiful natural resources and an abundance of food and water, and those living in marginal conditions where survival is highly dependent on the annual cycle of seasons and rainfall.

A Greater Understanding
Globally, scientists are seeking greater understanding of earth's living systems and their capacity to cope with these changes, to regenerate or maintain themselves under changed conditions. Toxic waste accumulation in the oceans, acid rain, species loss, land degradation and driftnet species are among the current range of concerns.

But this book is not about the health of the environment. Nor is it about the changes predicted in our surroundings in the future. There are many more scientists better qualified to write about the planet and its delicate species balance. This book concentrates instead on human health and the effects which the environment and change are having on the well-being of the human species. The present introduction briefly traces the history of the relationship between human health and the environment, focusing on public health and disease control.

Health and Environmental Chemicals

Attention to a possible link between human health and the environmental factors of a chemical rather than an infectious nature, has grown in recent time. About the middle of this century, scientists first began to take serious note of the increase in deaths from bronchitis, emphysema and diseases of the lung which coincided with the worst of the great black London smogs of the 1940s and 1950s.

A substantial increase in the number of people dying from heat exhaustion and stroke occurred in Athens at the time of extended heatwaves during three very hot summers in the late 1980s. The elderly and the very young proved to be particularly prone to cardiovascular problems and susceptible to heat stress. The weather and the level of atmospheric humidity were likely to be significant factors, but the difficulty lay in establishing a direct relationship between death and variable climate extremes and weather changes.

Though once seriously contested, the existence of a link between social conditions and health is now accepted as proved, thanks to a number of detailed studies based on extremely large populations.

Naturally, of greatest interest to humans is the effects which these environmental changes will have upon them, which are psychological, biological, physiological and social in nature (see Table 1.1). In common with all major health problems, they affect the individual's total health. And they are likely to be multiple, multi-faceted and probably compound one upon the other, making them all the more difficult to separate one from the other.

Environmental Health

By definition, environmental health encompasses the study of those things which affect the health and well-being of the community at large. The early work was mainly concerned with the control of infectious diseases and, as part of this, the living and working conditions of those particularly at risk. The water supply, food, housing, and hygiene first became important then and have remained so today, though different aspects are emphasized.

Now we have a body of knowledge on the environment and human health, backed by facts, science and logic. They constitute the material and issues for this book. The majority of the work referred to has appeared as published research reports and papers in scientific journals or specialized books. The sources are cited in references and additional reading lists at the end of each chapter. (At all times, the attempt has been made to attribute research and reference material to the correct source and to trace copyright, if applicable.

Sincere regrets are expressed if accidental errors, omissions or infringements have resulted.)

This book begins by identifying the problems and reviewing existing knowledge in each major area of concern. Often the evidence is circumstantial and frequently there is no established link between an imputed cause and an actual event. We might have observed that our hay fever coincides with the flowering of the privet hedge in the neighbour's garden, but unless specific allergy tests confirm that we are sensitive or allergic to privet, the issue remains unresolved.

Why study the environment and human health? We all know that the weather affects whether we feel hot or cold, whether we perspire or catch cold? But is the relationship always so clear? And, despite our experiences and our 'feelings', science already tells us that it is not the winter cold which causes pneumonia but a bacteria. How much do we really know, and can we influence our environment even if we do? These are questions and issues we will try to address.

A recent example is given by the following description of events. In the northern summer of 1988, record rains fell throughout Britain while prolonged heat-waves ruined the wheat harvest crops across the United States and withered those in India and central China. In other parts of rural and northern China, torrential rains triggered floods and landfalls leaving millions homeless and hectares of farmland waterlogged.

Yet these climate extremes are not confined to one year. During 1989, areas of Florida and California, accustomed to a balmy warm winter, were hit by blizzards and cold, and other areas of the United States had suffered similar aberrations in previous winters. In 1987, snows were reported in Venice, on the French Riviera and in subtropical Brazil.

Some of these changes have been blamed on the greenhouse effect, a global warming that could melt the polar icecaps and turn vast parts of the landscape into tropical jungles and arid deserts. Others have been attributed to changes in the nature and content of the fine mantle of gases which protect the earth from the damaging rays of solar radiation and sunlight.

The Future Planning Agenda
Unfortunately many of these symptoms and causes of global ill-health lie outside the traditional agenda for public health planners. This provides an easy excuse for many to avoid them, particularly where other priorities dominate. In the Third World, claims imposed by debt, basic education, poverty and lack of food and shelter often demand rapid growth and industrialisation. Costly controls and the use of more expensive alternatives are ignored where funds are scarce and pressures are overwhelming.

In this context, the maintaining and promoting of health take on added importance. An individual's health will only be maintained in circumstances in which there is clean air, a supply of clean and wholesome water, light (especially sunlight), adequate space for living and recreation, reasonably equable temperature without extremes, a plentiful supply of nutrients free from injurious contamination, access to quiet and peaceful surroundings in which to rest, and a congenial relationship with other co-habitants. Together these elements comprise the total environment. Each can be taken independently and examined for its contribution and the role it plays.

On the positive side, the World Health Organisation has adopted sections of a 1987 document called the Brundtland Report. It has convened a special Environmental Health Division and is organising it so as to 'promote the collective action of countries to respect and preserve the natural environment'. It is also addressing the difficult issue of educating reluctant Third World planners about the crucial long-term importance that environmentally sensitive development means for them and the rest of the world. Efforts are being made though they are slow to eventuate and there is considerable controversy. We will deal with this later in the book, especially in Section 6.0 on Interventions, but first we will look to a greater understanding of the problems and the issues.

TABLE 1.1: Effects on Humans

Psychological: crowding, decreased fertility, social tension, delinquency, crime, drugs, under-employment, unemployment

Biological: increased incidence of disease, change in nature of disease, e.g. bacteria resistant to current antibiotics; new viruses, e.g. HIV and agents that attack immune system

Physiological: increase in number of survivors with low productive potential, consuming disproportionate level health and social sup-port, e.g. permanently disabled (more low birthweight babies surviving); increase in number chronic disease sufferers since more people surviving to old age, e.g. arthritis, senility; increase in respiratory diseases due to environmental pollutants, e.g. asthma, bronchitis, and occupational (asbestosis, emphysema)

Social: higher costs through increase in number of people with chronic diseases, costs of potent medicines, proportion of community who are non-productive/have low productivity

References and additional reading

Brown M.H. *The Toxic Cloud*, Harper & Row, New York, 1987.
Canadian Cancer Society: Annual Report, 1987-8, Canadian Cancer Society, 1989.
Epstein, S. *The Politics of Cancer*, Sierra Club Books, 1978.
Haynes, V. and Bojcun, M. *The Chernobyl Disaster*, Hogarth Press, 1988.
McKeown, T. *The Role of Medicine: Dream, Mirage or Nemesis?*, Blackwell Scientific Publications, Oxford, 1979.
New Scientist, 'Code of Conduct May Keep Dangerous Pesticides At Home', 9 December 1989, no.1684, p.6.
Patterson, J. *The Dread Disease*, Patterson, New York, 1987.
Rowland, A.J. and Cooper, P. *Environment and Health*, Edward Arnold, London, 1983.
World Health Organisation Cancer Unit. *Cancer Control In Developing Countries*, WHO Publications, 1986.
World Health Organisation. *World Health In Developing Countries*, WHO Publications, 1987.

2.0 CAUSES

What components of the ecosystem affect human health? It is not hard to list the crucial ones — climate, geography, the atmosphere, living conditions, come to mind. They are subject to change and those changes must affect us too.

But the ecosystem is not simply a series of separate parts of a vast machine which functions reliably when the handle is cranked. It is more like a sensitive and delicate musical instrument; a violin, finely tuned so that the melody of the strings resounds against the timbre of the hollowed, polished sounding box of crafted wood to sing deep, rich notes. The individual parts act in unison to produce the whole.

The environment, too, forms a holistic system whose smooth function depends on the maintenance of balance between all of its components. Like a fine string instrument, to play at its best all its parts must work in unison. Before we can consider the effects human beings have upon this delicate instrument and the implications which this, in turn, has upon human health, we need to understand more about the basic components of the ecosystem. The first is climate, including the atmosphere, the changes occurring in it, and the effects they have on the weather experienced on earth by all living species.

Climate

Climate is of considerable importance to human health. It is the most obvious and easily recognised awareness individuals have of the environment. When humans communicate, they most often exchange initial greetings by talking about the weather, their personal experience of climate.

But climate is important for more reasons than these alone. Prevailing temperatures or humidity influence the occurrence and infectivity of a number of disease-causing organisms, insect vectors or infected individuals. Social conditions and levels of sanitation and hygiene can multiply these effects. These phenomena are particularly in evidence in poorly developed societies and communities such as the slums of Calcutta and Djakarta, where high temperatures, poor living conditions and endemic disease can result in the periodic outbreak of cholera, typhoid and dysentery.

At the same time, climate influences the clothes people wear, the food they eat, their housing and the conditions under which they work. On the whole, humans are well adapted to life in varying climatic conditions, but instances of maladaptation leading to ill-health can be recognised.

The Role of the Sun

Exposure to sunshine has a direct bearing on health. This can be positive or negative. Vitamin D is synthesized in the skin under the influence of ultraviolet light from the sun. At latitudes where exposure to sunlight is low and the sun's strength weak, such as in the Arctic Circle during the long winter, signs of vitamin D deficiency are in evidence. In these areas, a diet high in vitamin D and supplementation may even be necessary to prevent rickets.

An excess of sun in the form of prolonged exposure to high levels of ultraviolet radiation, results in an increased incidence of sunburn, heart stroke, skin cancers, cataracts and damage to the retina. Australians and other fair-skinned peoples living in latitudes close to the equator spend their leisure time outdoors, especially in the sun. As a result, they are increasingly likely to encounter skin problems, to be susceptible to disfiguring melanomas and to be at risk of other basal cell carcinomas.

Lack of sun can have equally adverse effects. Life in the higher latitudes means three months and more of life without sun, yet for six to eight weeks per year it shines weakly for a full twenty-four hours. There is certainly an effect on body hormone levels associated with these abnormal periods of light exposure, but their effects on body metabolism, behaviour patterns and psyche are extremely complex.

The Atmosphere

Changes in the atmosphere due to greenhouse gases or from depletion of the ozone layer will alter the amount of ultraviolet which reaches the earth's surface. Any increase in the amount of ultraviolet radiation and its intensity will directly increase the risk of damage to the skin and the incidence of skin cancers (Table 2.1). Already the recent decline in stratospheric ozone levels is claimed to be contributing to the large numbers of melanomas, cataracts and other skin cancers being observed in people from as early as late adolescence.

Heat leads to sweating and associated salt depletion; inadequate dietary replacement of salt can then lead to cramps. Physiological factors such as increased age can place individuals at greater risk from these disorders. If working conditions are associated with exposure to abnormally high temperatures, as in hot climates, sweat glands may become over-stressed and eventually fail in those who have not been able to acclimatize adequately.

Prickly heat, a skin irritation, is a manifestation of sweat gland failure. Interference with the control of body temperature caused by sweat gland failure or excessively humid conditions, which may prevent the efficient evaporation of water from the surface of the skin, can result in heat-stroke which may be life threatening.

Similarly, humid conditions, tight-fitting clothes and nylon socks and underwear can encourage tinea between toes of the foot and in the groin. Thus heat and the impact of associated clothing and lifestyle factors can encourage overgrowth of dermatophytes and fungi, their causative agents (Table 2.2).

In distinct contrast, extremes of cold are also potentially harmful. Local tissue damage is suffered in cold, dry conditions and at high altitudes. A generalized fall in body temperature accompanied by a loss of heat control (hypothermia) also occurs: the very young and the elderly are particularly prone. The effects of exposure to cold air are aggravated in wet conditions, especially if the air is moving, since the evaporation of the water increases the cooling effect.

Bad housing is often damp and draughty and thus very cold in the winter months. Among the aged, poor circulation and low levels of physical activity make cold conditions difficult. Shivering, hypothermia, chilblains and gangrene of the extremities are common among those at risk also of marginal nutritional deficiency. These factors contribute to undermining the health of those of lower socio-economic means, increasing their susceptibility to chronic respiratory conditions, viral and bacterial infections of the respiratory passages.

Atmospheric Changes

Human industrial activity leads to atmospheric pollution. In areas where winds are low and the air is still, pollutants tend to accumulate and their combined effect can be harmful to health, especially for those prone to chronic asthma and allergic rhinitis. For example, in parts of Britain and in the valleys of Wales as early as the first years of the Industrial Revolution, coal-burning industries were located on the plains adjacent to rivers for easy transport. While recognised for their benefits in providing employment, the local people blamed the factories and mines for making the air dirty and smelly, and for contributing to breathing difficulties and a huge increase in the number of lung diseases suffered by peoples living there.

In mining towns in Australia's northern Queensland, such as Mount Isa, a centre for copper, lead and zinc, winds heavily laden with yellow and grey dust regularly coat the town, the houses and cars in a film of fine particles of sulphur, lead and zinc. Residents complain of eye irritation, respiratory distress and allergy caused by the mining. Sophisticated modern equipment designed to restrict levels of atmospheric dust limit but fail to control the problem (Table 2.3).

As early as 1892, it was suggested that exposure to coal tar products might be responsible for a range of respiratory problems including cancer of the lungs and other internal organs. Yet, prior to 1938, the evidence linking lung cancer to coal tar exposure was limited to single case reports.

In the 1940s and 1950s in Britain, the combustion of bituminous coal in factory furnaces, and to a lesser extent in inefficient domestic fireplaces, was blamed for a series of pea-soup fogs and smogs which blanketed London and other urban centres. The severe urban atmospheric pollution provoked a public outrage against industrial pollution. The high incidence of bronchitis and lung conditions forced the passing of legislation which permitted the use of only coal-free fuels such as lignite for domestic purposes, and strictly monitored and substantially reduced factory emissions.

The serious nature of occupational health hazards to which coal and steel industry workers were exposed in the course of their work also came to attention. In an on-going University of Pittsburgh study commenced in 1962, it was demonstrated that, on average, coke oven workers die of lung cancer at a rate of 2.5 times that for all steel workers. The relative risks of lung cancer are 6.87 for men with five or more years employed at full topside, 3.22 for men with five or more years of mixed topside and side oven experience, 2.10 for men with five or more years' side oven experience only and 1.7 for all men with less than five years' experience. The study also indicated that coke oven workers have a 7.5-fold risk of dying from kidney cancers.

Other studies have suggested a higher incidence of cancers of the larynx, nasal sinuses, pancreas and stomach, and of leukaemia. In addition, coke oven workers have an increased risk of developing cancer of the urinary tract. Observations of animals and of human populations have shown that skin tumours can be induced by the products of coal combustion and distillation.

Chemical analysis of coke oven emissions reveals the presence of a large number of scientifically recognised carcinogens as well as several agents known to enhance the effect of chemical carcinogens, especially of the respiratory tract. In addition, workers show an elevated risk of non-malignant respiratory diseases such as bronchitis or emphysema.

Similarly, the interaction of nitrogen oxides from vehicle exhausts with hydrocarbons in the atmosphere occurs under the action of sunlight. The resulting photochemical smog contains oxidizing agents such as ozone and peroxyacetate nitrate, which irritate the eyes and respiratory passages. This kind of smog is particularly problematic in heavily industrialised Australian cities such as Sydney and Melbourne, where the level of ultraviolet radiation from the sun is sufficiently intense to create a high concentration of oxidants.

In traffic intersections in central Sydney, for example, levels of carbon monoxide are regularly recorded above 40 parts per million (ppm). Recorded levels of atmospheric lead and of nitrous oxides are also in excess of maximum levels for health, and are responsible for the asthma and eye irritation complaints suffered by city residents (Table 2.4).

Electromagnetic Radiation
Human evolution has taken place in an electromagnetic environment consisting primarily of the earth's magnetic field. Until the recent century, levels of electromagnetic radiation remained relatively constant and varied only with periodic energy emissions or activity cycles from the sun or other stars in space.

Electricity generation on a communal scale began at the turn of the century. About seventy years ago radio-transmission commenced, and during World War I radar came into use. By the 1950s, communities were criss-crossed with a series of wires, cables and emissions.

Today, the lowest level of electromagnetic radiation in a modern urban environment (50-60 hertz) is approximately 1000 times higher than pre-industrial levels (McMillan, 1990). Today there are micro-wave communication systems, a vast array of electrical appliances and power generation facilities, FM and CB radio waves, fibre optic cables, satellite and television signals exposing individuals to an unprecedented level of electromagnetic radiation — all invisible to the eye, and to our conscious senses. In homes near transmission lines carrying high-voltage power, the level of exposure jumps to 10,000 times greater than the norm, and 10 million times higher than natural background levels.

Power sources in the normal house environment run at about 240 volts, but high-voltage powerlines can carry up to 500,000 volts and have correspondingly large electromagnetic fields. In common with household power, the current in the powerlines is an alternating current varying from positive to negative 50 times per second, thus being referred to as 50 cycle current or 50 hertz.

A growing body of scientific evidence now suggests that most living beings, including humans, are attuned to the natural fields in the extremely low frequency range and the earth's static magnetic field (its geomagnetic field). It is becoming apparent that some biological processes have evolved in, and to an extent are dependent on, electromagnetic fields considerably lower than the current 50 hertz level experienced.

Concerns about levels of electromagnetic radiation first arose in the 1970s. An American epidemiologist, Mary Wertheimer, began investigating childhood leukaemia incidence in Denver, Colorado and noted that the houses of victims who had died in the previous twenty years between 1950 and 1969 were often located near pole-mounted transformers. The transformers stepped down 13,000 volt power to the 240 and 120 volt power used in homes. When Wertheimer began investigating the connection, she found a strong correlation between childhood leukaemia deaths and proximity to the high magnetic fields near the transformers and their associated distribution wires.

After thoroughly reviewing her work, Wertheimer and a colleague, Edward Leeper, published the findings in the respected *American Journal of Epidemiology* in 1979.

The articles caused little interest except among electricity authorities, who refused to accept the evidence of health effects from 50-60 hertz magnetic fields, and other health researchers who found strong evidence that the magnetic fields were associated with a greatly increased risk of cancer.

Eminent American epidemiologists, David Savitz and Genevieve Matanoski, have confirmed the fieldwork of Wertheimer and Leeper. This accumulation of evidence has convinced the British Columbia Hydro in Canada to offer to buy the homes of people living along its high-voltage powerline on Vancouver Island. By June 1990, fifty-five property owners had accepted the offer. Other electricity authorities have opted for placing street wiring underground or finding alternative routes for high-voltage transmission lines.

The debate over electromagnetic radiation came to Australia along with a 1983 proposal to build an eight-kilometre long, 220kv powerline between the Melbourne inner-city suburbs of Brunswick and Richmond. Despite protests from local residents and considerable opposition, the Victorian Government approved construction in August 1986. Even by that time, further evidence had accumulated concerning the effects of chronic exposure to 50-60 hertz fields.

Health and electricity authorities in Australia and in the United States have been put under more pressure to respond by a recent recommendation in a draft report on electromagnetic radiation prepared by the United States Environmental Protection Agency (EPA). After a two-year review of the literature, the EPA recommends that extremely low-frequency fields be classified as 'probable human carcinogens' and that radio frequency and microwave radiation be classed as 'possible carcinogen'. It agrees that high-voltage powerlines pose further and greater risks and that a number of common appliances are cause for concern among those chronically exposed to or making use of them.

Visual Display Units (VDUs) have recently attracted the attention of occupational health authorities associated with many trade unions representing clerical and secretarial workers. VDUs work on a similar principle to television and, in common with television, the image on the screen of a VDU is produced by shooting a stream of electrons onto a phosphorus screen which then glows. As the images on the screen are generated, various forms of radiation are emitted. The electron beam produces X-rays, most of which are absorbed by the cathode ray tube and are not considered harmful to human health. The activated screen, however, also emits visible light, ultraviolet light, infrared light, static electricity and radio waves. A significant proportion of the radiation is in the form of pulsed electromagnetic fields of 15 to 20 kHz and 50 kHz.

Concerns about VDUs centre on increased incidence of miscarriage, radiation-induced cataracts, skin rashes and stress symptoms, all cited as possible hazards of too great an exposure to radiation from VDUs. Work by Kaiser-Permanente, a health resources unit in California, found that of 1583 cases studied, the risk of both early (twelve weeks or earlier) and late (greater than 12 weeks) miscarriage increases by around 80 per cent for women working on VDUs for more than twenty hours per week, compared to women doing non-VDU work. The findings also reported a 100 per cent increase in miscarriages by VDU operators compared to non-working women.

Another appliance that gives cause for concern is the electric blanket. This is because users are exposed for long periods to electromagnetic fields which may be up to ten times stronger than background levels.

In May 1990, for example, a University of Carolina epidemiologist, David Savitz, reported to a meeting of scientists that children whose mothers used electric blankets during pregnancy had higher risks of brain tumours and leukaemia. Brain tumour rates increased two and a half times, incidence of leukaemia increased by 70 per cent and all cancers by 30 per cent. However, these studies await confirmation by further research.

Climate and Life
Climatic conditions also play a role in other less direct ways, thereby determining the characteristics of other life forms with which the human population must co-exist. Insects which may act as efficient transmitters of infection are particularly active in warmer climates. The classical example is the mosquito, certain species of which are specifically related to the transmission of several diseases. Climatic conditions must be appropriate if mosquitoes are to transmit infection. The same applies to transmission by other vectors and agents (Table 2.5).

Water
Water is essential for all living species, particularly human beings. Not only does the body contain in excess of 60 per cent water, but it is dependent on a regular and normal intake of quality water. However, with kidneys which are unable to handle salty and bracken waters, the human species requires a regular supply of pure water low in minerals and dissolved metals.

Excessive mineral or metal intake can be linked to cardiac failure, degenerative nervous complaints and mental deterioration, such as occurs when supplies become infiltrated by leachings from mines such as manganese or nickel, or despoiled by waste dumping. Mercury from the chemical industries surrounding Japan's Minimata Bay entered the food chain following dumping of chemical wastes into the Bay, a rich fishing ground for local people. Minimata disease, associated with major nervous system derangement, results from eating fish whose flesh has become infiltrated with mercury.

In parts of the world where rainfall is low or only occurs for part of the year, efficient methods for collecting and conserving water supplies are important. An adequate supply of water is necessary for good standards of hygiene as well as for drinking and cooking purposes. Modern methods of sewerage rely on water carriage, and where sanitation is inefficient there is an increased risk of the proliferation and transmission of infectious diseases, especially those affecting the gastrointestinal tract.

Temperature

Temperature is a function of sunlight energy, the atmosphere, latitude, and cloud cover permitting entry of heat and its exit or reflection from the earth's surface. The earth's temperature involves the complex interaction of earth's surface, the content and proportion of gases in the atmosphere, size and area occupied by water as oceans, seas and rivers, and the amount of vegetation. Water, the air, ground, clouds and vegetation absorb and reflect heat to varying degrees, and they influence the temperature as experienced by the individual.

Temperature is important as an indicator of climate and monitor of changes occurring in the local and global context. It involves the three basic questions which must be answered in forecasts of the climatic future: How much carbon dioxide and other greenhouse gases will be emitted? By how much will atmospheric levels of the gases increase in response to these emissions? What climatic effects will the resulting build-ups have after natural and human factors that might mitigate or amplify those effects are taken into account?

Human emissions of carbon dioxide depend on global consumption, population growth, rates of deforestation, and the policies and plans now being implemented world-wide to restrict industrial emissions into the atmosphere and develop alternative energy sources. Typical projections assume that global fossil-fuel consumption will continue to increase at about the current rate, yielding increases in carbon dioxide emissions of between 0.5 and 0.2 per cent a year for the next twenty years at least. This is a much slower rate of growth than that which occurred before the 1970s energy crisis. Other greenhouse gases such as methane, the chlorofluorocarbons (CFCs), oxides of nitrogen and low-level ozone together could contribute as much to global warming as carbon dioxide, due to their higher absorption of infrared radiation (Table 2.6).

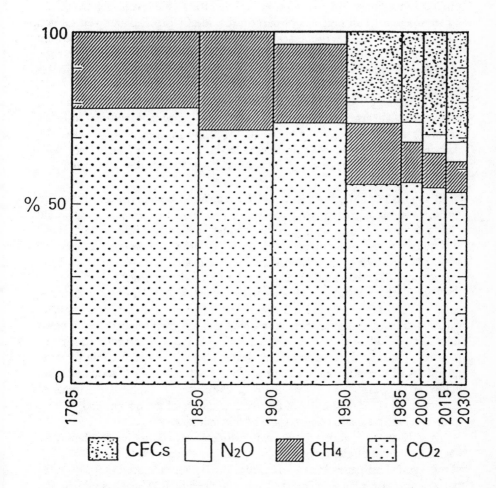

Fig. 2.1: Source: Wigley, T. 'Relative contributions of different trade gases to the greenhouse effect' from Climate Monitor, 1987, vol. 16.

Future predictions of these gases are even more complex than those for carbon dioxide emissions. The sources of some, such as methane, are not fully understood; the production of gases such as the CFCs and low-level ozone depends on technical developments and global politics. The increased carbon dioxide emissions and the factors which affect the levels achieved, are equated against constant use of carbon dioxide by green plants and natural chemical and biological processes. Final atmospheric concentration of carbon dioxide is the difference between these two amounts. Uptake of carbon dioxide is affected in turn by rises and declines in levels of carbon dioxide emissions.

It is possible that an increased concentration of carbon dioxide in the atmosphere will speed the uptake by plants, and increased photosynthesis will occur thereby counteracting some of the buildup. Similarly, because the carbon dioxide content of the oceans' surface waters stays roughly in equilibrium with that of the atmosphere, oceanic uptake will slow the buildup to some extent.

It is also possible, however, that an increased concentration of carbon dioxide and other greenhouse gases will trigger positive feedbacks that would add to the atmospheric burden. Rapid change in climate could disrupt forests and other ecosystems, reducing their ability to draw carbon dioxide from the atmosphere. Moreover, climatic warming could lead to rapid release of the vast amount of carbon held in the soil as dead organic matter.

This stock of carbon — at least twice as much as is stored in the atmosphere — is continuously being decomposed into carbon dioxide and methane by the action of soil microbes. A warmer climate might speed this process releasing additional carbon dioxide (from dry soils) and methane (from rice paddies, landfills and wetlands) to further enhance the warming. Large quantities of methane are also locked up in continental shelf sediments and below arctic permafrost in the form of clathrates — molecular lattices of methane and water. Warming of the shallow waters of the oceans and melting of the permafrost could release some of this methane.

In spite of these uncertainties, many workers expect uptake by plants and oceans to moderate the carbon dioxide buildup at least for the next 50 to 100 years. Typical estimates based on current or slightly increased emission rates put the fraction of newly injected carbon dioxide that will remain in the atmosphere at about one half. Under that assumption, the atmospheric concentration will reach 600 ppm, or about twice the level of the year 1600 by sometime around the year 2030 (see Fig.2.2). It needs to be borne in mind, however, that other greenhouse gases such as CFCs, methane and nitrous oxides are now building up at rates faster than carbon dioxide, which may have even more devastating effects.

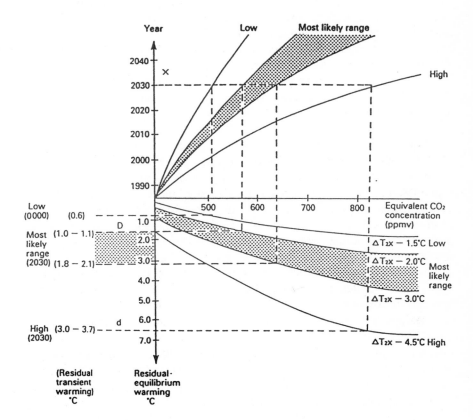

Fig. 2.2: Source: Warrick, R., Jones, P. and Russell, T. The Greenhouse Effect, Climatic Change and Sea Level: An Overview, *Climatic Research Unit, School of Environmental Sciences, University of East Anglia, 1988.*

Geology and Geography

The geology, geography and topography of a country affect the health status of the population and the range of diseases prevalent among its inhabitants. However, the relationship is not always clear. Within Scotland and other mountainous parts of Britain, there is no reason to link local cardiovascular disease rates to the softness of the public water supply. In Scotland, everyone's water is soft. The adverse effect becomes apparent only when the inquiry is extended to Britain as a whole.

Mortality and morbidity rates can be seen to vary with area and industries found there. In Great Britain, for example, respiratory illnesses and bronchitis are more common in areas where coal is mined. Similarly, the Hunter River Valley area of New South Wales, Australia — another coal mining region and one where heavy industry is concentrated — reports high prevalence rates for asthma, lung cancer and emphysema. High levels of the toxic mineral lead in the blood of children in Whyalla, South Australia have been blamed on the lead smelting and extractive industry, the area's principal employer.

It is logical that industries developed and factories were established where raw materials were near at hand, good communications and transport to market were assured. In Britain, metallic ores and coal deposits occurred near one another and the growth of smelting operations proved convenient, particularly where good communications existed and access to transport was possible. The ready availability of soft water facilitated the establishment of cotton and woollen industries in Yorkshire. In Wales, carboniferous deposits and limestone outcrops encouraged the growth of coal mining and associated minerals processing.

Geological factors can also have a direct influence on health because of variations in the occurrence of trace elements important to health. Residents of the Derbyshire Peak District in northern England have long been noted to be susceptible to goitre, a condition associated with a local soil deficiency of the trace element iodine. The same link can be drawn between the occurrence of goitre in the Eastern Congo, and the iodine-deficient soils of the areas of the New Guinea Highlands. Similarly it was observed that regardless of diet, a number of communities could be found where dental caries were rare. At the same time, the teeth of locals were mottled. This drew attention to a possible link between the level of fluoride in the water supply and the occurrence of tooth decay.

Living Environment

The dynamic interrelationships between various life-forms in a given habitat, the well-being or failure of one or another, may well have profound influences on the status and survival of others. The process hinges on the continual operation of the natural cycle which depends on producers, consumers and decomposers.

Often dismissed by non-scientists as undesirable, dangerous and unnecessary, bacteria play an important role in breaking down complex organic compounds into simple substances so that they may re-enter the natural cycle and be reincorporated into new living forms. Though one of the simplest forms of life, they complete the dynamic web of relationships which affects living species at lower and upper levels of complexity.

Other bacteria in the gut of humans and ruminant animals synthesize vitamin K, folic acid and components of the vitamin B complex. In animals such as rodents, where coprophagy (dung eating) is the norm, other vitamins may be absorbed from the dung.

The developing science of genetic engineering is opening up possibilities of utilising the manufacturing abilities of some bacteria for the production of medicinally important substances. In recent years, there has been increased exploitation of microscopic fungi which produce antibiotic substances as a natural system of defense against competitors.

Micro-organisms exist around, on and in all living species and for most of the time. This coexistence is harmonious even though some of them are capable of invading human tissue and causing harm in appropriate circumstances. This state of equilibrium depends partly on the maintenance of a dynamic balance between the vigour and invasiveness of the micro-organism and the state of the body's defences. If the defences fail an opportunity may be provided for ordinary inoffensive organisms to cause disease and even to become lethal. Less commonly, the organism itself may undergo a change which enables it to overcome normally adequate defence mechanisms, possibly by biochemical mechanisms, such as the use of other enzymatic breakdown paths, previously of minor value to the organism. Sometimes invading organisms are destroyed completely, sometimes a state of equilibrium is regained. Less often the organism prevails and disease ensues.

The visible animal life around is a more apparent and equally important part of the environment. Animals contribute to recreation and their presence gives aesthetic pleasure and enhances the quality of life. The work role of some species, though less important than in previous times, is still an important source for food. However, animals can also act as reservoirs of infectious disease agents. This means that although it is important to preserve wildlife, these animals, together with farm and domestic stock animals, need to be maintained in healthy conditions for human as well as their own well-being. Control of infectious diseases transmission ensures their health and well-being as much as that of their human masters.

The Social Environment

Humans have a liking for, and depend on, the company and mutual support of other members of the same species. Such a system of living is at once both mutually supportive and mutually destructive and achieving the right balance is a continual problem. While the human propensity to live in groups heightens the risk of competition for life factors, at the same time it provides social and psychological benefits whose subtle influences may determine how that person responds to disease, their attitude to life and life chances.

There is increasing recognition within industrialised nations such as Britain, the United States and Australia, of the close interrelationship between the health status of populations and discrete groups within them, and the social and economic circumstances of the lives led by these people. Statistics show that those at greatest disadvantage from the potentially harmful effects of environmental changes following the industrial revolution have tended to be the relatively unskilled, manual or industrial workers and their families, living in areas of considerable industrial development with their associated problems of close settlement, poor housing standards and atmospheric pollution. These people also tend to be low income earners, living in poor standard housing and existing on a diet of less than optimal nutritional value.

Thus, lifestyle tends to be a reflection of the social as much as the physical environment. Poverty, unemployment, poor living conditions, large families and a high level of crime are symptomatic of high levels of cardiovascular disease, cancer, respiratory disease and accidents, in the workplace and on the roads. Relative ignorance, incomplete or low levels of education, poor diets, lack of exercise, alcohol and drug abuse are risk factors for the major diseases, and these tend to be found in a significant proportion of those sufferers who come from the bottom end of the social hierarchy.

Differences in lifestyles, traditions, attitudes, culture, political and religious beliefs distinguish and separate groups which tend to be responsible for continual conflict and enmity. Territorial boundaries may represent not only spatial but also conceptual and philosophical differences, and it may be difficult and even dangerous to transgress them. This separation of social groupings is seen in its most extreme form in separate nations. However, social subgroups also exist within each nation, each identified by distinctive lifestyles and affiliations. At the same time, territorial boundaries tend to be less apparent and identity subsumed within the larger whole which comprises the nation. The Ainu of Japan, the Aborigines of Australia and American Mormon communities serve as examples of such separate ethnic or religious groups.

These subgroups are characterized by different sets of values, religious, ethical, moral and political; they have differing approaches to, and views about, fundamental aspects of daily life such as sexual behaviour, marriage, child-rearing and family life, education, diet and nutrition, the use of tobacco, alcohol and drugs, and all the myriad aspects of daily life. Each of these aspects of human life is becoming recognised as having a role to play in contributing to individuals' health.

Natural Selection: Its Role in Environmental Change
It was in the last century that Charles Darwin first described the process known as natural selection. Travel abroad and years of pondering the differences between living species led him to the conclusion that, in order to survive, each life form must come to terms with the environment in which it finds itself.

In other words, a state of equilibrium is reached between each life form and its environment which is such that those individuals least well adapted, die. Those which have sufficient variability of biological material, are able to adapt themselves to the changed environment. Over time, slight modifications in their form emerge and these improve their chances of survival over that of others.

In natural circumstances these evolutionary processes are extremely slow, although this tends to depend on the life-cycle of the particular organism. Thus, for some organisms such as bacteria, they may be relatively rapid. Bacteria reproduce over very short periods of time and new forms will be apparent within the human lifetime. The more sophisticated the organism, the more time it needs to adapt to a sudden change in the environment. Human evolution and that of the more complex species takes place over thousands of years.

The time during which changes occur in the environment and that required for evolutionary adaptation by means of natural selection is critical to their chances of survival in the changed environment. Hence, sudden or extreme changes may find that the majority of the population of individuals is unable to adapt. For example, the presence of a persistent new chemical in the atmosphere, which cannot be broken down by a living species, may result in the block of a metabolic pathway, inhibition of a critical enzyme system or the ability to produce an essential body hormone. This may compromise the life chances of the species members thus affected and possibly endanger the survival of the whole species. Such changes may then have a 'down-stream' or 'knock-on' ecological effect.

To give a more specific instance of this process: British scientists have found that the introduction of DDT as an insecticide led to its concentration in the food chain of predatory birds. As a result, they produced soft-shelled and infertile eggs, and their numbers began a dramatic decline. Had these birds been exterminated, there might have been changes in the populations of rodents which formed their prey.

Other examples exist in recent times. The accelerated spread of human communities into previously unsettled areas in Asia and South America has been responsible for the loss of many rare botanical species before adaptation has occurred. Substantial recent loss of rainforest plant species and biota, many of which have not been properly catalogued and described, has also occurred aided by large-scale commercial logging and clearing operations.

Changes in availability of natural food sources affects the health of bird species. Parakeets and indigenous parrot species in eastern Australia succumb to disease as a result of feeding on sugary biscuit and bread pickings scavenged in urban environments. Particularly in cleared and newly agricultural land, local streams and waterways are being forced to accept ever higher burdens of agricultural chemical runoff containing phosphate fertilisers and urea concentrates. Native fish species find the changed conditions often difficult to tolerate and a decline in numbers may occur. Others become susceptible to fish diseases or fail to reproduce.

As yet, we do not know the effects which our changed global climate and living conditions are having upon the human species. Is the accumulation of pesticides and organic agricultural chemicals in the food chain likely to impact upon our species survival? Will our offspring be less fit and able to survive the potent new viral diseases evolving because of our success in eliminating the previously deadly infectious diseases? These are all questions scientists are beginning to ponder and for which no answers can be provided — at this stage.

TABLE 2.1: Radiation

Ultra-Violet
* ocular cataracts/kerato-conjunctivitis
* increased skin cancers (squamous cell carcinoma, basal cell carcinoma, malignant melanoma)
* potentially shortened latency period for all cancers, possibly immune system suppression, increasing susceptibility to infection

Infra-Red
* ocular cataracts
* conjunctiva and skin erythema

TABLE 2.2: Effects of Heat and Humidity

* heat rash, fatigue, cramps, exhaustion, heat stroke, heat oedema
* heat rash and skin infections, e.g. fungal infection
* chronic heat fatigue
* adverse fetal outcomes

TABLE 2.3: Acid Rain

* depletes soil nutrients, plant skeletonizing/growth retardation
* aerosols damage respiratory tract and eye linings
* heavy metal toxicity from leaching of heavy metals from sediment, water-sheds and plumbing

TABLE 2.4: Noxious Gases

Radon (and daughter products) - respiratory tract carcinogen

CO - blood dysfunction (methaemoglobinaemia)
hypoxia (asphyxia)
- damage to central nervous system and other tissues/organs whose function already compromised by pre-existing disease

Dioxin - skin irritation (chloracne)
- suspected carcinogen

Formaldehyde - skin, eye, respiratory tract irritation (bronchospasm and sensitisation)
- suspected carcinogen

Toxic Hydrocarbons - carcinogens
- blood changes (including anaemia, bleeding disorders, leukaemia)
- damage to central nervous system

Oxides of Nitrogen - irritant to respiratory tract (bronchospasm), emphysema

Ozone - respiratory tract irritant (bronchospasm)

Oxides of Sulphur - asthma initiators

Vinyl Chloride - haemangiosarcoma of liver

Miscellaneous gases occurring in waste products
(NH_3, H_2S, CH_3SH, CH_4)
- malodorous
- asphyxiants
- irritants
- explosive

CFCs- stratospheric ozone is particularly affected by changes in the concentration of CFCs, CO_2, CH_4 and NO_2, the same gases that affect global temperatures and drive climate change. CFCs are released from pressure pack sprays, industrial cleaning processes, refrigeration systems and foam preparation. The resultant depletion of ozone is causing a greater proportion of biologically active UV radiation to reach the Earth, and is expected to increase the incidence of basal and squamous cell carcinoma and malignant melanoma. There is expected to also be an increase in ocular cataracts.

TABLE 2.5: Diseases Transmitted by Mosquitoes

Disease	Species
Dengue syndromes	
Onyognyong virus	*Anopheles*
West Nile dengue	*Culex*
Dengue viruses	*Aedeas aegypti*
Filiariasis	*Aedes,*
	Anopheles,
	Culex, Mansonia
Malaria	*Anopheles*
Virus encephalitides	
e.g. equine encephalitis,	
Japanese B and St. Louis	
encephalitis	*Aedes, Culex*
Yellow fever	
Urban	*Aedes aegypti*
Rural	*Aedes simpsoni*
	Aedes africanus
Ross River Fever	*Culex*
	various bush-tick
	spp.

TABLE 2.6: Heavy Metals and Radiation Poisoning

Lead (low level) - subtle damage to developing CNS in children
- gastro-intestinal tract
- induced anaemia
- induced encephalopathy
- induced nephropathy

Cadmium - induced kidney damage
- respiratory tract irritation (polyfume fever), emphysema
- suspected carcinogen (lung, prostate)

Uranium (natural) - induced nephropathy

Fuel cycle constituents - cancer/reduction in fertility and life-span

Mercury - irreversible neurological damage
- kidney and urinal tract damage
- reproductive effects

Chromium - lung and other cancers (6+)
- skin
- lung asthma
- nasal irritation/ulceration (polyfume fever)

Nickel - skin irritation
- respiratory tract irritation/suspected carcinogen

Arsenic - carcinogenic/skin/liver/lung
- neurological disease
- blood and liver dysfunction
- skin/mucous membrane irritation

Iron - siderosis of lung

Non-ferrous Metals - the above, also aluminium, antimony, barium, beryllium, bismuth, caesium, cobalt, copper, gallium, germanium, gold, hafnium, indium, the lanthanides, lithium, magnesium, manganese, molybdenum, niubium, the platinum group, rhenium, rubidium, selenium, silver, strontium, tantalum, tellurium, thallium, tin, titanium, tungsten, vanadium, zinc, zirconium.
- a range of health effects

TABLE 2.7: Reduction in Mortality since 1700 (England and Wales)		
Period	*Per cent Total Reduction Each Period*	*Per cent Reduction due to infection*
1700 to 1850	33	unknown
1850 to 1901	20	92
1901 to 1971	47	73
1700 to 1971	100	>100

Source: after McKeown (1979)

References and Additional Reading

Brodeur, P. *Outrageous Misconduct: The Asbestos Industry on Trial*, Pantheon, New York, 1985.

Brown, M.H. *The Toxic Cloud*, Harper & Row, New York, 1987.

Canadian Cancer Society: Annual Report, 1987-8, The Canadian Cancer Society, 1989.

Epstein, S. *The Politics of Cancer*, Sierra Club Books, 1978.

Forget, G. *Toxic Substances and Health*, International Development Research Centre, Ottawa, 1988.

Haynes, V. and Bojcun, M. *The Chernobyl Disaster*, Hogarth Press, 1988.

Lloyd, J.W., Lundin, F.E., Redmonds, C.K. *et al*. 'Long-term Mortality Study of Steelworkers. IV. Mortality by Work Area' from *Journal of Occupational Medicine*, vol.12, 1972.

MacMillan, I. *Electromagnetic Fields, Electric Power and Public Health*, Collingwood Community Health Centre, Melbourne, Australia, 1990.

McKeown, T. *The Role of Medicine: Dream, Mirage or Nemesis?*, Blackwell Scientific Publications, Oxford, 1979.

New Internationalist. 'Cancer — The Facts', August 1989, issue no.148, p.16, and extracted from Hester, J. *et al*, *Science for the People*, published May 1989.

Patterson, J. *The Dread Disease*, Patterson, New York, 1987.

Redmond, C.K., Strobins, B.R. and Cypress, R.H. 'Cancer Experience Among Coke By-Product Workers' from *New York Academy of Science Annual*, vol.271, 1976.

The Lancet, Editorial, 19 May 1984.

Wood, C.S. *Human Sickness and Health: A Biocultural View*, Mayfield Publishing Company, Palo Alto, California, 1979.

World Health Organisation. *World Health In Developing Countries*, WHO Publications, 1987.

World Health Organisation. *World Health*, WHO Publications, October 1980.

World Health Organisation Cancer Unit. *Cancer Control in Developing Countries*, WHO Publications, 1986.

3.0 CHANGES

3.1 Climate Essentials

The Prospect of Environmental Change
It seems that the human species and others residing with or dependent on human life on earth are facing a new battery of challenges, resulting principally from their own successes in colonising and dominating earth's surface. Nature seems to be responding, in its own way and in its own time. The first signs are apparent in major climate changes. However, controversy persists about whether the changes being experienced within decades constitute 'real' changes in the earth's climate, which is measured in thousands and millions of years rather than decades and centuries. To resolve these dilemmas and to better understand the factors involved and their role in human health, it is important to first understand the finer points of the science of atmospheric physics and weather, the 'climate essentials'.

Examining The Evidence
To better appreciate the problem, it may be useful to first examine some of the data on which claims of climate changes are based.

Climatologists at the Goddard Institute for Space Research in New York announced that the average global temperature in 1990 was 15.4°C. This makes it the warmest year since 1880, when records of the earth's surface temperature were first collected. The US research also indicates that the seven warmest years since 1880 all occurred in the last eleven years.
(*Sydney Morning Herald*, 11 January 1990)

At the British Meteorological Office and the University of East Anglia are records dating back to 1850. According to different estimates based on these readings, six of the seven warmest years have occurred since 1980. The six were 1980, 1981, 1983, 1987, 1988 and 1990. Some scientists there believe that the latest series of warm years could be the first indication that greenhouse effects are warming the globe.

Burning of vegetation, a common practice in the tropics, releases soot and several gases, particularly carbon dioxide, carbon monoxide, hydrocarbons, nitric oxide and nitrogen dioxide. This and other human activities — such as the burning of fossil fuels — account to a great extent for dramatic increases over the past two centuries in the atmospheric concentrations of trace gases. These increases are giving rise to such environmental perturbations as acid deposition, urban smog and depletion of the stratospheric ozone layer that absorbs damaging ultraviolet radiation. Warming of the planet is also expected from the build-up of greenhouse gases that trap infrared radiation.

Scientific findings tell us that the temperature and content of the atmosphere is increasing. Their investigations further reveal that this is not a new event. Large temperature changes mark earth's history: a series of ice ages, warm-intermediary periods followed by glaciation marked substantial changes in climate.

However, these changes took place over many hundreds of thousands of years. The current changes have been noted over less than one hundred years, a very, very small period in planetary time. The only other previous experiences of climate change followed major catastrophes in earth's history. Except for these chance misadventures, climate change has been gradual. Species had time to adapt, new species evolve, and older life-forms less 'fit' to cope with the altered conditions die out.

The human species has evolved during the latest 'warm period' in earth's history. Thus it has not been subject to a major climate change such as would usher in a new ice age. Scientists can only guess about the species' capacity to survive any such changes.

Extinctions
Extinctions can be thought of as one of the main processes in the evolution of life. They are the end result of a species being unable to adapt fast enough to new environmental circumstances. The majority of all species that have ever existed are now extinct. The background rate for major organisms has been estimated at 3-5 families or 180-300 species per million years. Some of the key factors that account for relatively gradual extinction are: predation, competition, habitat change and reproductive rate. The impact of the first three of these depends to some extent on the fourth.

Nonetheless there have been periods of abrupt extinctions, and these may have been the result of cataclysmic events such as the collision of asteroids with the earth or collisions of drifting continents. The fossil record indicates that there have been at least six periods of mass extinction on land and at least four in the oceans. Rapu and Sepkowski (1982) calculated that species disappeared at the rate of 20 families and up to 1200 species per million years during the mass extinctions in the oceans. Such mass extinctions occurred during the late Ordovician, Permian, Triassic and Cretaceous periods. There has been some recent speculation that mass extinctions occur about every 26 million years as a result of some kind of 'regular' asteroid shower.

There was an especially dramatic mass extinction at the end of the Cretaceous period 65 million years ago, when a very high proportion of the oceanic plankton suddenly disappeared along with the dinosaurs. It is now generally believed that this was caused by a collision between earth and an asteroid. This collision was believed responsible for a drastic decline in incident solar radiation, depleted the ozone layer and resulted in broad erratic weather patterns for hundreds of thousands of years.

Among the evidence for this hypothesis is the occurrence of an iridium-rich deposit at the boundary between the Cretaceous and Tertiary period, apparently laid down throughout the world at about the same time that the extinctions occurred. Iridium is relatively rare in the earth's crust, but relatively abundant in meteors.

The theory that the dinosaurs were wiped out by the impact of an asteroid has been challenged by an American astrophysicist and his wife, a biologist and a school student. These four people claim that dinosaurs met their death when nickel from a vaporised asteroid poisoned their plant foods.

Dr Thomas Wdowiak at the University of Alabama in Birmingham, USA claims that the theories invoking asteroid impacts to account for the extinction of the dinosaurs 65 million years ago do not adequately explain how the dinosaurs actually perished. He believes that the soot injected into the atmosphere by a disintegrated asteroid was laden with nickel which was soluble in water and very poisonous to plants.

The hypothesis accepted by most researchers proposes, instead, a two-staged effect. The initial impact is said to have resulted in giant explosions, fires and earthquakes across the globe. The theories say that the dinosaurs were dealt a final blow by the debris and soot which was blasted into the atmosphere and blocked sunlight for years. This would have brought major changes in climate, affecting all existing flora and fauna species. Wdowiak believes that nickel poisoning of plants provides a more satisfactory explanation for the eventual death of the animals.

In defence of his claims, Wdowiak contends that the type of mass killing which results from catastrophic events such as volcanic eruptions and earthquakes, produces major damage but does not result in species extinction. New plants emerge in the next growing season. However, with nickel poisoning, extinction is guaranteed in a manner Wdowiak compares with 'biblical salting of the earth'. The body that wiped out the dinosaurs would have been about 10 kilometres in diameter.

According to Wdowiak, such a body would have a density of 3000 kilograms per cubic metre. Spread over earth's surface, this would amount to 3 kilograms per square metre which corresponds to a layer a few millimetres thick. Wdowiak's figures mean that this fallout could contain between 130 and 1300 parts per million of nickel. The normal concentration of nickel in soil is 15 ppm: a level of 40 ppm is considered toxic.

On present evidence, life seems to have begun on earth barely 3500 million years ago and to have evolved only slowly until the present. The last 10 million years have been marked by significant changes in the climate which became more pronounced during the Quaternary and accelerated in the last 0.7 million years. Principally because of the impact caused by the human species, there have been further substantial changes in the parameters of global temperature, the content of the atmosphere, sunlight levels and oceanic plankton, etc., which govern climate and weather.

The Survival of Human Species
What effects will the increased rate of change have on earth and the survival of all living species? At one end, there are those who believe that the mass extinctions of whole species evident from past geological records have been triggered by brief, cataclysmic events such as the impact of a celestial body or periods of intense volcanic activity. Others argue that the extinction process is gradual and is brought on by environmental changes, which may be wrought by rapid tectonic, oceanic and climatic fluctuations. They are less able to explain the implications or the ramifications associated with the increasing pace of change now occurring.

Is this unique in earth's history? Has earth experienced a similar period of rapid change which has been induced or influenced by one of its dependent living species, rather than by a physical or solar accident? Many scientists maintain that the truth lies somewhere in between, a combination of earthly and extraterrestrial causes.

Will the human species survive? And if it does, in what form will that survival occur? Studies reveal that the continued existence of a species will depend on a number of factors. For example, widespread species generally seem to weather mass extinctions better than those endemic to small regions. Others at high risk are those that live in tropical climates and cannot readily adapt. The oceans, too, provide no refuge against extinctions, which hit marine creatures as much as land dwellers.

Size helped the dinosaurs to achieve dominance for 140 million years, only to then vanish from the fossil records. Other smaller mammals survived and have since diversified and themselves assumed a dominant role. Questions remain over why some groups, such as the trilobites and the ammonoids, survived several mass extinctions only to disappear entirely about 250 million years ago. Other species, such as the crinoids and the corals, came close to extinction only to reappear. Species like fish and mammals have proved to be remarkably successful in expanding and extending the number and range of their group. Both these species have dramatically increased in numbers since the Tertiary period began (approximately 65 million years ago).

Volcanoes, The Asteroid Shower and Other Theories
A rival to the Impact Theory favoured by many scientists, suggests that mass extinctions can occur directly as a result of events on earth. It argues for the role of volcanic upheavals and their resultant pollution of the atmosphere with vast quantities of those greenhouse gases which are responsible for global warming and temperature change, suggesting that intense volcanic activity also blocks sunlight. Periods between eras characterized by massive volcanic activity show the opposite environmental changes. Sea levels decline, ice ages and climate change allow for new species to evolve suited to the altered conditions. Such a theory would account for previous episodes of sudden, major climate change in terms of natural changes in earth's environment. Thus earth's fate and that of its species would seem to lie with nature and forces beyond human control.

Perhaps more controversial than the impact or volcanoes theories is the notion that asteroid showers regularly bombard the planet and instigate major environmental changes. Researchers Sepkoski and Raup from the University of Chicago have used fossil records to identify peaks of extinction about every 26 million years, more frequent therefore than those which mark the major mass extinctions.

The most obvious source for these regular calamities would seem to be the dense cloud of comets which astronomers believe surrounds the outer solar system. Periodic unsettling of that cloud could result in a barrage of the inner planets over several million years. Three mechanisms have been proposed but astrophysicists find problems with all three. Many argue that showers strike randomly not regularly. Others argue a frequency of 30 rather than 26 million years, and try to correlate frequencies with predictions about a possible cosmic timetable which includes one for the extinction of human and other related species.

Resolving The Dilemmas
Yet many scientists vigorously dispute the mass and small extinctions theories, taking a middle line maintaining that mass extinctions do occur through cataclysmic events from outer space, but that other factors have to do with local changes on earth. How else, they argue, could species disappear so rapidly and so completely? And they point out the role of technology and the changes wrought on the planet by the human species. In a very short space of time, the human species has changed the whole nature and form of the earth

and the conditions for all other living species.

Explaining the Current Weather Changes
In seeking explanations for current complaints about changing weather, unusually high temperatures and sudden weather extremes, we now need to pause and consider the factors which influence weather and climate. These are complex phenomena dependent on a series of finely attuned interrelationships we barely comprehend. They depend on atmosphere, temperature, the movement of air masses, and heat exchange physics. A basic understanding of the key components is essential.

The Nature of Climate
Patterns of weather occur in a regular sequence which constitutes the seasons. The regular cycle of dry and wet, hot and cold seasons forms the climate of a particular geographic region. Their character depends on the longitude, latitude, the topography, the atmosphere and the circulation of winds in that particular locale.

The amount of sunlight received by the soil and rocks, the water, plants and animals at the surface of the earth, and the temperature experienced, depend on the interaction of all these various factors. The resultant climate is unique to that region, and the species' response to it shows the close interdependence between all these components to the natural system. Essentially it depends on the complex movement of great masses of air over the differing regions of the earth's surface, where it is subject to the changing variables of solar emission (light, heat), global forces (convectional currents, eddies, turbulence and magnetic influences) and local regional factors (vegetation cover, damming of waterways). Often neglected, though important to include because of its size, is the ocean and its waterways which comprise more than two-thirds of earth's surface.

The earth's atmosphere, however, has never been free of change: its composition, temperature and self-cleansing ability have all varied throughout its history. However, the rate at which these changes has occurred has never been greater than those which have occurred in the past two hundred years.

Increasingly evident are the effects of ongoing changes including acid deposition by rain and other processes, corrosion of materials, urban smog, thinning of the stratospheric ozone mantle, and greenhouse gas-induced warming. However, these important modifications stem not from changes to the earth's major constituents, but from changes in the levels of its minor constituents, the trace gases.

Temperature

The temperature of the earth depends on the amount of sunlight received from space, the amount reflected by the upper atmospheric mantle and the earth's surfaces, and the extent to which the atmosphere retains heat. The heat absorptive capacity of atmospheric gases, surface rocks and soil, the oceans and the natural biota of forests, vegetation, etc., determine the earth's capacity to retain heat.

As described earlier, water vapour and gases such as carbon dioxide, methane and nitrous oxide are able to absorb and trap the sun's heat energy which penetrates the thin mantle protecting the earth. Their ability to retain this heat will govern the extent to which further heat is gained and lost by the earth.

The surface of the earth and the natural species living on it receive warmth and life-giving energy from the relatively small proportion of sunlight which eventually reaches the lower surface levels. Here most of this energy is either used or absorbed. The remainder is reflected back into space, or captured and retained by the same atmospheric greenhouses gases which limit the amount which has actually reached the surface of the earth. The result is a warming of the atmospheric blanket between the outer mantle of greenhouse gases and the earth's surface below. This blanket is divided into a series of layers — the troposphere, stratosphere, mesosphere and ionosphere.

The earth's heat balance is governed by the cloud mantle of water vapour, oxygen, carbon dioxide gases and minor gases in the first ten kilometres above the surface, and by the quantity and quality of the small ozone layer in the final outer layer. One of the most important of its gases is ozone, a highly unstable form of complex oxygen and constituting only 0.05 per cent of the mantle, located thirty kilometres above the surface.

The Oceans

Often ignored in considerations of climate, the oceans are a vast expanse occupying approximately two-thirds of earth's surface. They serve as a constant medium for the exchange of water vapour and other gases. To the climatologist, oceans are not only important for heat conservation and circulation effects associated with sea water, but because they serve as a 'sink' for greenhouse gases, in a way similar to the vegetation cover of vast tracts of tropical rainforest.

Gas Composition: The Greenhouse Gases

As early as the nineteenth century it was realised that carbon dioxide in the atmosphere gives rise to a greenhouse effect because it traps the heat from sunlight and captures the longer-wavelength infrared radiation released by earth. As seen from space, the earth radiates energy at wavelengths and intensities characteristic of a body at -18° C. The average temperature at the surface is some 33 degrees higher: heat is trapped between the surface and the level, high in the atmosphere, from which radiation escapes. There is little doubt among atmospheric scientists that increasing the concentration of carbon dioxide and other gases will increase the heat-trapping and warm the climate. The major gases involved in building and maintaining earth's heat balance are crucial for continued life in its present form.

Increases in population, together with greater demands for energy and on agriculture, increase and boost the production of greenhouse gases from ruminants, rice paddy cultivation, natural breakdown and decomposition. An annual increase of 1.5 per cent in carbon dioxide emissions has been linked with the burning of fossil fuels including coal, gas and oil products. Increased population and land under agriculture boost the carbon dioxide load a further one per cent. Chlorofluorocarbons (CFCs) introduced as refrigerants, solvents and propellants for aerosols, a persistent synthetic and non-biodegradable chemical, organic fertilisers and commercial carbon-based cleaners must also be acknowledged as contributing to greenhouse gas levels.

The rapid build-up of atmospheric carbon dioxide came to recent attention in the 1950s. Prior to this, the only substantial increase occurred during the eighteenth and nineteenth centuries with the expansion of industry and factory production which followed the Industrial Revolution. Worldwide pollution from carbon dioxide, the principal greenhouse gas, rose by 1.6 per cent in 1987 to 5.6 billion tonnes of carbon a year. (The figures are from the Carbon Dioxide Information Analysis Center at the Oak Ridge National Laboratory in the United States.)

Since 1983 global carbon dioxide emissions have risen by 10 per cent. The Toronto Conference on the Changing Atmosphere held in 1988 advocated a 20 per cent cut in these emissions by the year 2005, and rising to a cut of 50 per cent by 2020.

Climate Change Predictions Attributed to Greenhouse
Climate issues have been receiving increasing attention around the world since the mid 1970s, following a prolonged African drought and major shortfalls in grain harvests of Europe and North America which had accompanied a series of severe winters and unusual summers. Initially, speculation centred on the possible onset of a new ice age, a claim supported by satellite evidence of expanding snow cover in the northern hemisphere.

The physical, chemical and biological interactions which govern the behaviour of the global climate are complex and not yet fully understood. A massive international effort has been under way for several decades aimed at better understanding the mechanisms of the climate so as to be able, in future, to predict periods of flood and drought and to foresee and prepare for possible human influences on global and regional climate.

Since 1985, scientific predictions of long-term global warming due to greenhouse effects have triggered the preparation, for purposes of impact analysis, of scenarios of future regional climates as would be experienced in a warmer world. These scenarios have been rapidly sensationalised by the media so that the possible impact of the greenhouse phenomena has generated a widespread public perception of impending disasters for particular communities, the world and life as known today.

The relationships between greenhouse gases, climate cycles, weather and the levels of other atmospheric elements such as ozone and nitrous oxide are explored in the following chapter.

3.2 The Greenhouse Phenomena

What Does It Mean?
By the year 2030, it is predicted that as a result of increased levels of greenhouse gases in the atmosphere:

* global temperature will rise by between 2 and 3° C.
* rainfall during summer months will increase by up to 50 per cent and in winter decrease by 20 per cent: this will be particularly marked in countries located between the equator and approximately 40 degrees south.
* there will be increased tropical cyclone activity between 0 and 35 degrees latitude: the Tropics of Cancer and Capricorn are found at 23.5 degrees latitude north and south respectively from the Equator (0° latitude).
* more frequent weather extremes such as floods and droughts will be experienced on all continents.
* raising of the water table bringing increasing risk of salinity problems associated with irrigation and flood mitigation.
* increased warming will result in retreat of the snowlines, melting of a predicted additional 15 per cent of polar icecaps, effects which in turn will be followed by:
    (i) increase in water entering large riverine systems.
    (ii) rise in sea levels (up to 80 cm).

Climate Change
In 1979, a major international, interagency and interdisciplinary effort was mounted to establish a World Climate Programme. This Programme sought to coordinate scientific research and data on climate to provide the means of foreseeing possible future change. Its work has been further aided by that of the United Nations World Commission on Environment and Development (the Brundtland Commission).

Today, the prospect of major climate change within the lifetimes of the present generations, its possible impacts and the strategies through which nations might minimise and adapt to it, are now widely seen as among the most important issues facing humanity in the closing years of the twentieth century (Table 3.1).

Carbon Dioxide Emissions — The Latest Figures
To appreciate the extent of the problem of greenhouse gas emissions and the

comparative lack of progress thus far, the 1989 *Environment Digest* provides some sobering figures. In 1987, the Carbon Dioxide Information Analysis Center at Oak Ridge National Laboratory in the United States, reported that worldwide pollution from carbon dioxide rose by 1.6 per cent to 5.6 billion tonnes of carbon a year. This means that from 1983 to the present, carbon dioxide emission has risen by 10 per cent.

The United States, the Soviet Union (CIS) and China recorded the biggest increases, with China alone responsible for a rise of 4.8 per cent (Table 3.2). In contrast, France reduced emissions from 98 million tonnes to 95 (-3.2%) and western Germany from 185 to 181 million tonnes (-2.2%). These cuts were achieved as a result of more efficient use of energy and a greater reliance placed on nuclear power, particularly in France.

Despite the recommendations of the 1988 Toronto Conference, and in view of the severe winter recorded 1989-90, the first results suggest that the Americans have failed to cut back on continued increases in carbon dioxide emission levels.

It appears that a government committed to a policy of environmental protection and efficient energy use is necessary, as are legislation backed by sanctions and severe penalties against offenders, improved community education and individual consumer initiatives. The implications for human health are tremendous.

The Proof

Circumstantial evidence from the geologic and historical past confirms that there is a link between climate change and fluctuations in the levels of greenhouse gases. Between 3.5 and 4 billion years ago, the sun is believed to have been about 30 per cent fainter than it is today. However, the temperatures achieved within the atmosphere permitted sedimentary rock formation and the evolution of elementary life forms, although the surface temperature may have moved little above 0° C. Research workers have proposed that the early atmosphere contained as much as 1000 times the level of carbon dioxide detected in this century. These levels compensated for the weak solar radiation creating an enhanced greenhouse effect.

Increases in solar radiation during the Mesozoic eras are believed to have created an enhanced greenhouse effect which fossil evidence suggests was perhaps 10 to 15 degrees warmer than today. This was the age of the dinosaurs. At the time, 100 million years ago and more, the continents occupied different positions from those they do now, altering the circulation of the oceans and perhaps increasing the movement of heat from the tropics to higher latitudes. Calculations by Eric Barron of the Pennsylvania State University and others, however, suggest that paleocontinental geography can account for no more than half of the Mesozoic warming.

Russian scientists argue in support of Barron and his colleagues that increased carbon dioxide can readily explain the extra heating. They have constructed a geochemical model which suggests that carbon dioxide may have been released by unusually heavy volcanic activity on the mid-ocean ridges, where new ocean floor is created by upwelling magma.

Direct Evidence

Direct evidence, linking greenhouse gases with the dramatic climate changes of the ice ages, comes from bubbles of air trapped in the Antarctic ice sheet by the ancient snowfalls that built up to form the ice. A French team, headed by Claude Lorius of the Laboratory of Glaciology and Geophysics of the Environment near Grenoble, examined more than 2000 metres of ice cores recovered by a Russian drilling project at the Vostok Station in Antarctica. Laboratory analysis of the gases trapped in the core showed that carbon dioxide and methane levels in the ancient atmosphere varied in step with one another, and, more importantly, with the average local temperature as determined from the ratio of hydrogen isotopes in the ice water molecules.

During the current interglacial period of the past 10,000 years and the previous one around 130,000 years ago, the ice recorded a local temperature about 10 degrees warmer than at the height of the ice ages. At the same time, the atmosphere contained about 25 per cent more carbon dioxide and 100 per cent more methane than during the glacial periods. It is not clear whether the greenhouse-gas variations caused the climate changes or vice versa. However, scientists consider that other factors such as changes in the earth's orbital parameters and the dynamics of ice build-up and retreat, together with biological changes and shifts in ocean circulation, affected the level of trace gases in the atmosphere thereby amplifying any climate changes.

Formal climate records of greenhouse gases and global climates during the past one-hundred years indicate that a further 25 per cent increase in carbon dioxide has occurred above the interglacial level, and another doubling of atmospheric methane. However, attempts to construct models of global average temperatures during the past century are hampered by effects such as the 'urban heat island' phenomena. Readings from city centres tend to be skewed by heat released from machinery and industry, or stored by buildings, pavements and concreted areas. Corrections would add a further half-degree Celsius of unexplained 'real' warming. In keeping with the trend indicated, the 1980s appear to be the warmest decade on record.

It is tempting to suggest that these are the first signals of greenhouse warming, but the evidence is not sufficiently definitive. For one thing, instead of the steady warming anticipated from the progressive build-up of greenhouse gases, the record shows rapid warming until the end of World War II, a slight cooling through the mid-1970s and a second period of rapid warming since that time.

Global Climate Models
Predictions of the effect of increased build-up of carbon dioxide and other greenhouse gases on the climate are based on historical and geological records. These involve a complex interaction between atmosphere, oceans, land surfaces, vegetation, polar ice and events occurring within the solar system, including asteroid impact, sunflares and galactic electromagnetic disturbances. Early computer predictive climate models are becoming increasingly sophisticated.

The most complicated thus far are the global circulation models (GCMs). Developed at Princeton University's Geophysical Fluid Dynamics Laboratory, the National Center for Atmospheric Research (NCAR) and the Standard Institute for Space Studies in the United States, GCMs were intended for long-range weather forecasting. In these models, the atmosphere is represented as a three-dimensional grid with an average horizontal spacing of several hundred kilometers and an average vertical spacing of several kilometres. Climate is calculated only at the grid-line intersections.

The results of most GCMs are in rough agreement: a doubling of carbon dioxide or an equivalent increase in other trace gases would warm the earth's average surface temperature by between 3.0 and 5.5° C. A change of this size would match the five-degree warming which has occurred since the peak of the last ice age 18,000 years ago, but effectively has been achieved between 10 and 100 times faster. Unfortunately, the shortcomings of computer models limit the reliability of the GCM predictions. Many processes that affect global climate are simply too small to be incorporated and reflected within the models. Atmospheric turbulence, precipitation and cloud formation are all important climatic processes but occur on a relatively small scale which becomes swamped within the average humidity, temperature and cloudiness applying to the much larger grids of which they form components. A strategy called parameterization has been developed to effectively aggregate such small-scale phenomena that could act as feedbacks on climatic change, amplifying or moderating it, on the ground.

Clouds, for example, reflect sunlight back into space thereby tending to cool the climate. However, they also absorb infrared radiation from earth which creates a warmer effect. The influence which dominates at any one time depends on the cloud's brightness, height, distribution and depth. Recent satellite measurements suggest that clouds currently have a net cooling effect and that the earth as a whole would be much warmer under cloudless skies. Climatic change may have incremental impact on cloud character, altering the nature and amount of feedback. Present models, crudely reproducing only average cloudiness, fail to incorporate these and similar other feedback mechanisms.

Another deficiency with parameterization is apparent in their crude treatment of the oceans. The oceans exert powerful effects on the present climate and potentially influence future climate. Their enormous thermal acts as a sponge which slows any initial increase in global temperature while the oceans themselves warm up. The magnitude of the effect is likely to depend on ocean circulation which, in turn, may change as the earth warms. In principle, a climate model should couple a simulated atmosphere with oceans whose dynamics are simulated in equal detail.

The computational challenge, however, means that in most GCMs applied to greenhouse warming, the dynamics of the oceans are simplified, treated at coarse resolution or omitted entirely. This limits the reliability of any global forecasts and prevents the models from giving a definitive picture of how climate might change over time in specific regions. As long as the oceans are out of equilibrium with the atmosphere, their thermal effects will be felt differently at different places. An area in which there is little mixing between surface waters and cold, deep waters might warm quickly; high latitude regions where deep water is mixed up to the surface might warm more slowly. These thermal effects could in turn affect wind patterns, thereby altering other regional variables including humidity and rainfall.

These qualifications apart, climatologists tend to place considerable confidence in the global surface temperature predictions of their models. The skill of a model as a whole and its ability to account for relatively fast processes such as changes in atmospheric circulation or average cloudiness, can be verified by checking its ability to reproduce the seasonal cycle — a twice-yearly change in hemispheric climate that is larger than any projected greenhouse warming.

In spite of parameterization, most GCMs map the seasonal cycle of surface temperatures quite well, but their ability to simulate seasonal changes in other climatic variables, including precipitation and relative humidity, awaits further study.

During the course of decades there are slower processes than temperature increase occurring with global warming and they do not affect seasonal cycles. Changes in ocean currents, in the extent of glaciers and the circulation of air masses come into play. Simulations of past climates serve as a check on the long-term accuracy of climate models. To these are being added simulations of the climates of other planets such as Venus, where a dense greenhouse atmosphere maintains a surface temperature of about 450° C.

The records of the past one-hundred years provide the only direct test of the models' abilities to simulate the effects of ongoing greenhouse gas increase. When a climate model is run for an atmosphere with the composition of a century ago and then re-run for the historical 25 per cent increase in carbon dioxide and doubling of methane, models suggest temperature increase one degree higher than that recorded.

Explanations for the discrepancy suggest the heat capacity of the oceans might be larger than current models appreciate, that there has been a decline in the sun's output or that volcanic activity has injected greater amounts of dust into the stratosphere than was first recognised. The level of sulphur emissions into the atmosphere as sulphur dioxide from coal and oil-burning factories and power plants may be a further factor which limited greenhouse warming.

The Importance of Predictions
Changes in temperature and precipitation are required because they warn of future threats to natural ecosystems, affect agriculture production and have implications on human settlement patterns. For example, temperature has a substantial influence on forest types growing in a given geographical area. Ten thousand years ago, at the end of the last ice age, the spruce and fir forests covering much of the North American continent today were located further south and in the front of the vast ice sheet which covered present-day Canada. As the climate slowly warmed and the ice retreated, the forest belt migrated northward at a rate calculated to be around one kilometre per year. Forests probably could not sustain the much faster migration required by the projected warming, and many ecosystems were unable to migrate as easily; they existed only in preserves and niches which might become marooned in a newly inhospitable climatic zone.

The evaporation of moisture and reduced rainfall and runoff would have a major impact of human activities. A temperature increase of several degrees could decrease runoff substantially even if precipitation levels were maintained. Faster evaporation and a shortage of stream runoff water would place an increased demand on irrigation, further straining water supplies. At the same time, water quality would be likely to suffer as the same waste volume was diluted in lower stream volume.

Several climate models predict that in the United States summer rainfall is likely to decline, particularly in those areas of mid-continental USA which are the foci for grain and cereal production. American scientists estimate that the effects of a three-degree warming, combined with a 10 per cent drop in precipitation, increased crop water requirements and a reduction in available water for irrigation, could mean a fall in viable acreage of about one-third in arid regions of the western states and Great Plains.

Rising Sea Levels — A Present Reality
Early in 1987, exceptional flooding swamped many of the small islands which constitute the Indian Ocean nation of the Maldives. The Maldives consist of a double chain of twenty-six atolls. The large, ring-shaped reefs are studded with around 1300 tiny islands whose number fluctuates as sandy islets are washed away or become stable islands. In dimensions, these islands average 1 or 2 square kilometres and lie between 1 and 1.5 metres above mean sea level. The highest point in the chain is only 3.5 metres. The peoples of the Maldives tend to live in the zone between 0.8 and 2 metres above mean sea level.

The Maldives have become a symbol of the destructive power of global warming and a case study for those looking for ways to protect low-lying countries from rising seas. Scientists and politicians are still unclear about the gravity of the threat, but few dispute that sea-level rises will occur.

Since the end of the last ice age, 18,000 years ago, sea levels have been slowly rising, sometimes by as much as 10 to 20 millimetres per year. For the past century, the water has risen about 1.2 millimetres a year. There is general agreement that global warming should accelerate this rise. Some early models suggest a rise of about 65 centimetres by the year 2030, or almost 3.5 metres by the beginning of the twenty-second century. However, studies at the Climatic Research Unit at the University of East Anglia in Britain provide a best estimate of between 15 and 30 centimetres by the year 2030.

The results on the Maldives provided by the Centre for Tropical Coastal Management Studies at Newcastle upon Tyne, and those provided by Datum International Consultants for the Pacific nations of Tuvalu, Kiribati and Tonga, illustrate the problems and possible solutions for these countries. Both Kiribati and Tuvalu are as low as the Maldive Islands. Tonga is volcanic and hilly but Tuvalu totals 23 square kilometres only, comprising five atolls and four separate reef islands.

The peoples of these islands depend for their living on the seas, and until recently the lifestyle was subsistence in nature. Population pressures, particularly in the Maldives where annual growth is around 3.1 per cent, place heavy demands on land. At the same time, modern port and public facilities such as hospitals and schools place additional demands on the available land.

Scientists believe that as the climate warms, tropical storms will be of increasing intensity resulting in higher tides, deeper swells and bigger waves. These will erode more of the coastal areas and are expected to cause more serious flooding.

At the moment, the Maldives do not suffer directly from tropical cyclones but their shores receive the waves from distant cyclonic activity. In 1987, the swell from distant storms flooded the capital: a similar event flooded the western islands in 1988. While many of the coral islands are more used to coping with these emergencies, the islands of the Pacific are also made from rubble thrown up by repeated storms.

Gradually, this rubble weathers into a solid mass which is much more resilient than the loose sand that comprises much of the Maldives. Tropical cyclones may bring long-term benefits. For example, in 1972, cyclone Bebe swept across Tuvalu and overnight threw up a bank of rubble 19 kilometres long, creating new land behind (Wells and Edwards, 1989).

The increase in storm intensity contributes to problems associated with rises in sea level. Poorly planned coastal developments and damage to coral reefs further aggravate the situation. Wave action continually moves sand and re-shapes the small islands, and loss of beaches is a recurring problem faced by tourist resorts. The building of protective harbours, sea walls, groynes and jetties alters the natural movement of the water and results in erosion further along the coast.

Artificial breakways and causeways are expensive to build and repair. Coral reefs, vitally important as natural breakwaters, act as habitats for indigenous marine species and bring benefits to the tourist industry. At the same time, however, divers, snorkelers and pleasure boats damage and break up coral reefs. Builders break up and cart away the tops of reef for use in the construction industry. This is a particular problem for Male, the capital of the Maldives, where it has been anticipated that at the current rate of mining, the supply of coral from the inner reefs of North Male Atoll will be exhausted in thirty years. Thus, tourism, which is partly responsible for the erosion of the shores, will also suffer from it.

Rising sea levels also damage freshwater aquifers on the Pacific and Maldive Islands. Over-use of ground water has followed the boom in tourism. In islands such as the Maldives, fresh water is captured and held by the limestone and coral sands of the islands. These resources are shrinking rapidly on Male, and, as they are used, sea water penetrates the area formerly filled with fresh water, so that the supply of drinking water is an imminent problem.

Increasing salinity also affects agriculture. Poor soil on the islands means that there is little cultivated land. The Maldives, Kiribati and Tuvalu have already lost crops due to extra high tides and increased salinity. The basic staple, taro, is especially vulnerable to inundation and rising salt. Salt penetration also affects other root crops and fruit trees such as the mango. Increased importing of foods would greatly increase costs of living for the islanders, if crop loss became permanent through misuse of water or rising sea levels.

The Maldives government is aware of the problem and is implementing improvements. The use of rainwater is being promoted to improve supplies of drinking water, and to this end roads are being paved and drainage provided to soakaways. On the old, compacted earth roads, rainwater collected in puddles and was lost through evaporation. Desalination plants, though expensive, are being installed in tourist developments where costs can be passed on to the customers.

Islanders are also looking for new crops, tolerant of salt water and with shallow root systems to tap fresh rainwater. Both coconuts and breadfruits are relatively tolerant of salt. In Tuvalu, trials are being conducted on varieties of salt-tolerant sweet potato, and hydroponic gardening is under consideration.

Adaptation is also being called for within the fishing industry. Tuna is the main protein staple in the Maldives, and provides fifty-five per cent of the country's export income. The industry has expanded greatly in the past fifteen years with the use of mechanised boats, and the development of canning and freezing plants. While the Maldivians have little use for the fish dependent on the reefs, the small fish make good bait for tuna and tourists like eating exotic species. A rise in sea level would probably have little impact on big fisheries, but as the water warms, the movements of tuna are likely to change and the supplies of bait would be reduced by the death of drowned coral.

Most important of all, however, for the Maldives and many of the Pacific Islands is action to curb the growth of population. Measures are needed at once to ensure that the equilibrium between resources and population is restored here as in many other parts of the world.

Disputes About Greenhouse
But what if climate did not change and if earth and nature brought into play natural forces and mechanisms not fully appreciated by humans? Let it not be assumed that all climate scientists, or even the majority, concur about green-house gas and temperature change, climate patterns and global weather predictions. Dispute remains; and this is good, because it offers opportunities for developing new theories and measures. For example over temperature rises. To take one case in point: although many scientists predict global surface temperatures will rise by between 1.5 and 4 degrees within the next 30 to 50 years, some dispute these claims. Bryant, in the *Australian Geographic* (March 1989), says that present predictions are based on inadequate computer models, and that the evidence which scientists consider suggestive of global warming and sea level changes may simply be the result of natural climatic fluctuations. In support of his claims, Bryant argues that the majority of readings suggesting temperature increases have come from land-based city measuring stations. These are contradictory, he claims, because they result from population increase and concentration in the confined areas of these

large cities which generate heat and influence their own micro-climates.

Climate Change Response on a Global Scale

Beyond the local specific level as witnessed by the problems of sea level rise in small island states, global and international responses to climate change phenomena can be classed into three distinct types: technical, economic and preventive. These issues are the subject of more in-depth discussions in the concluding stages of this book when they are examined in the light of knowledge of the environmental effects on human health.

For example, the American Association for the Advancement of Science panel on climate change believes that action on water, an essential for life, must be immediate. It proposes that governments at all levels should re-examine the technical features of water systems and the economic and legal aspects of water supply management in order to increase the systems' efficiency and flexibility. As the climate warms and precipitation and runoff patterns change, water shortages will become more common and needs for regional transfers more complex. Even if climate did not change, more flexible water systems would make it easier to cope with the normal extremes of weather.

Update on Ozone Loss in Antarctica

In late 1989, the prestigious journal *Nature* reported that an ozone hole which reaches its largest extent over Antarctica in the spring was, in geographic terms, considerably larger than in 1988 and that it has appeared at an earlier date. The scientists based their conclusions on measurements made by an ER2 research aircraft on a flight between California and Chile. The measurements showed that at all southern latitudes down to 50 degrees, as much as 15 per cent of the stratospheric ozone was lost in August and 30 per cent in September. This is outside the region subject to polar temperatures.

Above Antarctica itself, chemical reactions that take place on the surfaces of ice crystals release chlorine in an active form which readily combines with available ozone. Outside the polar regions, however, there are few ice crystals. One explanation for ozone loss in these areas is that active chlorine is being released by similar reactions taking place on the surfaces of sulphuric acid droplets.

These effects are particularly important in countries like Australia, which presently has the highest rate of skin cancer in the world. Over 140,000 cases of skin cancer were diagnosed in 1981 alone and 80 per cent of these from melanoma. Medical experts predict that people face an increasing risk of damage, skin cancer and associated deaths due to increased levels of ultraviolet radiation reaching earth's surface.

TABLE 3.1: Health Implications of Climate Change Attributable to Greenhouse Effects

Health Effects	*Cause*
increase in skin cancer	increased UVR from ozone layer depletion
deaths from starvation and malnutrition	drought, floods
deaths from exposure	extremes of climate/weather
spread of vector-borne diseases	increased spread of vectors (mosquitoes, etc.)
increase in water-associated illnesses	increased pollution, reduction in quality and quantity of water, post-disaster flooding
increase in accidental deaths	breakdown of roads and other infrastructure
decreased ability to provide medical care	reduced financial resources for health due to other demands on economy
increased respiratory disease	increased air pollution outside (ozone) and inside: increased allergens & air-borne bacteria
deaths from heat exhaustion	increased temperature and humidity allied with air pollution
increase in crimes of violence	prolonged period of high temperature

Source: In Touch, *newsletter of the Public Health Association of Australia Inc., November 1989, Vol.6, No.4, p.4.*

TABLE 3.2: Annual Emissions of Carbon Dioxide
(millions of tonnes)

	1986	2030	Av.Annual Growth
World total	5,575 (100)	18,184 (100)	2.7
United States	1,299 (23)	3,257 (18)	2.1
Soviet Union	1,030 (18)	2,940 (16)	2.1
China	621 (11)	1,218 (7)	1.5
Developing countries	1,452 (26)	5,891 (32)	3.2
Japan	260 (5)	419 (2)	1.1
South-east Asia	146 (3)	532 (3)	3.0

Figures in parenthesis indicate percentage of world share. China is included as a 'developing country'.

Source: MITI, Japan

References and Additional Reading

Hsu, K.J. *et al.* 'Mass Mortality and Its Environmental and Evolutionary Consequences' from *Science*, 216, 1982, pp.249-56.

Graedel, T.E. & Crutzen, P.J. *Scientific American*, Sept. 1989, p.28.

Krause, F. & Bach, W. *Energy Policy in the Greenhouse*, Vol.I, International Project for Sustainable Energy Paths, El Cerrito, California, 1989.

Lewin, R. 'Extinctions and The History of Life' from *Science*, 221, 1989, pp.935-7.

New Scientist, no.1893, 2 September 1989, p.8.

Raup, D.M. & Sepkowski, J.J. 'Mass Extinctions in the Marine Fossil Record' from *Science*, 215, 1982, pp.1501-2.

Scientific America, September 1989, p.38.

Wdowiak, T. Special Paper no.247, 'Global Catastrophes', Proceedings of the Geological Society of America, 1990.

4.0 EFFECTS

4.1 General Effects

How do we recognize the first changes which occur in and to the body as a result of disease, infection, physical and environmental agents or disturbance to metabolism? This section contains chapters which describe these diseases. It begins by detailing the mechanisms by which biological, physical and chemical factors in the environment pose a risk to human health.

Five chapters discuss these risks and dangers under the headings of cardiovascular disease, cancers, the skin, the respiratory system and the senses. A sixth chapter, more general in nature, concludes the Section. It discusses a number of conditions often dismissed as psychosomatic but which are part of modern life, stress and society. Many of these, such as chronic fatigue syndrome and seasonal affective disorder, are not fully understood though they are the subject of research. The whole domain of environmental health and illness is only beginning to set its boundaries and, like much at the frontier of science, awaits more detailed findings, analysis, testing of theory and resolution of contradictions.

Disease Trends in Modern Societies
In the 1850s, a quarter of those people born in western populations were dead by the age of five years; another quarter died before reaching forty years. Today only one per cent die in childhood and 2 per cent before the age of forty; another 30 per cent die in 'middle age', and the rest reach their seventies and beyond. In the late 1980s and into the 1990s, the four big 'killers' are cancer, coronary heart disease, stroke and, in young adults, road traffic accidents.

Death represents the final outcome of a series of events marked by precipitants and causes. Deaths from infectious diseases, both acute and chronic, declined steadily in the first half of this century as a result of improvements in public, personal and clinical hygiene, and nutrition. Immunisation and antibiotics made a further contribution. In those times, the environment in which people lived included the pathogens, vectors and agents that caused disease and permitted it to flourish.

Today, the environment, in its myriad forms, plays a significant role as a precipitant and contributor to the end event. The diseases and illnesses of the majority reflect aspects of life in the late twentieth century: first, the greatly increased proportion of individuals who survive beyond middle adulthood; and second, the greater occurrence of degenerative disease processes caused by increased consumption and sedentary living. The diseases and illnesses of the minority who do not participate in or share the general affluence also include conditions directly related to unemployment, poverty and degradation, which in themselves weaken the life chances of these individuals.

Psychological Stress and Human Health

There is growing evidence that social and environmental stressors increase susceptibility to a wide range of diseases, and are common causes of both morbidity and mortality. Nevertheless, stress is only rarely considered to be a significant risk factor for disease in the public health arena. As a result, attempts to develop rational community-based strategies and policies to deal with stress are isolated and uncoordinated.

There are a number of reasons to explain this reluctance. First, the strength, validity and reliability of the data relating stress to many disease outcomes has been highly variable. Secondly, stress management programmes often focus on altering the individual response to stress, rather than on changing the social environment to either reduce stressor load or improve the range of coping resources available in the community. Often this type of activity is assumed to fall within the realm of curative policies. Third, stress is usually considered synonymous with major and not minor and recurring daily life events. The assumption is made that they are a necessary part of modern life. Consequently little is done to identify and remove them.

Stressors can be defined as sociological, psychological and physiological problems with which an individual has been burdened, with demands exceeding their potential ability for adaptation. An alternative definition identifies sources of stress as demands for which the individual has no readily available or automatic response, and which consequently disturb that person's emotional stability.

These two definitions highlight important facets of stress theory: that the stress-illness relationship is mediated not only by the presence, duration and intensity of environmental stressors but also by the individual's ability to cope with them. For a given level of exposure to a range of stressors, it is argued that adequacy of coping will be the major factor in determining whether an individual will proceed to develop an illness.

There is evidence that personality, exercise, social group, occupation, control over resources and coping style all play a role in mediating the stress-illness relationship. Social networks and social supports also appear to be important in relation to illness and health, but whether social supports have a direct influence on health or act only as buffers between stress and illness has yet to be established.

Other problems require the attention of scientists as well as psychologists: explaining how the immune system is undermined weakening normal body resistance to disease, the extent to which the social environment contributes to or undermines host resistance, and whether there are limits to individual stress management.

Biological Factors Affecting Human Health

The Pathological Process: There are many changes which occur in body cells and tissues prior to the development of detectable and specific symptoms and obvious signs of disease. It was the pioneering work of the early bacteriologists — Ehrlich, Koch, Pasteur — who, aided by the newly discovered and improved light microscope, first observed and recorded the series of pathological or degenerative changes which occur in the cell, as it passes from a normal healthy state to diseased, debilitated and eventually to die.

Cell Degeneration: Cell degeneration occurs when there is damage to cell structure or function which is, at least potentially, reversible. These changes are often first observed in the densely cellular, parenchymatous organs such as the heart, liver and kidneys. Understandably, due to the difficulty of taking a biopsy or sample of such tissue, detection of any changes or lesions in these internal tissues is particularly difficult. Modern medicine is forced to rely on blood and tissue samples which can be readily collected and examined.

There are a number of changes and processes associated with development of disease, within individual cells and in the whole human body. These range from the observed reactive processes such as inflammation (including the formation of vesicles, abscesses and granulomas), pyrexia and pyemia, to the degenerative processes of amyloidosis, emaciation and atrophy, to the broader systemic changes characterised by bacteraemia, septicaemia and toxaemia. Together with cell changes such as ulceration and cell necrosis (death), these processes describe the body's range of responses to challenge. This challenge may take the form either of invasion by micro-organisms and viruses or the presence of substances and conditions which affect the survival of human tissues.

Inflammation: Inflammation describes the dynamic processes by which living tissues react to injury, in particular the vascular and connective tissues. The inflamed area undergoes continuous change as the body repair processes start to heal and replace injured tissue. Thus, inflammation is a conservative process modified by whatever produces the reaction. It should not be confused with infection, an altogether different phenomenon, although one may arise from the other.

Within four primary categories the variety of agents which may cause inflammation can be grouped under the following headings:

Physical — trauma, heat or cold, radiation.

Chemical — simple chemical and organic poisons.

Infective — bacteria, viruses, parasites.

Immunological — antigen-antibody cell-mediated.

Outside of these any situation that leads to tissue damage must be considered. These may include vascular or hormonal disturbances.

The inflammatory reaction takes place in the surviving adjacent vascular and connective tissues. The initial stages are known as the acute inflammatory reaction. Where the process is prolonged the inflammation may be subacute or chronic. The classic signs and symptoms of acute inflammation are pain, redness, heat, swelling, loss of function and also headache, loss of appetite and a general feeling of discomfort. These gross signs are explained by changes occurring at microscopic levels.

In acute situations where exposure levels to a toxic agent occur, where infection is substantial or injury to the tissue high, death of local tissue results. In some circumstances, further damage might occur at other sites in the system and inflammatory processes there progress to cell and tissue death.

In tissues which are acutely inflamed and where cellular damage is relatively minimal and reversible, tissue death does not occur. If the causative agent is eliminated by the inflammatory reaction — for instance, the bacteria in an upper respiratory tract infection, contact with a allergic substance — or if the agent acts only once, such as in acute sunburn or exposure to nuclear irradiation, the acute inflammatory reaction is terminated by resolution. (Resolution is the term applied to the process by which the tissue returns to normal following acute inflammation.)

Tumours: Neoplasms or tumours arise when, for some reason, tissue cells multiply to produce an expanding, uncontrolled growth which lacks the normal limits which cells impose on their own rates and form of replication. In general, tumours are more common in older individuals, and in humans tend to be part of the ageing process occurring within cells and the body as a whole.

The original and precise meaning of the word tumour indicated any form of swelling, but in current usage it is equated with 'neoplasm' (i.e. new growth), and in lay terms is considered to be 'cancer'. Tumours are named according to the tissue of origin and whether the growth is judged benign or malignant.

The phenomenon of metastasis occurs when one or more tumour cells detach from the main mass, travel to a distant site and set up a secondary growth. There are two general groups of tumours, carcinomas and sarcomas, though the two terms are often used interchangeably.

Tumours can spread along various natural passageways within the body, but as a general rule carcinomas tend to spread via the lymphatic system, whereas sarcomas tend to spread via the blood stream. Common sites for metastases are organs with a large number of small capillaries, e.g. lungs, kidneys, liver and the brain.

Diseases from Bacteria and Viruses
Bacteria and the related organisms mycoplasmas, rickettsiae and chlamydia, are generally unicellular and many possess the capacity to cause disease in humans. This occurs by a number of ways: the production of either toxins or damaging enzymes, competition for vitamins and other essential body substrates and metabolites, and stimulation of self-destructive host responses. Bacterial toxins may be produced by pathogenic bacteria (exotoxins) and be present in food or water consumed by humans, or be components of the cell wall of invading bacteria (endotoxins).

Viral diseases occur in a wide range of living species including animals, fish, plants and insects which humans either eat or with which they interact (Table 4.1). Resistance of humans and animals to these diseases is compromised by stress, poor living standards and sub-optimal local weather conditions — in other words, by components of the physical, social and natural environment.

Fungal Diseases and Moulds
Fungal infection can result in systemic disease, local infestation or superficial contamination. Fungi thrive in warm, moist conditions and the systemic diseases, although usually sporadic in occurrence, are more common in tropical and subtropical areas. In temperate climates the main indication of fungi is mould growth on meat, due to post mortem contamination by fungal spores, and of particular significance on imported chilled or frozen meats. Because these moulds are moisture dependent, low temperatures alone are insufficient to contain their growth. Thus poor refrigeration with problems of condensation will predispose to mould growth, easily recognizable by a distinctive musty smell and taste. Fortunately most such mould growths are non-toxic.

Under certain conditions, fungi may elaborate toxic metabolites, called mycotoxins. The production of toxin depends on a number of factors including the growth substrate, microbial competition, relative humidity and environmental temperature. Serious outbreaks of mycotoxicosis tend to be more common in warm, damp climates, a significant problem in grains, grasses and other cereal products. Disease occurs as a result of ingestion of fungal toxins produced either within the fungus itself or the substance on which it is growing.

Mycotoxicosis: Many fungi are capable of producing toxins, of which the most important are those associated with *Aspergillus* spp. and *Penicillium* spp. The toxins can be roughly classified according to their main target organ, although many mycotoxins produce more than one type of disease syndrome and there is some species variation in susceptibility.

Liver toxins include aflatoxin, which has a spectrum of effects due to binding to nucleic acids and nucleoproteins in the cell so that there may be tumour formation, embryonic defects or inhibition of cell division. At high dose rates there can be pallor, enlargement, fibrosis and nodular hyperplasia of the liver. In severe cases, there may be bile-stained ascites and rarely there is massive hepatic necrosis.

The subclinical effects of mycotoxicosis have not been well defined. Prolonged exposure to low levels of toxins can be associated with reduced growth rates or infertility in cattle and fodder animals. It is possible that similar effects may occur in human populations dependent on locally harvested and stored cereal foods, such as poor farming communities in developing countries.

Factors Affecting the Growth of Micro-organisms
The conditions that affect metabolism and multiplication of micro-organisms include time, nutrients, temperature, moisture, pH, oxygen, radiation and inhibitory substances, either natural or artificial. In some bacteria the time between cell divisions may be as little as eight minutes, which could result in the production of millions of cells within a few hours. Usually, this rapid rate of growth cannot be sustained for more than a few hours because the cells start to die as a result of depletion of nutrients and the accumulation of toxic waste products. This follows the typical growth cycle which includes lag, log, stationary and decline phases.

To achieve energy for growth and metabolism, all micro-organisms require nutrients in the form of carbon-containing substances, nitrogen, vitamins and minerals. The availability of individual nutrients and the demand for specific minimal quantities of individual different components are limiting factors on growth.

Temperature also influences the rate of multiplication: each organism has its own minimum, maximum and optimum temperature range for life. Psychrophilic (low temperature), mesophilic and thermophilic (high temperature) organisms abound in relative profuseness, governed by day and night temperature norms on a seasonal or annual basis. Particularly in mountainous areas where glacial retreat is occurring slowly, a two-degree increase in global temperatures would be expected to favour mesophilic and thermophilic bacteria, at the expense of those thriving in low-temperature environments.

Potential human pathogens have an optimal range of 35-40° C. Psychrophiles, found in oceans and soil, can grow at very low temperatures and commonly spoil fish and other foods stored at refrigeration temperatures.

Moisture/Water
Water is the largest component of all living cells. It is essential to sustain metabolism, but in addition micro-organisms require an aqueous environment to transport food to the cell and waste products from it. For these functions, water must be chemically free and in liquid form. Water crystallised into ice or chemically bound in strong sugar or salt solutions is unavailable to micro-organisms. The proportion of available water in food or in solution can be described in terms of its water activity.

Changes in climate do not affect the individual cells, but may place a heavier demand on living organisms for water to maintain the normal tissue environment where there is heavy loss through sweat, perspiration and other body-cooling mechanisms.

The majority of bacteria are neutrophils; that is, they prefer living in environments near pH 7 and are unable to grow below pH 4.5 or above pH 9.5. Since acid is usually produced during metabolism, pH in a closed system usually falls until growth ceases. Highly acid or alkaline environments kill most micro-organisms. Despoliation of the environment has occurred in eastern Europe as a result of the discharge of raw factory and industrial wastes into the surrounding countryside or into local streams and waterways. The material discharged is often highly acid or alkaline and upsets the delicate natural cycle mediated by these bacterial species.

A similar situation can occur where extensive land clearance has laid bare the soil. Winds remove the surface top soil, rain, temperature extremes and time remove essential nutrient and particular matter and upset the acid-alkaline balance associated with vegetation and the presence of micro-organisms. A vicious cycle of salination and soil erosion further degrades the land.

Intestinal bacterial species alone have become highly resistant to the acid of the stomach and the alkali of the bile. Yeasts and moulds prefer a more acid environment than bacteria and have an optimum pH range of 4-6. They are often found as contaminants on acid foods, especially preserved meats.

Social Effects and Major Disease

Changes in Mortality From the Major Diseases: The health risks to western populations have radically changed from those of eighty years ago and it is now possible to point to the virtual elimination of the infectious diseases of diphtheria, polio, tetanus and tuberculosis which curtailed life and lessened total health status. Lifestyle and social environment, living conditions and standards of community cleanliness and knowledge continue to influence human health. These actions may be direct, through damage such as that caused by a poor diet, or indirect, such as by encouraging the growth of disease-causing micro-organisms. Alternatively, lifestyle may weaken the body's resistance to disease. Sun-bathing without protection, careless use of pesticide chemicals and handling of nuclear wastes weaken the body's natural healing mechanisms.

Physical Factors and Human Health

Oxygen: Oxygen is required by all living organisms and is obtained either directly from the air or as a result of chemical oxidation processes. Although some compounds can oxidize inorganic materials such as nitrate, sulphate or carbonate, most utilize simple carbohydrates — the sugars — by one of two basic methods, respiration or fermentation. Strict aerobes utilise only respiration and are unable to grow except in the presence of oxygen. All moulds and most common bacteria come into this category.

Anaerobes typically predominate in swampy, marshy areas, often the sites of waste and detrital material. Strict anaerobes are unable to grow in the presence of oxygen, whereas micro-aerophiles require only a small amount. Facultative organisms such as yeast and intestinal bacteria are adaptable and can use either respiration or fermentation as conditions allow.

Radiation: Micro-organisms can be inhibited by various forms of radiation. The effects of infrared and microwave radiation are primarily due to the heat generated during the process. Gamma irradiation is bactericidal if it is sufficiently intense and applied for an adequate time period. Ultraviolet rays can also be used as an effective anti-bacterial treatment but ultraviolet radiation does not penetrate despite the surface damage it induces in living tissues.

Radiation-induced lung cancer has a long history in European miners.

The problem persisted into the twentieth century amongst ore miners of nickel, arsenic and cobalt, all of which are carcinogens. A similar situation prevailed amongst miners of pitchblende at Joachimstahl in Germany, the very mines that provided Madame Curie with the pitchblende from which she first isolated radium.

Cancer is known to be a long-term result of the atomic bomb explosions at Hiroshima and Nagasaki, where the exposure to radiation was massive over a relatively short period of time. Concern has more recently focused on the possible cancer effects resulting from extended (chronic) low-level radiation exposure, the type of exposure frequently associated with a variety of occupations such as dental nursing and radio-pharmaceutical manufacture.

Ionising radiation, of the type associated with X-ray machines and exposure to radioactive (physically unstable) substances such as uranium and plutonium, involves exposure to very high energy particles. In the simplest case of X or gamma radiation, these particles are packets of energy known as photons emitted by excited atoms as they return to normal energy levels. In the case of radioactive substances, other particles (termed alpha and beta particles) are emitted as unstable nuclei decay to form more stable products, called daughter products. Alpha particles consist of helium atoms stripped of their electrons, whilst beta particles are simply radioactive substances which occur naturally in the earth, and which have largely been responsible for the radiation hazard associated with the mining, particularly underground, of uranium. The inhalation of radon and daughter products by such miners has resulted in elevated risks of lung cancer.

Despite the now very obvious cancer hazard associated with exposure to ionising radiation its history is littered with large-scale ignorance of the danger involved. The emergent use of X-rays as a diagnostic medical tool at the beginning of this century was associated with an appalling cancer toll amongst both treated patients and, in particular, the doctors and technicians using the technology. Indeed the pioneer of the X-ray apparatus, Wilhelm Roentgen, died of cancer as a result of his own massive exposure to ionising radiation. It took many years for the cancer hazard to become fully appreciated and several years after that before the first controls over exposure were introduced. Ionising radiation in workplaces is now one of the better controlled cancer agents.

Radionuclides: Fall-out resulting from the testing of nuclear weapons has caused concern in recent years, and the accident at Chernobyl nuclear power station in the CIS (formerly USSR) in 1986 has heightened public awareness of the potential problems. Radionuclides are the products of atomic fission, examples include strontium-89, strontium-90, iodine-123 and caesium-137.

The most important foods acting as vehicles of transmission of radionuclides to man are those from milk and plants.

Following the Chernobyl accident, the growing crops in all agricultural land along the direct path of wind traversed by the escaped fall-out was stripped and the soil ploughed, fallowed and re-ploughed. All farm animals were either killed and their products, milk and meat destroyed. Stoned fruits and herbs harvested in neighbouring countries were found to be so highly contaminated with beryllium as to be rendered unfit for human consumption.

Radiation can certainly cause cancer in animals and humans, and it produces mutations. The most penetrating kind is gamma radiation. X-rays are similar. Tissues and organs show considerable variability in their sensitivity to radiation. Cells which are actively growing tend to be the most sensitive — for example, blood cells and sites where they are formed, and the rapid progress of leukaemias affecting children.

Breast cancer, thyroid cancer (particularly in women), lung cancer and cancer of the digestive tract have also been attributed to the effects of radiation. However, there is considerable uncertainty about the nature of the dose-response relationship in radiation-induced cancer. Its effects are believed to be cumulative so that prolonged low-level exposure might have an equivalent effect to a shorter period of more intense radiation. Potential effects of relatively low but continuous exposure and the lengthy induction period of cancer mean that the nature of the relationship will not be easy to clarify.

Early studies of survivors of the atomic bomb explosions in Japan demonstrated an increased incidence of leukaemia in the most heavily irradiated people who were nearest the epicentre of the blasts. Early results from Chernobyl suggest that direct, high exposure results in a number of deaths within the first fourteen days. A second batch of deaths occurred in those exposed at high levels, but within five kilometres of the reactor.

In Japan, a latent period of several years was then noted, similar to that which has been found in experimentally induced leukaemia in mice. The peak incidence of leukaemia occurred in 1959, fourteen years after the bombs exploded. The same waiting period for the Chernobyl disaster will expire in the year 2000. Children exposed to the blast are being closely observed and regular health checks undertaken by Russian doctors and other international experts.

Persons exposed to X-radiation for diagnostic or therapeutic purposes have also been studied. In a controlled study over several years, an increased risk of cancer, especially leukaemias, has been demonstrated in those children who were exposed to radiation *in utero* as a result of pre-natal X-ray examinations of their mothers. Children exposed to this radiation have twice as much risk as unexposed children of dying of cancer before the age of ten years. Ultrasound investigation of pregnant women does not carry the same potential risks to either child or mother.

Between 1934 and 1954 it was common to treat conditions such as ankylosing spondylitis (a painful fusing of vertebrae of the spine) by X-ray therapy. It was subsequently noted that among 14,000 patients treated by this method, there was a ten-fold increase in the incidence of leukaemia. Increased susceptibility to thyroid cancer has also been noted among children treated with X-irradiation for enlargement of the thymus gland. Women who have had repeated X-rays for tuberculosis and those treated with X-rays for *post partum* mastitis have increased rates of breast cancer over other women.

The routine use of X-rays for tuberculosis screening no longer occurs in Britain or Australia as was the case prior to the mid 1950s. This is a safety measure: the low doses used then were and are generally regarded as safe. They stand in marked contrast to the medium to high levels of irradiation involved in the procedures described earlier. They, in turn, are higher than the background level of radiation in the atmosphere.

In most developed countries, the background level of radiation to which all members of the population are exposed will make up 70 per cent of the radiation dose to which each individual is exposed. The remaining 30 per cent will result from various medical procedures such as dental X-rays and chest X-rays. Accidents, however, do occur as a result of machinery failure causing leakage from a nuclear power plant into the atmosphere, or by damage to a canister of nuclear waste during transport to a storage site. These will result in chance exposure of nearby individuals to irradiation, with the resultant risks to their health. Constant monitoring, care and compliance with safety procedures minimise the chance occurrence of these accidents.

Metals

Metals and minerals such as iron and lead are sometimes taken into the body with food, inhaled into the lungs or absorbed through the skin. On most occasions, only small amounts are involved and they are eliminated with relative ease. However, on occasions, metals and the chemical complexes they form with other natural elements may be present constantly or in large amounts in the individual's environment or workplace. In these instances, chronic exposure permits the metals to accumulate in the body beyond the levels considered safe. At this stage, specific toxicity symptoms become evident and the individual's health is threatened (Table 4.2).

Naturally occurring minerals can have a similar effect; asbestos, for example, consists of six minerals which often occur with metals or metallic complexes. Of the metallic elements, nickel, chromium and their salts are associated with the risk of cancer of the nasal sinuses; iron as haematite is associated with lung cancer; and cadmium has been linked to cancer of the prostate.

Minerals and metals are important because they may also contaminate food if utensils, pots and pans of unsuitable material are used in food preparation and storage. Zinc and antimony poisoning have occurred following the storage of acid food in galvanized iron containers or poor quality enamelware. The time between consumption of the food and the appearance of symptoms (usually acute vomiting) ranges from a few minutes to a few hours, with the patient often complaining that the food has a metallic taste. Heavy metals such as cadmium, mercury and lead are also accumulated from the soil during growth by some food plants and animals. Knowledge of the effects of these substances and the extent to which they may be present is important to the health of communities in environments where there exists a risk of toxicity levels being exceeded.

As early as 1953, fishermen on Minimata Bay in Japan complained of numbed muscles in the arms and legs and then impaired vision and speech. In 1959, doctors identified the cause as methyl mercury discharged by the Chisso Corporation into the bay. The methyl mercury was absorbed into the fish which were eaten by the people surrounding Minimata Bay. More than fifty people have died of methyl mercury poisoning since 1953. Children of women who were poisoned have been born deformed and with mental defects.

Also in Japan, cadmium discharged into the Jintsu River by the Mitsui Mining and Smelting Company has been implicated in the development of diseases characterised by a tendency of the bones to snap under the slightest pressure. Again people ate fish which had absorbed large amounts of a heavy metal, in this case cadmium.

Chemical Factors with Potential Human Risk

Chemicals in the Environment
A number of naturally occurring metals, minerals and trace elements which are part of the earth's physical structures of rocks, soil and their break-down products enter the human body directly with food, or indirectly through contact with the skin and via the lungs. Most of them are relatively harmless in their mineral form: others are extremely toxic, such as lead, zinc and nickel. Those which are actively mined by humans can present a serious problem to the workers and those living in settlements near mines, factories and associated industries.

The comfortable little city of Hobart, Tasmania, nestles on the banks of the River Derwent. The city's biggest employer is the Pasminco-EZ zinc refinery, which first opened in 1917 to meet the demand for zinc for use in munitions during World War I. With proven zinc resources on the west coast of Tasmania and the promise of hydro-electric power from the Tasmanian rivers, the site was considered ideal for investigations into the embryonic electrolytic production of zinc. The site was chosen because it was close to the deepwater port of Hobart and there was an available source of labour.

Today the refinery produces 200,000 tonnes annually of zinc and zinc alloys, 40,000 tonnes of lead residue and 40 tonnes of silver, 400 tonnes of cadmium, 350,000 tonnes of sulphuric acid and 110,000 tonnes of phosphate fertilisers. However, nowadays it is seen more as an ugly, belching industrial dinosaur than the economic backbone of the small state of Tasmania, and its largest employer. Despite a nine-figure investment in environmental upgrading, the company and the plant is the centre of a heavy metals contamination scare. The local population is concerned about the results of a recent investigation into alleged 'insidious' contamination of hundreds of homes near the refinery, by cadmium and zinc borne on winds from open stockpiles over the past fifty years. The fear arises that a crisis predicted fifteen years earlier by Tasmanian scientist Professor Harry Bloom is coming true. Professor Bloom contends that airborne cadmium and zinc ingested in a mix of about 1:100 is detoxified in the kidney by a little-known protein, metallathonien. This protein promotes excretion of toxic heavy metals, such as cadmium, from the body. Despite Professor Bloom's warnings of the dangers, the company failed to cover its heavy metals stockpiles with plastic tarpaulins and merely agreed to dampening them to limit loss in winds.

In 1990, the Tasmanian State Government released a study, undertaken jointly by the Department of Environment and EZ, of soil from 65 sites in the suburbs around the plant. Immediately around the refinery and east across the Derwent River in the path of prevailing north-westerly winds, samples indicated a very high level of cadmium and zinc. Soil analysis showed that there was four times the level of cadmium and five times the level of zinc considered serious enough to warrant clean-up action and a management strategy. Entering the food chain at these levels, home-grown vegetables could be of potential danger if consumed.

Raw sewage, wood fibre from the Australian Newsprint Mills Plant further along the river, and organic discharges from the Cadbury plant at Claremont have combined with the EZ effluents to turn the River Derwent's water into a toxic metal and chemical mix, further endangering the lives of residents of the immediate areas.

Environmental and occupational scientists class cadmium and lead as systemic toxins which, if accumulated in the body for a long period, can lead to insidious effects, particularly upon the liver, other internal organs, the central nervous system and reproductive systems. High levels of cadmium affect the nervous system and have the potential to cause birth deformities. Given these implications, the local communities are becoming increasingly militant.

Remedial action from the companies has done little to improve the situation. Closing of the refinery would have a devastating effect on Tasmania's economy. A study by the University of Tasmania reveals that, had the plant closed down in 1985-6, there would have been a reduction in Tasmanian employment of about 2.3 per cent, representing a loss of some 4000 jobs. About 37 per cent of these jobs — approximately 1500 — would have been at the plant itself, while the electricity industry would have lost an additional 150 jobs. About 1900 jobs would be lost in tertiary occupations such as trade, transport, finance and community services.

If adequate environmental safeguards are not available, and economic implications are as major as these, small states and nations such as Tasmania are in an invidious position. Politicians are naturally wary of committing themselves and their governments to new industrial ventures, and it is understandable that potential solutions and closures are approached with caution.

Food Toxins and Chemicals
Food-borne infections and intoxications are defined by the World Health Organisation (WHO) as any disease of an infective or toxic nature caused by or believed to be caused by the consumption of food or water. Thus, food poisoning in human beings can be caused by a variety of food-borne agents which may be intrinsic or extrinsic to food. Intrinsic causes include food allergens and certain foods which are toxic in themselves. Extrinsic causes include chemicals, parasites and micro-organisms present in the food as contaminants. While bacteria are the principal causal agents of food intoxication and infection, mycotoxins, viruses, parasitic worms or protozoaimay also be implicated: similarly foods may be contaminated with their own naturally produced or introduced toxic substances.

Chemical Contaminants, Residues and Toxins
Toxic chemicals in food that are not of natural origin may include some food additives, agricultural chemicals (e.g. pesticides), and antibiotics and accidental contaminants (e.g. metals and radio-nuclides) (Tables 4.3, 4.4). The safety of food additives is a difficult exercise to resolve. It is almost impossible to provide absolute proof that an additive is non-toxic to all people under all conditions. Safety standards for the evaluation and re-evaluation of additives are becoming increasingly stringent, partly because of improved methods for determining very low levels of toxic substances. The legislative inconsistencies

are, however, difficult to explain. The flavour-enhancer monosodium gluta-mate, for example, can cause symptoms of poisoning in susceptible people, but under established safety guidelines and testing procedures does not constitute a danger to laboratory animals or human subjects.

There is an interesting story about the use of arsenic as a food flavour enhancer and its resultant effects on human health, told by author Robert Pritikin. Apparently people loved the taste of arsenic so much they used it in everything from breakfast cereals to hot drinks. Even when their hair started falling out, they lost their teeth, became nauseous, dizzy and suffered severe headaches, they did not stop using it and refused to consider the chance that arsenic use might be involved. Instead the pharmaceutical industry developed drugs to combat the symptoms, to which were added other drugs to mitigate their side effects. New hair-loss treatments were formulated and improved false teeth and dental aids marketed. So it continued until researchers confirmed that the common link in the illnesses was arsenic. People ignored the evidence and continued to use the enhancer because it made their food taste good. Food additives and flavourings are part of the modern marketing and manufacturing processes and little effort is made to monitor their long-term effects because of the small quantities in which they are used.

Food Sensitivity
Food sensitivity is an example of an adverse reaction to food. In recent years, an increasing proportion of the population have claimed to experience food-related experiences ranging from extreme anaphylactic shock to less life-threatening episodes such as local skin irritations, headaches, stomach upset, shortness of breath and a runny nose.

Scientific and medical opinion is divided over claims attaching the blame to certain foods and food chemicals. However, the fact remains that some individuals are more sensitive than others to unfamiliar chemicals and toxins. Whether more people are reporting adverse reactions or there have been changes in the quality, state or manner in which we prepare the food we eat remains at issue.

All foods can be broken down into their basic constituent chemicals. Organic molecules, which are very large and complex structures, include carbohydrates, proteins and fats. They are described as macronutrients and form the bulk of the diet. The balance of the diet is made up of micronutrients such as the minerals, trace elements, B vitamins, fat soluble vitamins A and E, vitamin C and other essential food factors such as beta-carotene, choline and taurine. They are found in food in company with these larger molecules. Together they supply all the essential substances the body requires for daily life.

There are also some other small molecular weight chemicals found in association with foods that are toxic or far from beneficial to the human body. These are naturally occurring food toxins and drugs such as alkaloids found in potatoes, salicylates in fruits and vegetables such as tomatoes, caffeine in tea and coffee, oxalic acid in rhubarb, theobromine in chocolate and amines such as histamine, tyramine, serotonin and phenylethylamine which occur in cheese, sausages, sauerkraut and fermented foods. Each of these chemicals can produce a pharmacological reaction in a sensitive person and if consumed in sufficient quantity. The amount required varies with the individual. Often the body will experience a toxic reaction in response to the presence of these and other naturally occurring food chemicals, but few symptoms are experienced. Indeed the vast majority of adverse food reactions are not serious enough to cause an upset and generally pass unnoticed.

Other adverse reactions to food have more to do with genetics than any special chemical principle of the plant or food. Phenylketonuria is caused by genetic deficiency of the enzyme to breakdown the amino acid phenylalanine. It is usually experienced when an individual lacks an essential metabolic enzyme to break down this component of protein foods. If not detected early in life, the child becomes increasingly impaired intellectually due to the accumulation of the toxic by-products. At the same time, many adults and young children may, for various reasons, be unable to digest either cow's or goat's milk and must rely on substitute dairy foods such as soy milk and tofu. This is a form of food sensitivity generally classed as an intolerance and will be described shortly.

There is a further problem. These days, the farming community responsible for providing fresh food such as fruit and vegetables, cereal crops and grains is producing food for large-scale commercial marketing concerns. Farmers use large amounts of chemical fertilisers to boost growth, and weedicides, fungicides and other pesticides to prevent crop loss and damage from insects. Despite best efforts, and on occasions intentionally, many of these chemicals remain on food reaching the consumer. If not removed prior to consumption, by washing, peeling or other forms of preparation, they too can be responsible for adverse reactions.

Manufactured Food Items

In recent years, food manufacturers and producers have sought to change the nature of food sold to the public, either to make it more attractive, increase its resistance to breakdown from prolonged storage or make it easier to use in cooking. They have developed new foods made from breakdown and re-synthesis of food chemicals. Modified starches, polyunsaturated vegetable oils, low-fat, low-cholesterol dairy products are all sold for their advantages of ease of use, health benefits or suitability. The food industry continues to work to develop new products targeted at the health- and diet-conscious consumer. These products are generally regarded as 'safe' and concerns should remain about the effect of life-time exposure to a diet based entirely on synthesized foods, low-cholesterol fats and other such items (Table 4.5).

However, some data is available concerning the inadequate nutrition of low-fat milk for babies and the potential vitamin deficiencies associated with use of lactose powder without supplementary vitamins and maltose. Long-term health effects of diets dependent solely on consumption of high protein and vitamin foods such as drink mixtures and meal substitutes have been observed in a number of reported case studies from young women. Clearly, the ethics of testing products such as these on volunteers prior to release onto the market is a problem which needs to be tackled by food manufacturers, medical scientists and ethics committees. The adequacy of consumer education on proper use also needs to be addressed.

Food Intolerance
There is a tendency to confuse food sensitivity with food intolerance. Strictly speaking, adverse responses classed as food intolerance are caused by an immunological response to one or more of its nutrients. Lactose intolerance, caused by a missing enzyme, will result in inadequate digestion of dairy foods with the result that medium to large molecular weight proteins, present in the intestines, can induce symptoms of food sensitivity. Similarly, in coeliac disease, a gluten intolerance develops as a result of an abnormal or missing enzyme.

In stimulating an inflammatory response to food, partially digested proteins from foods such as wheat and milk are absorbed systemically causing migraines, while others act on the brain in a manner similar to morphine-like hormones. These peptides are called exorphins and appear to be involved in some mental disorders, notably schizophrenia. Many people are sensitised to cow's milk and wheat foods as infants and perhaps even *in utero*, while others suddenly become sensitised in later life after some stressful event suddenly precipitates all kinds of sensitivities to foods, and also house dust, chemicals, pollen, moulds and any number of previously innocuous chemicals.

Other food peptides act as antigens and stimulate antibody production. Antigen-antibody complexes formed can cause inflammation and degeneration in the joints, bones and soft tissues. This occurs in rheumatoid arthritis and the collagen diseases, and in other so-called auto-immune disorders. These reactions may be classed as immediate hypersensitivity responses mediated by the inflammatory protein immunoglobulin-E, resulting in the symptoms of hay fever, asthma, or dermatitis and eczema. Alternatively, the allergic reaction may be delayed and involve other immunoglobulins, Ig M, A, D and G. Still others do not involve the immune system at any stage, but contain substances called prostaglandins and are related more to changes in fatty acid metabolism.

The expression of the most common types of food intolerance depends upon a multifactorial system of which the food itself constitutes only one part. The final allergic manifestations depend upon additional factors.

In order to manage food intolerance it is essential to recognise that most examples are far from simple and involve multiple food sensitivities, reactions which vary from day to day and which depend on the quantities consumed and the frequency of consumption of particular foods. Thus it is important to adopt a holistic approach to identifying causes. External environmental factors such as house dust, air-borne pollens, viruses, bacteria, temperature, weather change and humidity can influence the body's response at any one place and time when a particular food item is eaten. Similarly, the emotional and mental state of the individual, associated body stress levels and social influences play a part. At the same time positive anti-stressors such as aerobic exercise, nutrient intake, rest and active forms of mental relaxation are important mitigating influences.

Food Treatment Techniques

Traditional methods of preserving food, such as curing and smoking, may also result in the production of hazardous substances. The detection of known carcinogens such as 3,4-benzopyrene and other polynuclear aromatic compounds in wood smoke has led to concern over the safety of smoked foods. It has been suggested that the high incidence of stomach cancer in Iceland and Scandinavia is associated with the consumption of large quantities of smoked fish.

Nitrites used in the curing of meat may react with secondary and tertiary amines present in the gastro-intestinal tract to produce nitrosamines, substances found to be carcinogens in laboratory animals. Preserved hams, pork, pâtés and other organ meats contain high levels of nitrites together with other chemicals known to induce sensitivity reactions in susceptible individuals; pH and temperature are also important in their storage. A variety of moulds and other organisms which thrive in highly salty or acidic mediums can be responsible for outbreaks of food poisoning following ingestion of poorly handled meats.

Toxic Plants and Fungi

Natural toxins present in plant foods do not usually damage consumers' health because they occur in such small amounts. However, excessive or prolonged consumption of a particular food may result in an abnormal reaction. For example, some members of the Brassica family (e.g. cabbage, kale, broccoli) contain goitrogens which, if consumed in excess, may interfere with the body's ability to absorb iodine. Excessive consumption of carrots, pumpkin and other vegetables containing high levels of betocarotene may temporarily cause the palms and skin to appear yellow or orange. However, these same vegetables and the same chemicals are believed to exert a protective effect against intestinal cancers, by mechanisms thus far unexplained.

Agricultural Chemicals
Herbicides, fungicides, pesticides and growth regulators (mainly plant hormones) have been found at toxic levels after accidental contamination of foods. The consumption of food contaminated by organochlorine insecticides results in violent convulsions, while the organophosphate group of insecticides causes sweating, nausea and vomiting.

The group of pesticides which includes DDT, dieldrin and heptachlor accumulates reasonably uniformly in fat tissue such as breast milk, kidney and the liver. Among these, breast milk is the most usual fluid for monitoring levels because it is reasonably high in fat content (1-5 per cent), is readily available from a cross-section of a female population and does not involve invasive sampling techniques. Its major disadvantage is that it permits the survey of only a limited age-group of the female population. As well there remain questions about the levels of tissue pesticides which constitute a risk, are contributors to toxic effects and occur at minimal tissue levels. Difficulties are compounded by the fact that it is impossible to provide experimental evidence that anything is absolutely safe. Where chemicals are concerned, safety may be expressed in terms such as 'margin of safety' and 'acceptable daily intake'.

In addition, residues of organochlorine and other pesticides may remain in or on treated food products. Antibiotics administered to animals may result in residues in meat and milk products and there is concern that some bacteria, including pathogenic ones, may develop resistance to these antibiotics, rendering them ineffective in the treatment of human infections.

Industrial Products
Modern industry depends on countless complex chemical processes to manufacture the vast array of new plastics, fibres, teflons and materials which have become the stuff of normal daily life. Not a lot is known about the effects of many of the products themselves, or of their constituent chemicals and by-products, upon the workers who handle them in the workplace. Most firms

now carefully monitor the health of workers and have a team of scientists constantly reviewing research literature for information on the potential dangers of the chemicals they use. However, often it is years before the first signs of danger associated with use of a substance become known. Usually the evidence accumulates slowly, almost unnoticeable until it becomes overwhelming and irrefutable. Unfortunately it is often ignored — particularly initially.

On 22 January 1974, the rubber firm B.F. Goodrich of Louisville, Kentucky announced that three of its workforce exposed to vinyl chloride had died of angiosarcoma, a rare cancer of blood vessels of the liver. Within months, the US hygiene limit, which had only recently been reduced from 500 ppm was further lowered to 50 ppm. Almost immediately this was further reduced to 'no detectable level' and ended up at 1 ppm. Other countries were as quick to react, and in Sweden especially, drastic standards were passed limiting exposure to 'effectively zero' levels.

Large-scale production of vinyl chloride monomer (VCM) had begun in the 1930s, but production shot up dramatically after World War II when VCM became established as the foundation for a thriving plastic industry. Despite this enormous production and the widespread exposure of workers to polymerisation to form polyvinyl chloride (PVC), virtually nothing had been done to study the compound's toxic properties. It was not until 1970 that an Italian researcher named Viola noted, quite by accident, the carcinogenic effects of VCM after exposing rats to the gas at 30,000 ppm. This report was noted by a consortium of European chemical companies which commissioned more detailed studies from the Bologna Cancer Institute directed by Cesare Maltoni.

By late 1972, Maltoni had confirmed and extended Viola's results, finding a variety of tumours in several organs including angiosarcoma of the liver, at exposure levels as low as 50 ppm. These results were only released publicly with the announcements of the three angiosarcoma deaths of Goodrich workers in January 1974.

Since publication of Maltoni's findings several animal laboratory investigations have been performed reproducing the carcinogenic effects most notably in the liver but also other organs. Several epidemiological investigations of past VCM-exposed workers in a number of countries around the world have revealed an excess risk of cancer, the magnitude of which is related to the duration or the intensity of exposure. To date, no angiosarcomas have been diagnosed in Australian workers.

The Effects of Nuclear Disaster
Surviving a nuclear disaster has well-known physical effects. Less well known

are the psychological effects.

Of the million or more radiation survivors living in the United States, seven atomic veterans and their families were studied by a team of behavioural scientists led by Dr Bianca Cody Murphy, of the Department of Psychology at Wheaton College in Norton, Massachusetts. Their findings reveal that those exposed to low-level ionising radiation later experience high rates of physical and psychological illness in themselves and their families.

Health and behavioural scientists have often studied radiation survivors, people who live through a period of brief or intense exposure to possible contamination by ionising radiation. As a result of the Chernobyl disaster of 1986, the CIS has the largest number of radiation survivors, while atmospheric weapons testing in South Australia contaminated an unknown number of Australian military and civilians, including Maralinga Tjarutja Aborigines.

The Murphy team claims that after 1951, US atomic bomb tests were expanded to include the study of the effects on military personnel who were deliberately exposed. Soldiers were taken to desert test sites to watch the explosions. Their actual distance from the explosions was varied, with some approaching within three kilometres of the blast. After the explosion, the soldiers were ordered to approach the detonation site. On occasions they wore radiation badges to measure the amount of gamma radiation to which they were exposed, not to ensure their limited exposure. In addition, no protective measures were taken to guard against the gamma and beta radiation that covered their clothes and entered their skin and lungs.

On return to their units, the specially selected personnel were under orders not to discuss their experiences. This wall of silence was maintained until 1979 both by the government and the veterans. In 1979, however, three of the veterans who had been involved in Pacific tests, three from the Nevada tests and one who had helped in the Hiroshima clean-up broke their silence. All had at least one serious physical problem and many had more. Four had some form of skin cancer, two had serious skin diseases, two had severe thyroid problems and one had major respiratory difficulties.

The research team discovered that four of the six families who had borne children reported birth defects, cancer or other serious health problems in at least some of the children. However, they also found that all of the atomic veterans and their families shared common experiences and coped with their health and behavioural problems in nearly identical ways. In all, four themes emerged.

First, for atomic veterans, there was the 'invalidation of their experience' and 'institutional denial' on the part of the US Government, veterans' organisations and medical providers. This had a demoralising psychological effect causing the atomic veterans to take measures to re-affirm the reality of their experiences.

Second, the veterans and their wives who had children experienced psychological anxiety about the transmission of genetic defects and birth problems to their offspring. Feelings of guilt and stress about the health of their children extended to include concern for the well-being of future generations. The veterans whose children suffered birth defects experienced intense feelings of guilt which affected their abilities socially and in the workplace.

Third, the veterans and their wives also went to great length to protect their children and possibly themselves, from the disturbing feelings about radiation exposure. Often a shroud of silence characterised all aspects of family interaction. Silence served as one means by which the parents sought to protect their children.

Finally, the atomic veterans, as a coping response, expressed a great desire to leave a record of their experiences of radiation exposure in order to help prevent the suffering of others. Their perceived need to disclose their knowledge and experiences stemmed from an inner belief that their lives were likely to draw to a sudden and abrupt close at any moment.

Are these feelings any different from those of others who have been forced to come to terms with the knowledge that their lives may be shorter than otherwise expected? Cancer, AIDS and genetic diseases which limit life-span impose similar psychological stresses. The issue of post-nuclear trauma of a psychological nature thus needs to be resolved in the context of life experiences not quite as unique as the veterans themselves might wish to indicate.

TABLE 4.1: Pathogens and Health Effects
(vector-borne)

INSECTS: *Mosquitoes* — malaria
— Ross River fever
— dengue
— Australian encephalitis

Fleas/Ticks — tularaemia (rabbit fever), skin/lymphatic lesions, lung, gastro-intestinal tract
— peste (plague):
Fleas on rodents — lymphatic, pulmonary, blood, meningeal infection

Cockroaches/Flies — diarrhoeal diseases (e.g. shigella, faecal/oral)
— trachoma, kerato-conjunctivitis, blindness
— typhoid fever (via human patients and carriers, foods), systemic illness
— yaws, caused by treponema pestenue
— skin papillomas (lumps), proceeding to ulceration and degeneration

SNAILS — schistosomiasis (snail fever), liver filuosis/portal hypertension
— paragonimiasis (lung fluke disease), respiratory tract degeneration
— clonorchiasis (oriental liver fluke disease) invasion of bile ducts and liver cells by fluke

WATER — gardiasis, malabsorption, diarrhoea
— hepatitis A
— leptospirosis
— typhoid fever
— rotaviral diarrhoea
— shigella, staphylococcal diarrhoea, bacillary dysentery

FOOD — salmonellosis, (rats, pets) contaminated hands, acute enterocolitis (diarrhoea, nausea, abdominal pain)
— staphylococcal food poisoning, humans, animals, nausea, vomiting, cramps, prostration
— bacillus cereus food poisoning, diarrhoea
— clostridium botulinum (botulinism), central nervous system depression, respiratory failure

TABLE 4.2: Wastes

Agriculture: crop residues (chaff, stubble, spoiled food, washings), animal excrement (faecal, urinary)

Forestry: wood dust/copper chrome arsenate/pulp washings/chlorine/dioxin, skin sensitisers/chloracne and potential immunosuppresant

Fisheries: excrement from intensive fish-farming, increasing BOD, increase in pathogen/contamination of fishery

Mining: heavy metals, acids, leachates, radioactive isotopes, dust tailings

Energy production: heavy metals, thermal pollution, fly ash, coal dust and other particulates, COx, NOx, SOx and hydro carbons, oil spills, alienation of land and chemical waste

Manufacturing: COx, SOx and NOx, industrial chemicals and gases, asbestos, particulates, heavy metals, ultra-violet and infra-red radiation, radioactive isotopes, waste organic and inorganic materials, etc.

Transport: COx, SOx, NOx, particulates, heavy metal, chemical and oil spills

Energy use: COx, SOx, NOx, particulates, oil, heat

Tourism: COx, SOx, NOx, particulates, heavy metals, oil, litter and effluent

TABLE 4.3: Pesticides/Herbicides

ORGANOCHLORINES: Aldrin, Dieldrin, Chlordane, Heptachlor
— central nervous system stimulation/depression, blood disorders, liver/kidney/GIT disorders, respiratory tract irritation, suspected carcinogens. Reproductive effect, blood dyscrasias.

ORGANOPHOSPHATES: Parathion, Malathion, Diazinon
— cholinesterase inhibitors, nervous system stimulation/depression, progressive failure of body systems; allergy

CARBAMATES: Carbaryl, Dithiocarbamates, Phenylthiazine
— renal, reproductive teratogen

NICOTINE:
— chlorpheroxy acids, dinidtropherol, dermatose, (herbicides) respiratory irritants

PCBs/PBBs
— teratogens/mutagens

TABLE 4.4: Organochlorines

Substance	Possible Health Effects
H.C.B.	Carcinogen
Dieldrin Heptachlor Chlordane Lindane B.H.C.	Known or suspected carcinogens; neurotoxic effects; chloracne and other skin diseases.
PCBs	Neurological, liver and skin disorders; tumour promoter, carcinogen
Dioxin	Teratogen. Chloracne. Liver damage. Extremely toxic.

Source: US Office of Technology Assessment, Wastes in Marine Environments, *1987, reproduced in S. Beder* Toxic Fish and Sewer Surfing, *Allen & Unwin 1989.*

TABLE 4.5: Nutrition

Inadequate supply — starvation

Poor quality — lack of proteins/carbohydrates/minerals/vitamins/
   trace elements, etc.: e.g. lack of iron in plants — anaemia;
   excess odium — disturbs metabolic balance; shortage of iodine
   — goitre; magnesium/zinc — impaired body function, especially nervous system; cobalt shortage — pernicious anaemia.
   — vitamin deficiencies: e.g. beriberi, scurvy, pellagra.

Contamination of food — refer to TABLE 4.1: Pathogens
   — excess leads to low calcium absorption.

References and Additional Reading

Brodeur, P. *Outrageous Misconduct: The Asbestos Industry on Trial*, Pantheon, New York, 1985.

Brown, M.H. *The Toxic Cloud*, Harper & Row, New York, 1987.

Cassel, J. 'The Contribution of the Social Environment to Host Resistance' from *American Journal of Epidemiology*, 1976, vol.104, pp.107-23.

Epstein, S. *The Politics of Cancer*, Sierra Club Books, 1978.

Forget, G. *Toxic Substances and Health*, International Development Research Centre, Ottawa, 1988.

Graham, N.H. 'Psychological Stress as a Public Health Problem: How Much Do We Know?' from *Community Health Studies*, 1988, vol.XII, no.2, pp.151-60.

Haenszel, W. and Hillhouse, M. 'Uterine Cancer Mortality in New York City and Its Relation to the Pattern of Regional Variation within the United States' from *Journal of the National Cancer Institute*, 1959, vol.22, p.1157.

Haynes, V. and Bojcun, M. *The Chernobyl Disaster*, Hogarth Press, 1988.

Hester, J. *et al. Science for the People*, May 1989.

Murphy, B.C. *et al. American Journal of Orthopsychiatry*, 1990.

New Internationalist.'Cancer: The Facts', Aug. 1989, no.148, p.16.

New Scientist. 'Code of Conduct May Keep Dangerous Pesticides At Home', 9 December 1989, no.1684, p.6.

Newhouse, M.L. and Thompson, H. 'Mesothelioma of Pleural and Peritoneum Following Exposure to Asbestos in the London area' from *British Journal of Industrial Medicine*, 1965, vol.22, p.261.

Patterson, J. *The Dread Disease*, Patterson, New York, 1987.

Rawls, W.E., Tompkins, W.A.F., Figouerora, M.E. & Melnick, J.C. 'Herpesvirus Type 2: Association with Carcinoma of the Cervix' from *Science*, 161, 1968, pp.1255-8.

Rowland, A.J. & Cooper, P. *Environment and Health*, Edward Arnold, London, 1983.

Wood, C.S. *Human Sickness and Health: A Biocultural View*, Mayfield Publishing Company, Palo Alto, California, 1979.

World Health Organisation. *World Health In Developing Countries*, WHO Publications, 1987.

World Health Organisation Cancer Unit. *Cancer Control In Developing Countries*, WHO Publications, 1986.

4.2 Cardiovascular Disease

This chapter is the first of those dealing with the health-related outcomes of environmental processes. It covers cardiovascular issues: hypertension, atherosclerosis, coronary heart disease and stroke. Mortality from stroke has declined steadily since 1950, presumably reflecting in part advances in the treatment of high blood pressure. Coronary or ischaemic heart disease, which was the main cause of death in Australia for many decades and even now accounts for about 30 per cent of all deaths in men and 25 per cent in women, reached a peak in the late 1960s and has declined steadily since, as it has in the United States and New Zealand, and as has only recently begun to occur in the United Kingdom. A reduction in the actual number of people developing the disease, and improved treatment (and therefore survival) of heart attack patients may have contributed to this decline. Recent evidence suggests that the former factor has been substantially more important.

The Nature of Coronary Heart and Artery Diseases
Arterial diseases, specifically coronary artery disease, is a problem of major importance in Australia, the United Kingdom and America. It precludes and contributes to coronary heart attacks and myocardial ischaemia, the principal causes of death among adults. The capacity of arteries to deliver a consistent and smooth flow of blood to all parts of the body depends on environmental factors.

The coronary arteries are comparatively small vessels which traverse the front and back of the heart, conveying aortic blood to the heart. If a coronary artery or one of its branches becomes obstructed, the related area of the heart muscle will be starved of oxygen and essential nutrients. The muscle may go into spasm, it will weaken and some of the cells may die. This area of the heart is permanently damaged, and during exertion or stress the extra demands for blood placed on the heart will not be met. Shooting cramps and chest pain (angina) will result. Clinically a condition called myocardial ischaemia occurs and this tends to worsen with age.

Coronary arteries become blocked usually because of local arterial plaque, sections of vessel damaged by the infiltration of fatty fibrous materials and subsequent localised calcification. This disturbs smooth blood flow and small clots or thrombi may form in the 'eddies' in the vicinity of the plaqued section. This process called atherosclerosis has been the subject of considerable research.

In modern affluent societies, the formation of plaque or atheroma seems to commence relatively early in life, in childhood, and the resulting atheroma become increasingly extensive with the ageing process. Diet and levels of consumption of foods high in cholesterol and saturated fats (triglycerides) govern the extent of deposition of atheromatous plaque and the age at which it begins. Genetic factors also play a role. Familial hypercholesterolaemia is an inherited tendency to raised levels of cholesterol in the blood serum. It is accompanied by increased risk of coronary disease.

Diabetics are also more prone to coronary disease than non-diabetics. A family history of early death from heart attack, congestive heart failure and cardiovascular abnormalities increase risk. In one study, men under the age of 55 years with a male relative who had died of heart problems before age 55, were calculated to be at a five-fold risk of a similar fate. Women with a comparable family history were 2.5 times more vulnerable.

Causes and Risks

Although there is no recent truly international information on risk factor prevalence in cardiovascular diseases, the results of a comprehensive Australian study, conducted in 1983 by the National Heart Foundation, detail the results of a survey sample of approximately 6000 adults aged 25-64 years and confirm data gathered in a similar study in 1980. Diet, smoking, alcohol consumption, body weight, physical fitness and levels of stress are confirmed as the principal causes.

Thus food and eating play a major role in determining the levels of cardiovascular health and disease. On average, 17.2 per cent of Australian men surveyed and 22.8 per cent of women said they were following a special diet, whether for weight loss, on medical advice or through vegetarianism. Two and a half times as many women (7.4%) as men (2.8%) were on a weight control diet. More men than women always put salt on their food after it was cooked (34% versus 23%) while more women than men regularly took vitamin, mineral or other supplements (34% versus 20%).

Survey results from western countries indicate that the type of diet typically eaten by urban dwellers in affluent societies often falls short of national dietary goals despite the ready availability of money, choice and information about correct eating habits for health. By way of illustration, Fig. 4.1 compares the major nutrient composition of Australian and American diets with that recommended by the United States Government. It indicates that people either do not know or choose to ignore the advice given them by their doctors.

Cigarette smoking decreases cardiovascular health as well as having implications for respiratory function and cancer. The 1983 Australian National Heart Foundation survey found that 32 per cent men and 25 per cent women regularly smoked cigarettes. Average consumption was twenty-one per day for men and sixteen per day for women; 80 per cent males and 60 per cent female smokers smoked more than ten per day, and 40 per cent and 21 per cent respectively smoked more than twenty a day. Further studies suggest these rates may be low and that approximately one-third of adults were smokers.

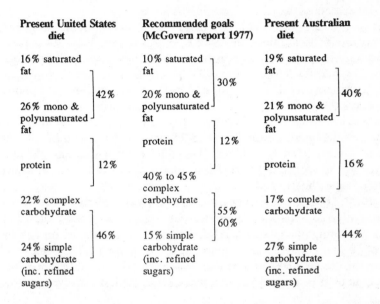

Fig. 4.1: Contributions of major nutrients (excluding alcohol) to total dietary energy intake: comparison of actual diets in United States and Australia with United States dietary goals.

A particularly disturbing feature was the high rate of smoking in the youngest age groups and particularly among young women aged 16-24 years. Recent surveys have shown that an increasing number of teenagers are smokers and the increase has been greater among girls than boys. Among school-children aged 12-15 years, as many as 10-25 per cent are regular smokers.

The influence of diet on the incidence and prevalence of coronary heart diseases is exerted principally through the role of dietary fats. Diet choice determines the circulating blood levels of cholesterol and triglycerides, significant causes of coronary heart disease and atherosclerosis. Only 30 per cent of Australians have been recorded with levels below 5 mmol per litre of blood cholesterol and at reduced risk of these diseases.

In contrast, approximately one-in-five Australian men and women have levels above 6.5 mmol/L, a level considered to be unacceptably high. The average blood cholesterol concentration in men rises with age, and levels at around 45 years of age. In women, however, it starts lower than men, rising steadily and exceeding that of men around the age of fifty.

Epidemiological evidence indicates that lipoproteins are better predictors of coronary artery disease than clinical or electrocardiographic results. There are three important forms of lipoproteins: very low density lipoproteins (VLDL), low density lipoproteins (LDL) and high density lipoproteins (HDL). Partitioning total cholesterol into different proportions of lipoproteins improves the ability to predict heart disease because those lipoprotein fractions which transport cholesterol and triglycerides behave differently in atheroma development. Normally 60 to 75 per cent of the total plasma cholesterol is transported as LDL; and plasma LDL concentrations, like total cholesterol, are atherogenic. The two have approximately the same predictive power for coronary heart disease.

In contrast, HDL are inversely correlated to total cholesterol and to the occurrence of arteriosclerotic plaque. Recent studies have found that of all lipids and lipoproteins measured, HDL have the largest impact on risk for those over 50 years of age. Furthermore, in adults, plasma HDL cholesterol levels are consistently lower in men than in women. This sex difference does not appear until at or around puberty, when HDL cholesterol levels in boys decline.

Dietary Fats: Recent measurements of blood cholesterol in Australian children aged 5-18 years showed levels 50 per cent higher than those recommended by the World Health Organisation, and 10 per cent higher than recommended in the United States.

Body weight, also related to diet, is a further component in coronary heart disease and cardiovascular disease, and it is often found to occur in individuals with high levels of circulating cholesterol and/or transport lipoproteins. The average weight of both men and women in western societies tends to peak in the 50-54 year age group and is reflected in the proportion of overweight people. In Australia, 6.4 per cent of men and 8.7 per cent of women are classed as obese (body mass index or BMI exceeds 30).

Significant levels of obesity and overweight in Australian, British and American school children have raised concerns. These are further heightened by physical fitness levels and lifestyle surveys suggesting that as these children

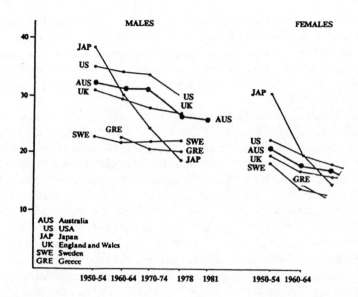

Fig. 4.2: *Premature Mortality: proportion of 15-year-olds dying before age 65, Australia, 1950-81.*

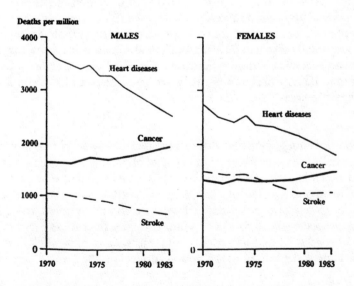

Fig. 4.3: *Age-standardised death rates from major causes of death, Australia, 1970-83.*

pass into adolescence they often tend towards being overweight, become less physically fit and are more sedentary. Obesity among Australian adolescents is the subject of preventive programmes aimed at reducing weight and increasing regular exercise. These programmes are aimed especially at the high proportion of teenage girls and boys aged 16-18 who are obese (9.5% and 7.0% respectively).

Studies have demonstrated that populations with relatively high levels of serum cholesterol and lipoproteins tend to be the same ones that show high levels of mortality from ischaemic heart disease. Information available from WHO for some 65 countries indicates that mortality from ischaemic heart disease is highest in Finland, South Africa, the United States, Scotland, Australia, Northern Ireland, New Zealand, Canada, Iceland and Israel. Death rates are consistently low in Thailand, Honduras, Guatemala, El Salvador, Taiwan, Ecuador, Paraguay, Jordan, Peru, the Philippines and Hong Kong. They are rising in Czechoslovakia, France, Netherlands, England and Wales.

It is now considered that total energy content of the diet, particularly the ratio of simple, refined to complex carbohydrates is also important to cardiovascular disease. Yudkin is given credit for establishing this link although other researchers have since failed to confirm his work. Nonetheless, clinically, excess body weight, the result of a high energy, low roughage diet condemned by Yudkin, has been found to be associated with increased blood pressure, declining physical fitness and subsequent increased risk of coronary disease.

The fact that less refined diets, high in complex carbohydrates and roughage are usually associated with less affluent standards of living suggests that, as socio-economic conditions improve and people become more affluent in lifestyle and dietary preferences, they are more at risk of atheroma and coronary heart disease (Table 4.6).

Alcohol

Alcohol consumption is an equally disturbing issue. Sixty-six per cent of men and 41 per cent of women report drinking alcohol on at least one day of each week; and there is little difference among the age groups. However, the proportion of Australian men who drank on at least five days each week rose steadily from 15 per cent of those aged between 25 and 29 years, to 40 per cent for those aged between 55 and 59 years, dropping back to 33 per cent for those aged between 60 and 64 years. While Australians, as a whole, record almost the highest per capita consumption of alcohol, the British fare little better.

A similar pattern for alcohol consumption emerges for women. The proportion rises from 7 to 18 per cent with age, declining slightly to 16 per cent for those aged 60-64 years. Moreover, on the basis of previous research conducted by the Australian Bureau of Statistics, these responses are almost certainly an underestimate of the true levels of consumption, particularly in respect of the over-60 age category.

In general, excessive use of alcohol is associated with increased risk of high blood pressure, overweight, accidents, cirrhosis of the liver, brain and heart damage, as well as various types of cancer. The severity of this risk is determined by the number of days per week on which alcohol is consumed and the average number of drinks on those days. Around one-in-ten men and one-in-seventeen women were in medium, high or very high risk groups. The age groups with the greatest risk were 35-39 and 50-59 years.

The Effects of Social Change
The relationship between standard of living, nutritional status, proportions of saturated and unsaturated fats in the diet, serum cholesterol levels and cardiovascular disease is demonstrated in work from South Africa. Comparing populations from separate ethnic and social groups, living under the same geographical and climatic conditions, Bronte-Stewart and a team of doctors showed that the very poor Bantu fared better than the middle-class Cape Coloured population, who in turn were less likely to suffer heart disease than the affluent Europeans.

As standards of living improved so too did the total fat content of the diet and the proportion of saturated animal fat to unsaturated vegetable fat. Blood lipid levels were highest in the Europeans, intermediate in the Cape Coloureds and lowest in the Bantu. These levels directly correlated with their standards of living and the fat content of their diets.

In Europe, changes in the social class incidence of coronary heart disease have been apparent since the early part of the twentieth century. Around the 1930s, the first indicated rise in the incidence of heart disease was observed in Britain and occurred predominantly in social class I. (Members of social class I include judges, political leaders, managers of large corporations and specialist professionals such as medical doctors.) This observation contributed to the perception that stress, particularly that associated with the type of decision-making, is required of the occupations of members of social class I, and is responsible for placing them at increased risk of heart disease.

Twenty years after the first rise in coronary disease was noted, the social class gradient in standardized mortality ratios upwards from the lowest socio-economic groups was still unmistakable, although the gradient was less than before. By this time, there had been a considerable rise in the mortality rates attributable to coronary heart disease in all social classes. When the male social class mortality was examined by age group, the ratios showed no particular trend in men aged 20-24, but at 25-34 years they indicated an upward mortality gradient from class I to class V (the lowest socio-economic group). This contrasted with strong downward gradients from I to V at 45-54 and 55-64 years.

British figures published in 1961 demonstrated a further change. The greatest mortality then lay in social class V for men and married women aged 20 to 64 years. Little difference has been apparent in the 1971 and 1981 data.

These changes have occurred in parallel with considerable social changes in the United Kingdom. While social welfare measures have resulted in a certain amount of social levelling, particularly in terms of nutrition, education and living standards, sedentary life has increased and modern appliances minimise active work in the home. Changes in alcohol consumption, diet and smoking have accompanied increased affluence. Members of social class V would appear to be seduced more by these lifestyle changes and less informed of, or prepared to take, the preventive action necessary to minimise risk of disease. Recent figures from other industrialised countries confirm these trends.

High Blood Pressure
High blood pressure is one of the major risk factors in coronary heart disease and is the most important risk-factor for stroke. High blood pressure or hypertension is defined as a diastolic blood pressure of 95mm Hg or more, a systolic blood pressure of 160mm Hg or more, or both. A raised blood pressure accelerates atherosclerosis, so blocked and ruptured blood vessels occur some twenty years earlier than in people with normal blood pressure. The pressure itself appears to be directly involved in this process and, as a result, the higher the pressure, the greater the vessel damage. It is also likely that high blood pressure helps precipitate the clinical event as well as accelerating atherosclerosis.

Smoking further exacerbates cardiovascular risk, with or without high blood pressure. The greater the number of smokes per day and the earlier age at which smoking begins increase the potential risk of coronary heart attack and ischaemia. For an individual with existing high blood pressure, smoking adds an additional risk in the order of 1.8 to 3.

The specific causes of high blood pressure require further investigation, but environmental behaviours associated with affluent lifestyles and diet are prominent. High sodium intake (salt in the diet), overweight, obesity and alcohol intake are important. Blood pressure levels increase with weight and reported alcohol consumption and can, in time, be lowered by reducing weight, controlling alcohol and salt intake. Exercise appears to benefit both weight control and hypertension.

High blood pressure appears to run in families and a family history often helps identify individuals at risk. Social class and stress, at home and in the workplace, also relate to high blood pressure.

The Physical Environment
Cardiovascular diseases are also influenced in each area by the unique physical environmental features of that environment. Important to human health directly and indirectly are the water supply and the presence or absence of ground metals and minerals, in the water or in the ground where food for human consumption is grown.

In the first instance the content and quality of water consumed by humans seem to play an unexplained role in the aetiology of cardiovascular disease. It is considered likely that minerals such as calcium, sodium and magnesium are naturally present in local water supplies as dissolved salts and may exert a protective effect on cardiac muscle cells, though the findings in this regard are by no means consistent. Bore water drawn from the Artesian Basin, the vast underground water supply in parts of inland Australia, is generally very hard and high in calcium salts.

Evidence from the United Kingdom suggests that whereas 'hard water' may exert a beneficial effect, natural or deliberate softening of the water supply in an area can be accompanied by an increase in the death rate from circulatory disease. Results are based on long-term studies on populations throughout the country and provide circumstantial if not substantive evidence (Rowlands and Cooper, 1986). In contrast, foods with high levels of potassium and grown in areas where water contains high levels of the dissolved mineral tend to exert a protective effect.

It seems that all circulatory diseases are affected to some degree by a 'water factor', which operates through increasing proportionally the incidence of the most commonly occurring form of disease in that country. Thus, in Japan, where cerebrovascular accidents are significant contributors to high death rates from cardiovascular diseases, an increase in dissolved calcium minerals is accompanied by increased reports of stroke-related deaths.

Individual minerals have differing effects on cardiovascular health. Trace minerals such as cadmium and manganese can enter groundwater through leaching from industrial tailings dams when sites are mined. These metals are directly toxic to cardiac muscle cells.

Other environmental agents also act on cardiovascular mechanisms. Phosphate fertiliser and pesticide residues which enter the water supply from runoff through agricultural catchment areas, serve as a potential source of concern for country people in particular. For city people, recent studies have shown that carbon monoxide, a common urban pollutant associated with motor vehicle emissions, increases the occurrence of irregular heart beats thus posing an increased risk to heart patients. The multifactorial and complex interrelationships between these and other factors which affect human health leave many potential areas for future research.

TABLE 4.6: Characteristics of Population Screened for Cardiovascular Risk Factors		
	Number	*Percentage*
Sex		
male	4832	40%
female	7235	60%
Total	12,067	
Age range	18-98	
Mean age	49	
Cholesterol (mmol/L)		
<5.5	6862	57%
5.5-6.5	3237	27%
>6.5	1968	16%
Mean cholesterol	5.42	
BMI>25	5551	46%
Mean BMI	25.08	
High Blood Pressure (treated)	2655	22%
Family history of heart disease	4465	37%
Previous cholesterol test	3620	30%
Current smokers	1689	14%
Number per day (mean)	16	
Exercise (never/rarely)	7964	66%

Source: James et al, Community Health Studies.

References and Additional Reading

Australian Government. *Looking Forward To Better Health*, Better Health Commission Report, Vol.1, AGPS, Canberra, 1986.

Bronte-Stewart, B., Keys, A., Brock, J.F., Moodie, A.D., Antonis, A. & Keys, M.H. *Lancet*, 1955, vol.ii, p.1105.

Dwyer, T. & Hetzel, B.S. 'A Comparison of Trends of Coronary Heart Disease Mortality in Australia, USA and England and Wales, with reference to three major risk factors — hypertension, cigarette smoking and diet' from *International Journal of Epidemiology*, vol.9, 1980, pp.65-70.

Marmot, M.G. *et al.* 'Employment Grade and Coronary Heart Disease in British Civil Servants' from *Journal of Epidemiology and Community Health*, 1978, vol.32, pp.244-8.

Rowland, A.J. & Cooper, P. *Environment and Health*, Edward Arnold, London. 1983.

Truswell, A.S. 'End of Static Decade for Coronary Disease?' from *British Medical Journal*, 1984, vol.289, pp.509-10.

Winslow, R. 'Air Pollution by Carbon Monoxide Poses Risk to Heart Patients' from *Wall Street Journal*, 4 September 1990, p.B4.

Yudkin, J. 'Dietary Fat and Dietary Sugar in Relation to Ischaemic Heart Disease and Diabetes' from *Lancet*, 1964, vol.ii, p.4.

4.3 Cancers

Cancer is one of the main causes of death worldwide, especially in the developed countries where the risk of death from communicable disease has been largely eliminated. Originally believed to be a natural process signifying ageing and an accompanying deterioration, the vast majority of cancers are now believed to result from human behaviour, diet and lifestyles. Many cancers take years to develop, insidiously appearing only around middle age. While serving to speed up the ageing process and bringing forward the age of death, some forms can actually bring an early, painful and sudden death.

Cancer is primarily classified according to the system of the body which is affected, the particular organ within that system and the kind of tissue from which the growth is derived. There is also a differential incidence apparent between the sexes and in relation to the occurrence of each type of common cancer. Table 4.8 highlights these differences. In most body systems, the death rates of males exceed those of females. This relationship is reversed only in the case of cancers affecting the reproductive system and the breast.

Cancer of the lung in men is widespread, but there is now a steady increase in the number of women contracting and dying from lung cancer. The increase in the number of women, particularly adolescent girls, who smoke cigarettes has been claimed to be responsible. The role of passive smoking has also been raised in recent years and current research suggests that this is an important factor, too long ignored by medical researchers.

The incidence of skin cancers in countries with a warm, sunny climate such as Australia has also shown a sharp increase in recent years. Environmental degradation and depletion of ozone in the earth's stratosphere are said to contribute to this alarming trend.

Cancer, Occurrence and Spread
With the exception of lung cancer, deaths from which are increasing worldwide at a prodigious rate, the increase in cancer deaths is only slow. The global mortality rate is 4.3 million annually, of which 2.3 million occur in the Third World (Table 4.8).[1]

Although cancer is by no means confined to older age groups and certain kinds are a major cause of death in children, the incidence of the disease and its tendency to increase with age confirms observations that exposure to carcinogenic agents needs to continue for a period of time. It is relatively uncommon to get cancer before the age of forty. However, there are some cancers which are more common in children than adults. For example, acute leukaemia accounts for approximately 30 per cent of all childhood cancers in the USA, of which 80 per cent are acute lymphoblastic leukaemias. Chronic lymphocytic leukaemia is almost never seen before mid-adulthood and is most common above the age of fifty.

Carcinogenesis

The way that cancers start is not yet fully understood although there is general agreement on some aspects of carcinogenesis. It can take years between exposure to a carcinogen and first diagnosis of the cancer. During this time the process of carcinogenesis, involving many steps, is occurring. Tissues appear to go through recognizable pre-cancerous growth changes before any definitive lesion becomes apparent.

It has been suggested that there are at least three or four steps involved in the production of leukaemia and at least six or seven steps for the generation of a carcinoma. The principal event that eventually gives rise to a cancerous cell is some form of damage to the genetic material, deoxyribonucleic acid (DNA). Some carcinogens can directly cause mutations in DNA. When it gets into a cell of the body, the carcinogen chemically reacts with DNA producing a mutation or change in the sequence of DNA bases in the chromosomal material of the cell. Other carcinogens do not themselves chemically react with DNA, but when taken into the body are chemically altered by natural body processes to form a chemical which does react with DNA. This process involving 'pro-carcinogens' may explain why some chemical carcinogens cause cancer in specific sites of the body but not in others.

Since DNA damage may occur in any part of the cell DNA, there is a vast range of potential mutations which can occur. Nonetheless, many are 'silent' and do not appear to have an identifiable effect. On the other hand, if the mutation occurs in parts of the DNA coding for a crucial enzyme or one responsible for cell reproduction and growth, digestive processes, foetal development or hormonal production could all be affected.

Oncogenes

The discovery of parts of DNA called oncogenes has produced a flurry of scientific research. It is believed that they normally play a crucial role in cell growth and differentiation. Oncogenes can be 'switched on' in a number of different ways, all involving DNA damage. It is also conceivable that some of these oncogenes should be 'switched on' for controlling cell growth, but if DNA damage switches them off, cell growth patterns are affected. But no one knows which oncogenes play a role in carcinogenesis and when.

It is now known that there are probably several of these oncogenes which have to be affected before a cancer is seen. The discovery of oncogenes opens a wider spectrum of action for substances like the chemically inert fibre asbestos. It is currently believed that asbestos fibres in a certain range of sizes can enter the cell. While they do not combine with the DNA or its mutation, they seem to be responsible for breakages of the chromosomal DNA strands during cell replication. Depending on the extent of breakage that occurs, the strand will either re-join or join incorrectly, resulting in the normal replication process or a fault. The extent of breakage may also mean that the cell dies.

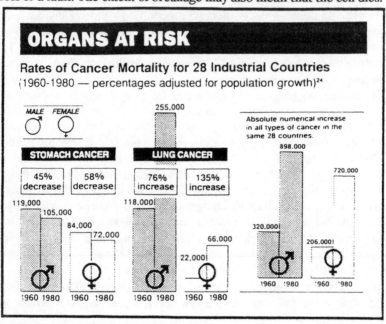

Fig. 4.4: Source: WHO Cancer Unit, 1986. Rates of cancer mortality for 28 industrialised countries.

The WHO Cancer Unit, following a comprehensive review of cancer occurrence and control in developing countries, provided the data described in Table 4.8. Figures for a twenty-year period in the industrialised countries can be described by the histograms in Figure 4.4 (percentages adjusted for population growth).

The most easily accessible data on cancer concerns mortality, in this instance the cancer fatality rate. This is governed, however, by the accuracy of diagnosis, the time in life when this develops into or is identified as cancerous, and the efficiency of the consequent treatment. International comparisons tend to be influenced by the form and quality of health services, practitioner knowledge and population access to health services. Apparently comparable data reflect these factors as well as real environmental differences between countries and their populations. Time between exposure and first signs of disease, and the duration and intensity of exposure are all important in determining the extent, the form and the future prognosis of the cancer.

Time

The fact that cancers develop over time means that it is often difficult to establish a definitive link between tumour growth and an imputed cause. In addition, changes in the incidence of various types of cancer in a community over time accompany or follow changes in exposure to environmental conditions. For these reasons, suspicions remain unproved about possible links between air-borne agricultural pesticides or defoliants such as Agent Orange, waste chemical dumps, tobacco smoke or mineral residues in industrial tailings dams and the development of tumours within the body.

Geographic Distribution

The incidence of cancer differs from place to place. Just as differences in occurrence may be observed on an international country-to-country basis, occurrence rates and their levels for different parts of the body can be observed between rural and urban populations. Variations of a geographic nature may arise from genetic variation between the people living in different parts of the world, or they may reflect a difference in environmental factors. Genetic and environmental causes can be distinguished by observing the effects of migration on comparable racial or ethnic groups.

Where a true geographical and environmentally linked association exists, the high incidence rates occur in all inhabitants of the region whatever their racial or ethnic origin; similar high frequencies cannot be observed in comparable population groups living elsewhere. Healthy people who move into the affected area will soon acquire the same disease prevalence rates of the inhabiting population. Those who subsequently return to their previous home will, in time, experience a return to the previous disease prevalence applying there.

These relationships have been observed in diseases such as hypertension and atherosclerosis, diabetes and with certain kinds of cancer. However, while the data suggests that environment, of a physical or social nature, plays a role, explaining and proving the association is difficult. As a result, most are imputed and remain suspect and unproven. For example, cancer of the oesophagus is common in Iran and among Africans living in South Africa (one per thousand per annum). However, lower levels of oesophagal cancer (25 per million) are evident in Africans living elsewhere on the African continent.

Similarly, WHO (1980) describe a geographic distribution in the incidence of stomach cancer. Crude mortality suggests the highest occurrence of the disease is in Japan (48 per 100,000 population), Eastern Europe, principally Hungary (43 per 100,000), and East Germany: the United States records a low incidence of stomach cancer but a particularly high mortality for lung cancer, cancer of the colon and cancer of the breast.

The liver is often identified as a site for cancer, yet in reality it is more often the site for secondary tumour growth. Primary liver cancer is not among the more common cancers in the United Kingdom but is a focus for cancer in Africa where ground nuts such as the peanut are a major food source. Aflatoxin is a known carcinogen: ground nuts, stored under damp conditions, become mouldy and develop high levels of aflatoxin. There is also an increased risk of developing primary liver cancer after an attack of serum hepatitis, particularly if chronic infection occurs as is the case in poor, rural communities in Africa and even in disadvantaged groups in developed countries, such as among Australian aboriginal fringe town dwellers.

Cancer of the breast tends to be found most often in western countries and there is an imputed link between a diet high in saturated fats and cholesterol and the development of breast cancer. The United States, Australia and, to a lesser extent, Britain show a high incidence of breast cancer. Britain (particularly Scotland), the United States, Australasia and northern Europe show a high fatality rate for lung cancer, with figures climbing in newly developing nations such as China and other parts of South-East Asia.

Cancer and Locality
Small local changes in environmental conditions may also be significant in the aetiology of human cancers. For example, there have been conflicting reports concerning the effect of high-voltage power lines on animals and humans. Cows grazing within 200 metres of high-tension wires have been claimed to produce less milk than in other paddocks away from the wires. Protesters disputing sitings of such wires on their properties are voicing their concerns. Furthermore, new US research suggests that electromagnetic fields from power lines can cause leukaemia and other cancers, particularly in children.

In order to resolve the dilemma, researchers have recruited 800 children to study links between electrical power lines and cancer. The children, half of whom have leukaemia, will wear a device to measure the electromagnetic fields they live in. It is hoped that the evidence obtained will resolve the issue.

Cancer and Occupation
The first cancer recognized as being occupation-related was cancer of the

scrotum, originally described by Dr Percival Pott in London chimney sweeps in 1775. Chimney soot, the inadequately combusted carbon from coal fires, was blamed. Pott's observations formed the basis for the first preventive measures against occupationally induced cancer, as three years later the Danish Chimney Sweepers' Guild urged its members to take daily baths. A century later, Butlin, another English physician, concluded that the much lower incidence of scrotal cancer among chimney sweeps in northern Europe compared to their English counterparts was related to the hygiene measures and protective clothing, instituted by the former group. It is now known that the cancer effects are the result of continual exposure to the chimney soot and that soot owes its cancer-causing property to the presence of some polycyclic aromatic hydrocarbons (PAHs).

The first observed cancer effects, other than those involving the skin, attributable to occupation were probably those of Dehn in the latter part of the nineteenth century.[2] He reported on the development of cancer of the urinary bladder in three workers in a so-called aniline dye factory in Germany. These reports led to subsequent observations in many countries on the association between bladder cancer and work involving exposure to 2-napthylamine (b-napthylamine), benzidine or 4-aminobiphenyl, an important class of organic chemical amine that rose to prominence in German dyestuff production during the 1880s and 1890s. A subsequent epidemiological survey conducted in Germany and Switzerland in 1912 confirmed that bladder cancer was 83 times more common among workers in the aniline and aromatic amine dye production industries than among the general population.

For many years these findings were ignored in English-speaking countries. During the First World War, sizeable dyestuff industries were established in Britain and the United States. By 1920, aromatic amines were being produced and used extensively on both sides of the Atlantic. To forestall a repetition of the German experience of bladder cancer, the International Labour Office published a report in 1921 entitled *Cancer of the Bladder Among Workers in Aniline Factories*. The report concluded that there was insufficient evidence to incriminate either benzidine or beta-napthylamine, but advised that there be rigorous attention to hygiene to minimise risks. This advice was also ignored. The dyestuffs industry in Britain and the US continued large-scale production based on the use of the suspect aromatic amines with little regard to controlling occupational exposure. Within ten years, instances of bladder cancer among workers were being reported daily: a problem that could have been avoided and thousands of lives saved.

Social Class
The association between occupation and socio-economic class affects cancer

rates as much as it does the development of other less serious diseases. Differential social class distribution also indicates the effects of varying physical environments or of socially determined behaviours, particularly in cancers of the lung, the colon and the liver. Here, factors such as cigarette smoking, poor diet, high alcohol intake and living conditions play a significant role. Among the poor, late detection tends to contribute to higher death rates than among better educated, more privileged sectors of a population.

Occupation or Lifestyle?
The proportion of the total number of cancers which are caused by occupational exposure is controversial and is often boiled down to a distinction between 'lifestyle' and 'work' factors. The issue is important because the nature and scope of cancer prevention programmes funded by public moneys and donations is strongly influenced by the contribution that occupation makes to all cancers (Table 4.7).

Perhaps the greatest obstacle to obtaining general agreement on the question of how much cancer is occupation related, is that the majority of cancers are multi-factorial in origin; that is, two or more factors may be responsible for cancer development. For example, lung cancer in a heavy smoker with a known history of exposure to an occupational carcinogen may be due to the smoking, or to the industrial exposure, or to a combination of both. The difficult question that must be resolved is to what degree each factor is likely to be responsible. This type of answer can to some extent be obtained by epidemiological investigation (study of disease distributions in human populations), where certain factors are controlled while variation in others is looked for. Alternatively it is possible to attribute cancer to all of the most likely causes. This has the drawback that when added together the total figure exceeds 100 per cent.

The approach that has most often been taken to arrive at an estimate of the occupational contribution to cancer involves estimating the number of cancer deaths directly attributable to prior industrial exposure to known occupational cancer agents, and from this exercise, producing forward estimates. Such attempts have unfortunately produced widely varying estimates ranging from 2 to 40 per cent. This range of figures results from the differing methodologies employed by the various investigators and from the different sets of databases that have been used.

A most widely quoted document addressing itself to this issue is the NCI/NIOSH/NIEHS Report (commonly referred to as the Bridbord Report). This report estimates that occupationally related cancers may comprise as much as 20 per cent, and the data relating to five other industrial carcinogens — arsenic, benzene, chromium, nickel and petroleum fractions — suggest at least 10 to 20 per cent more. Added together, this makes 23-38 per cent of the total, to which the effects of ionising radiations and other known carcinogens must still be added. The methodology employed by this investigation involved estimating for each of the above listed industrial carcinogens the current excess risk and the exposed populations. When these are multiplied by the population-wide cancer incidence for the carcinogens' target organ, an overall estimate of occupational cancer is obtained.

The report's estimates have been criticised most notably by Doll and Peto (1981). They contend that the lack of attention paid to the duration and degree of exposure to these industrial carcinogens results in risk estimates that are grossly exaggerated. In effect, Doll and Peto argue that the absolute worst excess risk estimates that are derived from heavily exposed workers were applied across the board to other working populations with trivial or minimal exposures.

While it is certainly likely that the NCI/NIOSH/NIEHS report results suggest a greater than real risk, traditional estimates which they challenge are likely to be highly conservative. For instance, most such estimates suffer from the fallacy of a 'one effect — one cause' explanation which, because of the likely multifactorial element in cancer, is inappropriate. Further, the development of conservative estimates suggesting perhaps 1-2 per cent of cancer being attributable to occupation can be ambiguous. Often it is the practice to ascribe proportions to all other likely known causes, then attributing the remainder to occupation. A sounder approach is to review the extent of occupational contribution to cancer at various sites such as the lung, bladder, brain, etc., based on the known effects of individual industrial carcinogens, and from this obtain an overall estimate.

Using this approach, Higginson (1981) estimates that around 6 per cent of all cancer is due to occupational factors, a figure that coincides with the findings of Doll and Peto. It represents a substantial increase from earlier suggestions by Higginson of 1-2 per cent using a similar approach. This indicates the dependency of this approach on the degree of information available concerning the number and effect of industrial carcinogens in use. Clearly, as more substances are proven to be carcinogenic so too will our occupational incidence estimates rise.

Viruses and Cancer
Viruses are among those biological factors which seem to play a role in the development of cancers. However, whether they directly stimulate tumour growth, undermine the integrity of the body's natural immune system or simply weaken the body's total defences against infection so that it becomes subject to reproduction errors in the healing process, remains a matter of

speculation.

It was in 1911 that Rous, experimenting with chicken viruses, discovered that they were capable of causing animal cancers. Both DNA and RNA (ribonucleic acid) viruses are involved, and in some cases the infection can be transmitted from one animal to another.

It has proved more difficult to identify viruses in human cancers though circumstantial evidence exists, but transmission-type experiments cannot be undertaken for ethical reasons. The longevity of the human species also makes difficult any tracing of possible links between cases, and the variability of biological responses and their interactions complicate any interpretation of findings. Furthermore, if viruses are found in cancer cells they may only be 'passengers'. If found more frequently than expected, this may simply be because the cancerous cells, being different, are more prone to infection.

Lymphomas are among those cancers which may be due to an infectious agent. The lymphomas are a group of cancers related to leukaemia, but differing from it in important respects. Whereas leukaemia in early stages attacks the bone marrow or the blood, the lymphomas usually start in lymph nodes or organs rich in lymphatic tissue, such as the spleen. They can occur in almost any organ of the body.

Lymphomas are uncommon cancers (in Australia there are fewer than 800 cases a year) occurring in men and women of all ages but rarely in children. There are two kinds of lymphomas. One is Hodgkin's disease, named after Thomas Hodgkin, a nineteenth-century English Quaker physician who first described cases of the disease. The other is a group of several types known jointly as the 'non-Hodgkin's' lymphomas. Both require different forms of treatment: that for Hodgkin's disease has been highly successful.

While the causes of lymphomas are unknown except for that of Burkitt's tumour, which is chiefly found in tropical areas, infection with a virus is now known to be a factor. However, no one knows whether the virus causes the cancer or simply makes the tissues susceptible to it.

Transmission of a herpes virus by mosquitoes is believed to be involved in the development of a human lymphoma that occurs in African children. It has been called the Epstein-Barr virus (EBV) after the two workers who first isolated and described it. In laboratory cell cultures, the virus brings about changes in lymphocyte cells similar to those observed when the cells become malignant. Also, high concentrations of antibodies to the virus are always found in patients with Burkitt's lymphoma, a situation which occurs only when a foreign intruder such as an infection has stimulated an allergen-antibody response.

EBV is now known to be the cause of infectious mononucleosis (glandular fever) which often afflicts adolescents. However, while the appearances of the mononuclear white cells of the blood are very similar to those seen in some cases of leukaemia, the condition is always self-limiting.

Cancer of the cervix is strongly believed to be associated with genital infection with herpes virus. Human papilloma virus (HPV) is believed to be included also. Genital warts, a benign growth caused by a common wart virus, are often found also in the same group of patients. All three diseases — genital warts, genital herpes and cervical cancer — seem to occur most often in women who have had multiple sexual encounters, have a history of poor genital hygiene and whose sexual experiences began at an early age. The possible link between them remains ill-defined.

An etiological role for the following viruses in the development of human cancer has been suggested: hepatitis B virus (HBV), Epstein-Barr virus, human papilloma virus and human T lymphocyte virus 1 (HTLV-1). Experimental and epidemiological evidence shows that viruses are implicated in about 20 per cent of female and about 10 per cent male cancers worldwide (Zur Hausen, 1986). In women, the high figure is related to the incidence of cervical cancer in which human papilloma-virus is believed to play a key role.

Generally there is a long latency period between the primary viral infection and tumour development. The exceptions are Burkitt's lymphoma (BL) where the latent period can be from three to twelve years only, and cervical cancer, which may develop within five years of an HPV infection.

For a virus to be considered oncogenic, there needs to be clear evidence of association of the virus with the tumour, best proved by presence in the tumour cells. However, the virus may become integrated into the host cell genome and therefore not easily detected. Ideally, presence of antibodies to viral antigens, induction of tumours in animals by related oncogenic viruses, and transformation of human cells *in vitro* by the virus should also be apparent (Meekin, 1990).

Carcinogenesis is recognised as a multistep process that may take many years to complete. Zur Hausen has put forward a theory of 'failing intracellular surveillance' to explain viral involvement in carcinogenesis. This theory views the development of cancer as a three-stage process occurring over many years, which ultimately causes host cell control of viral genes to fail. Modification of host genes coding for an intracellular suppressor protein — cell interfering factor — causes inactivation of the viral repressor mechanism in the host cell. Subsequent viral activity leads to transformation of the cell to malignancy. The hypothesis accounts for latency, the monoclonality of tumours and the fact that only a small proportion of infected persons may develop cancer following infection with an 'oncogenic' virus.

Viruses may vary in their oncogenic potential. It is generally accepted that viruses often need co-factors to involve cell oncogenicity. These factors include immunosuppression, carcinogenic chemicals, action of human oncogenes, environmental influences and presence of other infections. For example, it is now widely accepted that HPV alone does not cause cervical cancer. Co-factors are necessary for malignancy to result. Smoking, chemical carcinogens, immunosuppression, sexual practice, hygiene, hormonal changes, and presence of other sexually transmitted diseases such as herpes simplex virus type II.

Similarly hepatocellular carcinoma (HCC) is one of the most common cancers worldwide, and 75 to 90 per cent of primary liver cancer is attributed to HBV. The role of HBV in liver cancer was first suggested in the 1970s, since confirmed by epidemiological studies. A geographic correlation exists between the incidence of HBV chronic carrier state and that of patients with liver cancer. Supportive evidence also links related hepatic viruses and liver cancer in animals. However, co-factors are not thought to be of major importance in progression of this cancer.

The oncogenic potential of Epstein-Barr virus was first noticed in Burkitt's lymphoma. In addition, geographic distribution of both malaria and the lymphoma suggests that malaria either stimulates EBV-infected cells or alters the sensitivity of B lymphocyte cells to the virus. Nasopharyngeal cancer is a comparable example. This form of cancer is most common in South-East Asia where the incidence is 10-20 cases per 100,000 population per year (Galloway, 1989). It is more common in men and the peak age group is 45 to 55 years. The ethnic distribution implies that a genetic factor may operate. No *in vitro* or *in vivo* models exist but the virus is believed to act in a similar way to its role in Burkitt's lymphoma. However, in this instance, chemical co-factors are also believed to have a role. Carcinogenic substances used in herbal medicines and foods, including chemicals extracted from *Euphorbia* plants, activate EBV in human cells.

The most convincing evidence for viral oncogenicity is anticipated to follow results on the use of antiviral vaccines. Immunisation programmes against HBV are being carried out presently in a number of countries under the WHO. In twenty to thirty years the incidence measures of nasopharyngeal and liver cancer, and Burkitt's lymphoma could provide the necessary evidence to confirm viral oncogenicity.

Cancer Causes
Despite the above qualifications, the vast majority of cancers are now believed to be caused by lifestyle and environmental factors of a social, behavioural or chemical nature. Because they are associated with a particular lifestyle or result from customary practices in various countries, it is important to deal with them here.

There are believed to be many factors which contribute to cancer ranging from poor nutrition, a diet high in cholesterol and fats or a low-fibre diet, to tobacco and betel nut chewing, smoking, nuclear radiation, working with industrial products such as benzene, asbestos and coal, and prolonged exposure to the ultraviolet radiation from the sun. There is a vast amount now known about these conditions, methods of treatment and prevention but no cure has yet been found. Here we consider only those cancers caused by or associated with the environment and the effects which changed environmental conditions may have upon their incidence, prevalence and forms as suffered by humans.

Direct or indirect causes of cancer by the environment are believed to include the following:

Employment: Statistics indicate that 5 per cent of cancers are due to workplace hazard, amounting to a minimum of 25,000 annually in the US and 13,000 in the UK.
* in carcinogenic industries, e.g. nuclear industry, in radiography (exposure to high doses of X-rays)
* with chemicals, e.g. petroleum-based products, plastics
* passive smoking
* ray emissions; accidental discharges from equipment
* hazardous mining, e.g. asbestos, coal. One study estimates that some 270,000 Americans will die from exposure to asbestos between 1980 and 2009.[3]

Air-borne: Toxic natural and industrial material is released daily into the atmosphere. Health problems and cancers associated with atmospheric pollutants have been increasing steadily since the nineteenth century. US estimates of the proportion of cancer due to pollution of the environment vary from 5 to 14 per cent.[4] Sources of air-borne carcinogens include:
* nuclear emissions into the atmosphere. The official Soviet report on the Chernobyl accident predicted 49,000 cancer deaths as a result of the nuclear catastrophe.[5] Other estimates vary between 5 and 10,000.
* industrial gases
* volcanic gases
* passive smoking in confined spaces

Agriculture: Cancer-related deaths arising from chemicals used in agriculture are increasing annually, particularly in the developing world where inadvertent use, poor education and unsafe practices are responsible for excess quantities being used and carelessness in regard to safety precautions. J.

Forget of the International Development Centre in Ottawa provides evidence that some two million people are poisoned by pesticides per year, resulting in 40,000 cancer-related fatalities.[6] *New Scientist* in 1989, basing its information on surveys by the UN Environmental Programme and the latest WHO figures, places the number of deaths from pesticide misuse alone at close to one million annually.[7]

At home: Sources of risk in the home include:
* life-long exposure to low levels of radon gas arising from household appliances. The US Environmental Protection Agency estimates 20,000 lung cancer deaths per year may be due to high levels of household radon exposure.[8]
* long-term exposure to pesticides used in termite and rodent eradication (regular treatment of buildings and under homes).

Diet: Epidemiological studies suggest that a significant number of cancers are associated with eating habits. For example, a low-fibre diet, high in saturated fats and meat proteins has been linked to bowel cancer: a taste for raw fish and birds' nest soup by the Japanese seems to be linked with a high incidence of stomach cancer. It has been estimated that better diets could cut cancer rates by up to 35 per cent in industrialised countries alone.[9] A 20 per cent initial decline in the incidence of oesophagus cancer in France is predicted simply by decreased consumption of brandy in the area around Calvados, a prolific producer region.[10]

The environment: Mercury poisoning such as that which occurred at Minimata in Japan was responsible for deformities in the newborn, for nerve system derangement and mental damage. The high levels of minerals in the water, in agriculture crops grown around Minimata and in the fish harvested from the Bay waters were implicated in the abnormally high level of cancer deaths and birth abnormalities (previously discussed).

Chemical Agents
A range of environmental factors of a chemical nature have the potential to affect human health. Basically these can be divided into two main groups: inorganic and organic molecules. The latter may be further subdivided into those of use in or by industry where the hazard is mainly occupational, and those which are biological in nature, used for medical or pharmaceutical purposes. Inorganic chemicals may be minerals, elements or their inorganic compounds.

Chemicals may act as initiators, promotors or both, and they exhibit all grades of potency. Their potential to interact with one another and with other factors influencing human health, and the biology of other ecosystem species, combine with the variable induction periods to make it almost impossible to unravel the web of causation of cancer.

While studies in animals assist in identifying possible carcinogens and their potencies, these same chemicals may have a completely different effect in humans, targeting different organs and possessing different levels of potency. At the same time, the degree of risk associated with use of the substance, either by industry or the community in general, has to be balanced against the safeguards which can be implemented to minimise exposure together with its benefits.

Special Environmental Cancers
Just as the environment is multifaceted in its nature, so too are the roles played by its physical, chemical and social components and the ways they interact in the development of cancers in animals and humans.

Discussed below are those cancers in which some aspect of the environment plays either a major or a minor role. Significantly they include the major causes of death among the middle-aged and elderly in the population of the developed, western countries.

Lung Cancer
Lung cancer is the major single contributor to cancer mortality in the United Kingdom and Australia, currently accounting for some 6 per cent of all deaths in England and Wales: 9 per cent of all lung-cancer deaths occurs in males. The 1984 data on major causes of deaths in Australia show that in the age cohort 45-64 years, 11 per cent male deaths were attributed to lung cancer. Women with lung cancer accounted for 6 per cent of all female deaths in this age group. Between 65 and 74 years of age, lung cancer continued to claim 10 per cent of all deaths, with chronic lung diseases accounting for a further 8 per cent.

Mortality from lung cancer as measured by SMRs is greater in urban than rural areas. This is particularly the case in the industrial towns of the English Midlands and the southwest, where the likelihood is that a number of physical environmental factors act synergistically in bringing about the increase in urban mortality. Thus the general pollution of the atmosphere, due to emission from and locations of industry, further jeopardises the life chance of workers exposed to potential carcinogenic agents as a result of their occupation. Pollution of the atmosphere, especially by smoke generated by the inefficient combustion of coal, is accompanied by the presence of known carcinogens such as 3,4 benz-pyrene in the atmosphere.

During the twentieth century, there has been a substantial rise in the incidence of lung cancer. There is little accurate information on lung cancer deaths prior to the 1920s because of confusion in diagnoses with other lung afflictions, principally pneumonia and tuberculosis.

An estimated 40 million total deaths in 1926 had by 1988 risen to almost 700 million, on a global basis. Tobacco and the consumption of cigarettes has been blamed for much of this tremendous loss of life. The epidemiological data confirm the increased death rates with differential consumption patterns for the sexes and the gradual spread of cigarettes throughout the world.

Tobacco and Lung Cancer: Tobacco has been estimated to be responsible for the deaths of 2.5 million people annually resulting from cancer of the lung, cardiovascular disease and chronic respiratory diseases (e.g. emphysema, bronchitis and asthma). As a result of warnings and health education, tobacco consumption in the industrialised countries is falling annually by about 1.1 per cent.[11] However, this decline is greater among men than women. Recent surveys indicate that among women, especially those less than twenty-five years old, tobacco consumption has increased in some countries in excess of 20 per cent in recent years.[12]

In Third World countries and particularly among the young, tobacco use has also increased, on the whole in excess of 2.1 per cent per year.[13] On a per capita basis, Figure 4.5 shows the percentage change in cigarette smoking which occurred in the ten-year period between 1975 and 1984 inclusive. Mouth cancers, mostly caused by chewing betel nut and tobacco, occur mainly in Asia and Oceania.[14] Currently, China is the world's leading consumer of tobacco products. In Shanghai, lung cancer is the leading cause of death.[15]

Tobacco use is so widespread and so much a part of society that it is often overlooked as being dangerous. A poll conducted in 1986 for a major national Australian newspaper found that 47 per cent people rated alcohol as the drug most harmful to society, 36 per cent nominated heroin, and tobacco came in third at 5 per cent. The magnitude of the tobacco problem is recognised by few and is confirmed in a study conducted by the Anti-Cancer Council of Victoria in 1983, which showed that among smokers there is much ignorance about illnesses and cancers caused by smoking. Only 58 per cent current smokers linked lung cancer with smoking, 8 per cent linked smoking with stroke and 3 per cent knew that smoking could cause complications in pregnancy. Almost a quarter of those surveyed did not believe smoking to be at all harmful to health.

In the USA and Japan, mortality from lung cancer is lower than would be expected from their levels of cigarette tobacco consumption. This anomaly may be due to the fact that Americans often smoke less of the cigarette than consumers in other countries. A comparison made by different researchers suggests that, on average, Americans smoke only about half of their cigarette whereas in the United Kingdom only 25 per cent is thrown away.

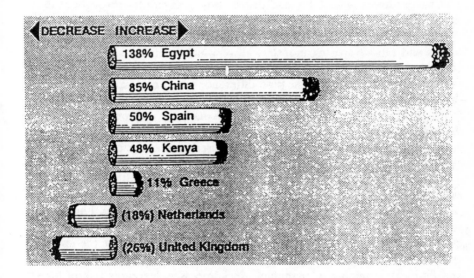

Fig. 4.5: Percentage changes in cigarette smoking consumption per person. Source: Multinational Monitor, *July/August 1987.*

Animal experiments have shown that the tar distillates from tobacco contain carcinogenic substances. When smoking cigarettes there is a greater chance of the distillates reaching the bronchi because of the temperature of combustion of the tobacco and the steady consumption of the length of the cigarette, in which some of the earlier distillate will have condensed. For pipe smokers, much of the tars distil out in the stem and bowl of the pipe. Their risk of developing lung cancer is significantly lower than that of cigarette smokers.

Risk of developing lung cancer is also substantially greater in those cigarette smokers who inhale the smoke. Different brands have varying levels of nicotine, tars and other potential carcinogens. A further contributing factor is the presence or absence of filters.

Aside from lung cancer, tobacco use causes several other respiratory diseases with high incidence in both industrialised and developing countries. It is associated with cancer of the mouth, lips, oesophagus, larynx and bladder. Since the 1920s, however, the population death rate from lung cancer has risen from about 40 million to nearly 700 million. This rise has occurred differentially in the sexes, starting first in men around 1926 but not appearing to any significant extent in women until the 1950s.

Changes in the availability and distribution of cigarettes throughout the world were responsible for the initial increase in sales. With the Second World War came changes in community attitudes and standards, fashion trends and the advertising industry, all of which contributed to increased consumption. It was at this time, too, that large numbers of women began to smoke, and it is this thirty-year time difference which can explain the lag in the onset of the lung cancer epidemic between men and women. During the 1970s the incidence of female lung cancer has made the major contribution to the continuing rise in overall cancer mortality. Furthermore, the age at which women are commencing to smoke has lowered, and those reporting early signs of the disease are now often in their late thirties.

Certain occupations are also known to be associated with an increased risk of developing lung cancer. Some are well known, such as that associated with exposure to fumes from the combustion of coal in coke ovens. Less often recognised is the increased risk from metal smelting, especially nickel and chromium. The effects of irradiation by radon are believed to be the important factor in uranium mining. Exposure to asbestos, particularly to the dust and fibres as in mining or manufacturing processes, increases risk of lung cancer — a risk which is potentiated if there is concomitant exposure to cigarette smoke.

Also associated with exposure to asbestos is mesothelioma, affecting the pleural membrane surrounding the lung. Although less common than cancer affecting the bronchi of the lungs, it appears to be on the increase. The Australian Navy is currently facing a plethora of claims of negligence associated with asbestos insulating materials, from men who served on its post World War II warships and aircraft carriers.

Cape Province, South Africa, was the site where the first definitive evidence of a relationship between asbestos and mesothelioma of the pleura of the lung was made. Wagner, Slegg and Marchand in 1960 described a total of 47 cases of mesothelioma, 45 of which were subsequently found to be a direct result of involvement in crocidolite asbestos mining.

The relationship was confirmed by the work of Newhouse and Thompson (1965) who followed 83 Londoners with mesothelioma and found that over half of them gave a history of occupational or domestic contact with asbestos.

Of those with no such history, one-third lived within half a mile of an asbestos factory. The interval between first exposure and the development of the terminal illness ranged between 16 and 55 years. Asbestos, or asbestos bodies associated with a local reaction to the presence of asbestos, were present in biopsied material in each case.

The incidence of lung cancer differs with social class and is highest among those of lower socio-economic groups in all industrialised countries. This differential risk can be explained by two factors. First, there are occupational risks which would affect specific social groups. Second, a large proportion of those in lower social classes live in urban industrial areas where they are more exposed to atmospheric pollution. A third and more controversial explanation is the different lifestyle, attitudes and levels of education found when comparisons are made with those in higher socio-economic groups which record lower levels of lung cancer. These better-off individuals are also more likely to be screened early in the course of the disease and to take preventive action to control and reduce their risk of eventual death from lung cancer.

The Cancers Of Women

Profiles of female cancers prove as alarming as those for lung cancer. The mortality data focus on cervical, breast, ovarian and vaginal cancers. In England, approximately 2000 women die each year from cancer of the cervix. In Australia, the figures suggest an SMR between 1 and 2 per cent of female deaths from cancers in all ages. The majority of these cases of cervical cancers occur in the 25-44 age group. It has been suggested that frequent sexual partners, early age of sexual experience, infection with venereal disease, low socio-economic status, smoking and poor sexual hygiene are all likely causal factors, acting either alone or in synergy.

While not significant as a cause of death from a numerical viewpoint, for women in industrialised countries cervical cancer is important as an example of a cancer which is susceptible, not only to secondary prevention but also, in theory, to primary prevention. There is substantial evidence of the role of environmental factors in its occurrence and spread.

Breast Cancer: Breast cancer is a cause of even greater concern for many women. In 1984, 11 per cent all female deaths in age groups 25-44 and 45-64 in both Britain and Australia were attributed to breast cancers. Thus a woman's lifetime risk of developing breast cancer has been calculated at one-in-15. Worldwide, deaths from breast cancers continue to rise though data suggest that Australia and the United States show the highest incidence of this form of cancer. In a more positive vein, breast cancers under 2cm in diameter when diagnosed and treated are likely to have little effect on life expectancy. With these small cancers, there is a possibility of limited surgery that largely preserves the breast, instead of mastectomy (removal of a breast).

Though data is inconclusive, studies suggest a possible link between breast cancer and lifestyle, particularly the consumption of a diet high in saturated fats such as the triglycerides. The link with a high-fat diet is based on studies of women living in Asia where the incidence of breast cancer is low and coincides with a low-fat diet. However, when Asian women move to western countries and change their eating habits, their risk of breast cancer increases. In Japan, an increase in breast cancer has been associated with the rising consumption of animal fats since World War II.

Also significant appears to be the age of child-bearing and whether or not breast feeding was practised by the mother. The higher fat content of diets consumed by women in the wealthier industrialised countries, and their tendency to marry and bear children at a later date, together with better health care and lower levels of mortality for other diseases, mean that this form of cancer tends to be the most common cause of early death.

Cervical Cancer: In Third World developing countries, cervical cancer has been blamed for premature deaths. Between 1960 and 1980, WHO surveys confirm that the incidence of cervical cancers in the industrialised countries fell by 30 per cent, largely as a result of widespread screening techniques. About 75 per cent of all cervical cancer deaths in these countries occur because of inadequate screening programmes and infrequent medical check-ups. Age of commencing sexual activities and sexual practice are also important.

Sexual behaviour has recently been strongly implicated in the causation of cancer of the uterine cervix via the transmission of the carcinogenic virus which causes genital warts. It would seem that cancer of the cervix is a sexually transmissible disease which could be prevented through reduction in casual sex, the use of condoms and improved standards of sexual hygiene among 'at risk' groups.

Identification of potential causal factors has come from studies of the differential incidence of cancer of the cervix between ethnic or religious groups. In studies undertaken in the United States, Haenszel and Hillhouse (1959) reported an incidence of 4.7 per 100,000 Jewish women per annum in New York, compared with an overall rate of 18.3 per 100,000. In similar studies, Moslem women have also been found to have a relatively low incidence. Although male circumcision was at one time thought to be a possible common factor, further studies on uncircumcised Jewish and Moslem men have failed to confirm the link.[1 6]

Data from the early 1950s of studies conducted on Canadian nuns adds weight to conclusions that the activities associated with sexual intercourse itself may be responsible, possibly through tissue damage or the introduction of some foreign substance. An examination of the medical records of 13,000 nuns over twenty years identified 130 cancers, among which were 53 cases of breast cancer but not a single instance of cancer of the cervix.

A similar search of statistical records by Rawls *et al*[1][7] led to the hypothesis that sexual transmission of an infectious agent, possibly the herpes simplex virus, may play a role. Traces of herpes simplex virus nucleic acids have been detected in cancer cells by some researchers, and other laboratories report herpes viral antibodies in the serum of cervical cancer patients, while absent in that of a control group of patients who did not have cancer. While appearing to lend support to a causal association between herpes virus and cervical cancer, it is possible that the virus may only manifest in patients at a time when cancer cells are present. These cancer cells are less able to resist infection by the virus than normal healthy cervical tissue.

However, if sexually transmitted viruses are the cause of cancer of the cervix, it might be expected that the use of barrier methods of contraception (such as the condom or the diaphragm) would interfere with transmission and reduce the incidence of the disease. Unfortunately, increased use of artificial measures of contraception has tended to coincide with a more relaxed attitude towards sex, particularly in western societies, together with lower age of first experience and increased number of sexual partners. This further confuses the data on cervical cancer. The only conclusions that can be made equally implicate all causal factors and agents, social and environmental, behavioural and biological.

Skin Cancer

Skin cancers are mostly caused by over-exposure to the sun. There are three major types: the two most common are basal cell carcinoma (BCC) and squamous cell carcinoma (SCC). These rarely spread to other parts of the body and are seldom fatal. On the other hand, they can be very disfiguring if not treated early.

BCC is also known as rodent ulcer and is perhaps the most common of all human malignancies. These tumours grow slowly and seldom metastasize. They are easily removed. SCCs are malignant tumours of the prickle cells in the epidermis. They invade surrounding tissue and will metastasize. An early lesion may resemble a solar keratosis.

Malignant melanomas, the third type of skin cancer, can behave like an internal cancer. If not detected early and treated properly, they can spread to other parts of the body. A melanoma usually appears initially like a mole, unusual freckle or birthmark. Technically, it is a neoplasm of the melanocyte which is extremely dangerous if left untreated for even a short period. Malignant melanomas can occur at any age, but are rare before puberty. Whether they are related to excessive sun-exposure or not is still a subject for debate as the evidence is not conclusive. While they do occur more often in sun-exposed areas, this is not always the case as they can develop on parts of the body rarely if ever exposed to the sun.

In Australia, the country with the highest rates of skin cancer in the world, approximately 145,000 new cases are reported each year. Of the cancer fatality rate of 1000 per annum, 80 per cent are due to melanomas.

Normally the skin protects the body from the dangers of the sun. Squamous cells are constantly being sloughed off and replaced by new cells. The melanocytes produce more dark pigment called melanin, which partly protects the skin from excess ultraviolet radiation in sunlight and from exposure in certain industrial processes. Most Australians and people of Caucasian descent have skin which produces inadequate amounts of melanin to ensure protection against strong sunlight.

Those at greatest risk of injury are people with fair skins, especially if they do not adequately protect the skin by sunscreens and clothing, including hats. People who work in the sun, those who burn easily and do not tan, and those with a family history of skin cancer or who were born or live in countries with a climate like Australia are most likely to be susceptible to skin cancers. Recommended precautions against skin cancer are given on page 126.

Cancer of the Stomach

Cancer of the stomach, together with cancer of the bowel, is among those cancers which are not easy to detect. Because of its relatively low risk, screening is not undertaken as a routine procedure in either Britain or Australia. However, in Japan, where the disease has a relatively high incidence, screening techniques have reduced mortality rates. Researchers link its frequency there with the popularity for pickled and highly salted foods.

For stomach cancer there is also a remarkable but unexplained geographical variation in incidence both globally and within individual countries. A gradient in mortality can be observed by tracing death rates as one progresses westwards from Japan across Asia and the European continent, or eastwards across the Pacific. The lowest rates occur in the United States. Migration studies have tended to discount genetic factors in the relatively high Japanese incidence because Japanese migrants to the USA show a reduced rate. In addition, the reductions are related to the length of residence in the United States.

However, the picture is not clear because genetic factors seem to play a role. American studies demonstrate that US-born citizens with Japanese ancestry suffered only one-third the stomach cancer mortality of Japanese living in Japan, and that Japanese residing in America who had been born in Japan had just over half the mortality rate experienced in their own country. Japanese migrants to Hawaii show rates intermediate between the Japanese and US figures.

Average mortality rates for a country also tend to mask local variations. For example, in the north of England, SMRs for stomach cancer are relatively high: similarly in north-west Wales. Studies of garden soil from properties in randomly selected counties in the United Kingdom where people had died from stomach cancer after fifteen or more years of residence, recorded higher concentrations of zinc and chromium than garden soils in the same area where a person had died of a non-malignant condition, or had died of cancer after less than two years' residence (Howe, 1970).

Other research teams have linked zinc and copper ratios in garden soil with stomach cancer deaths after prolonged residence, and have further extended the work to focus on the zinc:copper ratios in potatoes grown in such soils. In Welsh districts with a relatively high rate of stomach cancer mortality, the incidence after 10-19 years residence is excessive where the soil has a high content of inorganic carbon, such as in areas where peat is significant.

Thus, the soil mineral content or proportion appears to be important in itself. So too is the mineral and chemical content of the foods grown in such soils. More complex dietary relationships and dietary habits are also significant. For example, parts of northern Australia show a disproportionate level of stomach cancer. This is most marked among Aborigines whose general health and low socio-economic status tend to be consistent with a poor dietary history often characterised by alcohol abuse, smoking and substance abuse (e.g. petrol sniffing, intoxication with methylated spirits).

Then again, food chemicals naturally present or added to the finished product for prolonging life or improving flavour may play a role. Chile, Costa Rica and Argentina show relatively high mortality rates for stomach cancer when compared with the rest of the American continent. In some cases, for example in Chile, a high dietary intake of nitrates has been identified, usually from the use of nitrogen-bearing fertilizers and the consequent contamination of water supplies. N-nitroso compounds (N-nitrosamines and N-nitrosamides) are powerful carcinogens in animals, and are formed as a by-product of stomach digestive processes. A high dietary intake of inorganic nitrogen, or a diet involving use of tinned and preserved meats and sausages where nitrites act as preservatives, can, theoretically, lead to high concentrations of the N-nitroso breakdown products in stomach acid. The latter is highly likely in western communities where preserved meats and hams are increasing in popularity and are being eaten by a larger cross-section of the population.

Cancer of the Colon

The geographical distribution of cancer of the colon tends to show a recipro-
cal relationship to that of the stomach. Globally, it occurs less commonly in
Japan than in the United States. In general terms, cancer of the colon tends to
be a disease of industrialised western countries. Australia has one of the
highest rates of bowel cancer in the world. People over age 40 years are at
greatest risk, but the survival rate of those diagnosed is much better than
commonly realised, particularly if treatment is initiated early.

In nations like Britain and Australia, about one-in-25 people will develop
bowel cancer — one of the highest incidences in the world. One hypothesis
which has gained support in recent years targets for blame the high-fat,
low-fibre diet consumed in affluent western countries. Researchers consider
that a high-fibre, low-fat diet is likely to carry reduced risk of bowel cancers.
A similar diet is also recommended to minimise risk of stomach, rectal and
prostate cancers.

Dietary fibre is the indigestible portion of the diet. Raw vegetables, fruit,
plant-based foods, grains and cereals, nuts and seeds all contain substantial
dietary fibre. It includes a number of different compounds such as some
proteins, fats and carbohydrates as well as other less well known substances
like phytates. Each has its own particular effects within the gastrointestinal
tract.

Fibre originates mainly from cell walls and seed coats (bran) of cereals
and other plants. A generally accepted definition of dietary fibre is the plant
polysaccharides and lignin (a non-carbohydrate cell-wall polymer), all of
which are resistant to hydrolysis by human digestive enzymes. The polysaccharides include cellulose and hemicellulose, polymers of simple five and six
carbon sugars, and various pectic substances comprised of polymers of other
small sugar molecules and which are capable of forming gels with water. The
properties of the polysaccharides are such as to make them resistant to
digestion.

Lignin is a complex polymer which is not based on sugar but on an organic
compound of polymerised and chemically modified phenylpropane, a
medium chain carbon. It is a tough and extremely intractable material. Other
minor fibre components are indigestible gums and mucilages, such as agar,
carrageenan and gum guar which are natural components of some plant foods
and are used extensively in processed foods.

The value of fibre in the diet depends on two essential properties which
are related to the chemical composition of the food components listed above.
Cereal grains and similar foods containing large amounts of these polysaccha-

rides are the most concentrated form of fibre. They contain relatively large amounts of pentosan hemicelluloses, molecules which retain water and add bulk to the gut contents. At the same time, these carbohydrates are only digested slowly and parts of the molecular structures are resistant to gastric enzymes. As a result, foods with large amounts of these molecules pass through the gastric tract and proceed to the bowel, where they can be broken down, in part by the bacterial flora. At this stage the fecal matter still contains a large amount of water with the result that the final stool passed will be more bulky than for an individual consuming a different diet.

A number of researchers postulate that colonic fermentation of fibre into relatively simple organic acids, called volatile fatty acids, may play an important role in maintaining health of the lower colon, and reduce the risk of colon cancer. Flatulence (or wind) is produced by these reactions with the release of methane and other similar small molecular weight gases. These gases assist passing of the stool.

**TABLE 4.7: Principal Occupational Hazards
to Human Health**

PHYSICAL — noise
— vibration
— temperature
— radiation
— atmospheric pressure

CHEMICAL — metals (lead, cadmium, mercury, zinc)
— inorganic chemicals (arsenic, carbon disulphide)
— organic chemicals (benzene, xylene, phosgene 2.4.5.-T,
organophosphates)
— gases (carbon monoxide, nitrous oxide, hydrogen cyanide,
hydrogen sulphide, nickel carbonyl)
— acids
— alkalis

SOCIAL AND PSYCHOLOGICAL — performance requirements
— repetitious work
— lighting
— conflict at international level
— work organisation
— smoking (passive smoking)

TABLE 4.8: Frequency of Occurrence of Common Cancers in Men and Women (international comparisons)

Region	Sex	1		2		3	
		Cancer	Total cases	Cancer	Total cases	Cancer	Total cases
Latin America	M	Stomach	28,000	Lung	25,000	Prostate	19,000
	F	Breast	49,000	Cervix	44,000	Stomach	17,000
Africa	M	Liver	43,000	Lymphatic	20,000	Mouth	13,000
	F	Cervix	37,000	Breast	27,000	Lymphatic	12,000
China	M	Stomach	128,000	Esophagus	109,000	Liver	81,000
	F	Cervix	137,000	Stomach	68,000	Esophagus	59,000
India and other Asia	M	Mouth	97,000	Lung	62,000	Stomach	43,000
	F	Cervix	141,000	Breast	95,000	Mouth	48,000
UK*	M	Lung	30,000	Prostate	10,000	Bladder	7,000
	F	Breast	24,000	Lung	11,000	Colon	9,000

* UK figures (1984) included for comparison to indicate the pattern typical of industrial countries.

Source: WHO Cancer Unit, 1986.

Notes

(1) Editorial, *The Lancet*, 19 May 1984.
(2) Cited in *New Internationalist*, 'Cancer: The Facts'. August 1989, issue no.148, p.16, and extracted from Hester, J. *et al*, *Science for the People*, published May 1989.
(3) Forget, G. *Toxic Substances and Health*, International Development Research Centre, Ottawa, 1988.
(4) Brown, M.H. *The Toxic Cloud*, Harper & Row, New York, 1987.
(5) Haynes, V. and Bojcun, M. *The Chernobyl Disaster*, Hogarth Press, 1988.
(6) Forget, 1988, op.cit.
(7) *New Scientist*, 'Code of Conduct May Keep Dangerous Pesticides At Home', 9 December 1989, no.1684, p.6.
(8) 'Indoor Radon and Lung Cancer' from *New English Journal of Medicine*, March 1989.
(9) Patterson, J. *The Dread Disease*, Patterson, New York, 1987.
(10) Epstein, S. *The Politics of Cancer*, Sierra Club Books, 1978.
(11) *The Lancet*, 19 May 1984.
(12) NHMRC/Westmead Hospital, December 1989/January survey reported in SMH.
(13) *The Lancet*, 19 May 1984.
(14) WHO Cancer Unit. *Cancer Control In Developing Countries*, WHO Publications, 1986.
(15) WHO. *World Health In Developing Countries*, WHO Publications, 1987.
(16) Haenszel, W. and Hillhouse, M. 'Uterine Cancer Mortality in New York City and Its Relation to the Pattern of Regional Variation within the United States' from *Journal of the National Cancer Institute*, 1959, vol.22, p.1157.
(17) Rawls, W.E., Tompkins, W.A.F., Figuerora, M.E. and Melnick, J.C. 'Herpes-virus Type 2: Association with Carcinoma of the Cervix' from *Science*, 161, 1968, pp.1255-8.

References and Additional Reading

Canadian Cancer Society. *Annual Report, 1987-8*, The Canadian Cancer Society, 1989.
Bridbord Report. *Estimates of the Fraction of Cancer in the United States Related to Occupational Factors*, 1978, US NCI/NIOSH/NIEHS.

Brodeur, P. *Outrageous Misconduct: The Asbestos Industry on Trial*, Pantheon, New York, 1985.

Doll, R. & Peto, R. *The Causes of Cancer: Quantitative Estimates of Avoidable Risks of Cancer in the United States Today*, Oxford Medical Publications, 1981.

Galloway, J. 'Epstein-Barr Virus: Chance and Felicity' from *Nature*, 1989, vol.338, p.463.

Higginson, J. 'Proportions of Cancers due to Occupation' from *Preventive Medicine*, 1989, vol.9, pp.36-45.

Howe, J. (ed). *National Atlas of Disease Mortality in the United Kingdom*, Thomas Nelson & Sons Ltd, 1970.

McKeown, T. *The Role of Medicine: Dream, Mirage or Nemesis?*, Blackwell Scientific Publications, Oxford, 1979.

Meekin, G. 'Viruses and Cancer' from *Patient Management*, 1990, vol.14, no.4, pp.85-91.

Miller, E.C. 'Some Current Perspectives on Chemical Carcinogens in Humans and Experimental Animals' from *Cancer Research*, 1978, vol.38, pp.1479-96.

Newhouse, M.L. & Thompson, H. 'Mesothelioma of Pleural and Peritoneum Following Exposure to Asbestos in the London area' from *British Journal of Industrial Medicine*, 1965, vol.22, p.261.

Rowland, A.J. & Cooper, P. *Environment and Health*, Edward Arnold, London, 1983.

Wagner, D., Slegg, T. & Marchand, J. 'Diffuse Pleural Mesothelioma and Asbestos Exposure in the North Western Cape Province' from *British Journal of Industrial Medicine*, 1960, vol.17, p.260.

World Health Organisation. *World Health*, WHO, October 1980.

World Health Organisation Cancer Unit. *Cancer Control in Developing Countries*, WHO, 1986.

Zur Hausen, H. 'Intracellular Surveillance of Persisting Viral Infections' from *Lancet 2*, 1986, p.489.

4.4 The Skin

The skin is our barrier against the outside environment. It shields us from the wind, the cold, the sun, invading bacteria, dirt, dust, metal objects, plant hairs, thorns and fungal spores. It is the first line of outer defences against infection. Being the largest body organ directly in touch with the environment, the skin comes in contact with a wide range of plants, animals and micro-organisms, is subject to the vagaries of climate and weather, and suffers injuries, traumas and insults through our daily ablutions, at work and during leisure play. It is therefore required to be remarkably resilient and adaptive, and to be capable of rapid repair. At the same time, it must be sufficiently sensitive to detect minute changes of potential danger.

Physiologically, the skin is made up of three layers of cells: the outer epidermis, the reticular layer and the subcutaneous layer. The dermis is a tough, flexible and elastic layer of dense, irregularly arranged connective tissue. It is made up of cells older than other dermal layers below. They regularly slough off with age, wear and tear, and as a result of damage or injury. They are replaced by younger cells underneath.

Immediately beneath the dermis layer is the reticular layer. It contains numerous blood vessels arranged in two layers. In the deep portion of the reticular layer, many sinuous arteries are found. New cells form in this region and gradually move upward through the dermal layers as each top layer is sloughed off.

The subcutaneous layer is composed of loose connective tissue typically containing much fatty tissue. There is no clear line between it and the dermis. The tissue is loose enough to accommodate significant volumes of fluid and it is into this layer that subcutaneous injections are made.

Premature Ageing and the Skin
Ageing is a universal phenomenon which proceeds in a relentless fashion. All organs and tissues age at different rates but there is also considerable variation in the rate of ageing between races and between individuals. The skin, if properly cared for together with the body and emotions, will show few signs of ageing. Those areas of the skin which have received little or no exposure to ultraviolet radiation will often appear to maintain the subtlety, smoothness and texture of infant skin into old age. In contrast, those areas of the skin which have been continually exposed to sunlight will show signs of flaccidity, looseness, wrinkling, thinning, dryness and areas of keratolytic thickening.

All of these signs of aged skin are more evident in those who have been exposed to sunlight for prolonged periods or who live in areas where the intensity of the light is high. In contrast to people in Europe, those in a country like Australia who experience continual exposure to sunlight will show evidence of accelerated ageing of the skin. This damage begins in infancy and progresses over the years. Studies have shown that by age twenty, more than 80 per cent of all people examined had some evidence of skin damage. By age forty, all individuals show damage in sun-exposed areas.

All the signs of ageing are due to the thinning of the epidermis. There is also a change in the ability of the melanocytes to transfer pigment to the keratinocytes. Major changes in the appearance of the skin, however, are produced as a result of changes in the collagen and elastin fibres of the dermis. Studies have shown that soluble collagen decreases and the insoluble collagen increases as a function of the duration of sun exposure. These changes, together with the degeneration of the elastin fibres, produce the general condition known as actinic elastosis. The skin develops a furrowed, leathery appearance with numerous wrinkles. This is often called sailor's or farmer's skin. Cutaneous blood vessels rupture easily because of the decreased support by the dermal connective tissue. As a result, areas of purpura appear.

Plants and the Skin

Plants play an important part in the lives of all species and make a welcome contribution to the visual environment. Besides playing a vital role in the carbon, oxygen and nitrogen cycles essential for all other living species, they provide food, shelter, moisture, and many are a pleasure to behold. Chemicals essential for other animals and of medicinal value to humans are present in plant tissues; their perfumes and fragrances attract and repel pollinators and predators.

Plants come in contact with humans intentionally, as a result of farming, in cooking, timber-working, herbal and pharmaceutical manufacture. Even the most committed urban-dwelling westerner may tend a home garden, raise a few kitchen vegetables, visit a botanical garden, go hiking in a recreation reserve or camp in the wild during summer.

Apart from being stung or bitten by passing insects, developing tetanus from implanted rose thorns and sporotrichosis from cactus spines, and receiving forefoot amputations from rotary mowers, the gardener is exposed to a myriad of potential plant allergens and irritants. On all these occasions, transient casual contact with a plant can also occur which results in irritant dermatitis or a contact urticaria.

Plants or plant products may interfere with the skin in a variety of ways:

Irritant hairs and spines: Many plants produce spurs, burrs or trichomes (hairs) which enable plants or seeds to adhere very effectively to animal hides and to clothes. This permits wider dispersal of seeds in the environment so that the species becomes distributed to new sites. However, these same accessory parts may produce transient irritation of the skin.

While most people are aware of the impressive spines produced by the larger cacti or succulents, the smaller, more insidious, softer spines of species such as *Opuntia* can produce severe pruritus and even granulomata. These soft spines, called glochids, are easily shed into the skin after light contact.

A condition called 'Sabra dermatitis' is particularly prevalent in Israel in pickers of prickly pears (*Opuntia ficus-indica*) and can closely mimic scabies. The glochids are best removed by gentle application of adhesive tape to strip off the affected areas of the skin. Irritant crystals occur in the stems or leaves in many plants, particularly rhubarb, and also in bulbs such as daffodil, tulip and garlic.

Stinging Emergences: Members of several families, especially the *Urticacea* (the stinging nettle family), defend themselves from passing browsing animals by stinging emergences on the leaves and stems. Each emergence carries a small tip of oxalate crystals which is easily sheared off after light contact, revealing a sharp hypodermic needle loaded with inflammatory mediators, including histamine and acetylcholine; the resultant wheal and flame from stinging nettle contact is familiar to all. However, there are some singularly unpleasant species of stinging nettle, one of which, *Urtica ferox*, a native of New Zealand, has caused death from extensive contact.

Phototoxicity: Several families, notably the carrot (*Umbelliferae*), *rue* (*Rutaceae*) and fig (*Moraceae*), produce psoralens, substances which induce photosensitivity to long-wave ultraviolet radiation (UVA) and which are used in the therapy of skin disorders such as psoriasis or mycosis fungoides. It is interesting that the ancient Egyptians used leaves from a member of the carrot family (*Ammi majus*) in the treatment of psoriasis. Unfortunately these plants can produce toxic blistering eruptions if contact is followed by strong sunlight.

Contact Urticaria: Several plants produce contact urticaria, a localised irritation with or without an accompanying rash, in a small group of susceptible individuals. This is not due to direct injection of the histamine, as in the case of the stinging nettle, but rather to a glycoprotein on the surface of the plant. Many grass pollens which can be responsible for hay fever can also cause contact urticaria. Simply peeling fruit or vegetables such as apples or tomatoes (common culprits) may be responsible.

Contact Dermatitis: Several house and garden plants and weeds can cause florid contact dermatitis which may be either irritant or allergic, or a mixture of both. Thus, garlic handlers may develop irritant changes from the sap or a true allergic dermatitis due to an allergen, Tulipalin A, which is also present in other bulbs. Susceptible garlic handlers, such as chefs, will develop a fingertip contact dermatitis, and a similar rash can occur in professional bulb growers during the planting season (tulip finger, hyacinth scabies).

In the United States, the most common cause of contact plant dermatitis is sensitivity to members of the poison ivy family (*Toxicodendron* spp.). This family, which also includes the poison oak, is very commonly found in North American woodlands and waste grounds and can cause a very severe, often blistering eruption after slight contact. Some people develop a generalised erythroderma (severe reddening of the skin of the whole body) and may require hospital admission and high-dose steroids for control. Fortunately the toxic members of this family are not generally found except in botanic gardens, where they tend to be screened behind wires and cages.

Sun-induced dermatoses: Some rare congenital disorders may produce a rash resembling severe sunburn on exposure of the skin to the sun's rays. A group of conditions known as porphyrias can be responsible for blisters and erythema in the sunlight: also certain drugs such as tetracyclines, sulphonamides, phenothiazines.

The Sun and The Skin

Human beings have long associated sunlight with health and good fortune. Worship of the sun, not only as a source of power, heat and light but also as a supreme deity, can be traced back to the days of the early Egyptians. Persian, Greek and Roman writings are also full of references showing that the sun was an object of awe and wonder.

As far as the skin is concerned, the sun is not a benevolent god but a dangerous enemy. Over-indulgence in its pleasures frequently results in severe and irreversible damage. The only known beneficial effect of sunlight is its role in the production of vitamin D. The ultraviolet component of the sun's spectrum converts 7-dehydrocholesterol into vitamin D3. On the other hand, knowledge of the number of skin diseases in which sunlight plays a direct or contributory role is being expanded on a daily basis.

The term 'photosensitivity' is used to describe adverse reactions to sunlight. Table 4.10 shows that there are five separate categories of photosensitivity reaction. These include a total of thirty-six separate disease states that are either photoxic reactions or squamous cell carcinoma, or are conditions which are aggravated by exposure to sunlight, such as porphyrias.

Sunlight Induced Disorders and Protection

The sun, particularly its ultraviolet rays, is responsible for substantial acute and chronic damage to the skin, and for a number of sunlight-induced disorders (see overleaf, Fig. 4.6).

Acute effects: Sunburn varies in intensity from a mild erythema to severe vesicles and bullae formation. Symptoms include heat, burning feelings, rigors, nausea, vomiting and prostration.

Chronic effects: Premature ageing of the skin, senile elastosis, solar keratosis, basal and squamous cell carcinoma, melanomas and actinic cheilitis are among the many chronic effects associated with exposure to the sun over a prolonged period of time.

The Sun and Exposure

The degree of sunburn or erythema which develops in a sensitive skin will depend both on the duration and intensity of that exposure. An individual with normal skin sensitivity to sun (someone who burns moderately and tans gradually) will show the initial signs of a sunburn with only a twelve-minute exposure at noon in mid-summer (Table 4.9). The longer the duration of exposure, the greater the extent of the damage. A two-hour exposure of unprotected skin can result in permanent damage and might even have fatal consequences.

Unfortunately sunburn is a delayed phenomenon and as a result of over-exposure can occur readily. It is an inflammatory reaction whose first signs appear approximately four to six hours after exposure. These symptoms reach their peak in sixteen to twenty-four hours: they begin to fade, under normal circumstances, in about seventy-two hours.

The amount of sunburn-producing radiation in the atmosphere varies with the factors listed in Table 4.11. All of these factors can be related to the distance the radiation has to travel through the atmosphere. On any one day the most dangerous period of exposure is between 9am and 3pm, when about 80 per cent of all sunburning radiation is received.

Sensitive or even semi-sensitive individuals can still sunburn on a moderately cloudy day. The warmth of the sun stems from its infrared, not burning radiation. On cloudy days, this radiation is more effectively blocked and the period of exposure is often longer because no heat discomfort is felt. At the same time, the intensity of the ultraviolet radiation is increased because of the scattering effect of the clouds.

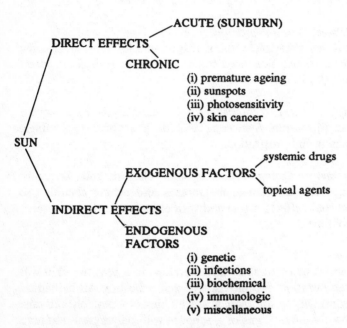

Fig. 4.6: Sun-induced Disorders of the Skin

The Nature of Ultraviolet Radiation

Ultraviolet radiation which causes sunburn and other forms of skin damage is actually made up of two components, direct sunlight and indirect skylight. The latter is the radiation which is scattered by the clouds, dust, moisture, etc., in the atmosphere. In summer these two components make contributions which are of about equal proportion. Together they are known as total global radiation, a concept which explains, in part, the ease with which people burn when on the beach or the golf course. Unlike the garden at home, there is nothing present to block skylight and so the intensity of radiation striking the skin tends to be greater.

The radiation emitted by the sun (solar radiation) is a form of electromagnetic radiation which can be pictured as a series of waves containing particles called photons. It is ordered into bands of waves (wavelength), measured in nanometres (nm). Skin reactions to radiation from the sun are largely concerned with the ultraviolet radiation band, which is the invisible region of the spectrum between 200 and 400 nm. Radiation in the ultraviolet band below 295 nm (UV-C) is all absorbed or scattered in the ozone layer of the stratosphere: it does not reach sea level.

The very short-wave gamma and X-rays which have the ability to penetrate deeply into tissue and cause extensive damage are also not present in natural sunlight. The main area of concern with regard to sunburn, therefore, ranges between 295 and 400 nm. The radiation between 280 and 320 nm (UV-B) is the erythema or sunburn region of natural sunlight. Radiation in this area of the spectrum has damaging effects upon the DNA of the epithelial cells, calling for subsequent enzymatic excision and repair. Failure of the repair mechanism could lead to mutations, so that ultraviolet light is potentially mutagenic.

The UV-B band radiation is also responsible for initiating melanogenesis. Most of the other damage associated with excessive exposure is caused by this band of radiation. This includes solar keratosis, a localized thickening of the skin. Specifically the most damaging is the 305 nm wavelength. Animal experiments also suggest that the carcinogenic effects of UV-B light are cumulative and that the incidence of cancer is directly proportional to the square root of the annual dose.

The radiation between 320 and 400 nm (UV-A) is responsible for the darkening of melanin already present in the skin. This is the phenomenon known as immediate pigment darkening, a tan which appears during or shortly after exposure and which lasts for only a short period.

Ozone and Skin Cancer

For details of skin cancer see Table 4.12.

The 1989 report on stratospheric ozone over the polar regions indicated that the ozone hole which reaches its largest extent over Antarctica in the spring was, in geographic terms, worsening. Measurements showed that at all southern latitudes down to 50 degrees, as much as 15 per cent of the stratospheric ozone was lost regularly in August and 30 per cent in September. This is outside the region subject to polar temperatures.

Over 140,000 cases of skin cancer were diagnosed in Australia in 1981 alone and treated at a conservative cost, then, of $91 million. Since then the numbers exceed 200,000 cases and costs have tripled. Medical experts believe that if the continued depletion of the ozone layer proves as predicted, people face an increasing risk of damage because the amount of UV-B light reaching the earth's surface will increase substantially. The effect will mean a considerable further increase in deaths from skin cancer and associated skin problems.

Protection Against Skin Cancer
Cancer councils, government agencies and skin specialists have set out a number of minimal precautions which individuals can take to minimise the cancer of sun-related damage to the skin and associated skin cancers. Briefly these are generally inclusive of the following:

* keep out of the sun in summer months, particularly between the hours of 11am and 3pm (GMT 1100 and 1500 hours)

* wear protective covering clothing on the body and cover the face and head, ideally with a broad-brimmed hat

* use zinc cream for total protection of vulnerable areas such as the nose, the lips and the backs of the hands

* use 15-plus broad spectrum sunscreen for maximum protection of areas such as arms, back, legs, etc.: particular care is needed for children and those with sensitive and fair skins

* apply sunscreen creams before and during time spent in the sun and especially if bathing (water-resistant creams)

Proposals for an international framework to reduce emissions of CFCs, put forward in 1976 by the anthropologist Margaret Mead and William W. Kellogg in the USA, would keep emissions of carbon dioxide below a global standard by assigning polluting rights to each nation. However, their potential success in reducing the incidence of skin cancer resulting from increased ultraviolet irradiation from ozone depletion of the stratosphere will be difficult to establish and prove.

Miscellaneous Skin Allergies

Atopic Dermatitis: A skin irritation of unknown origin and which is characterised by itching and scratching. There may be allergic, hereditary or psychological components. In about 70 per cent of all cases there is a family history of the disease and those who suffer from it are classed as being hypersensitive. About 3 per cent of all infants have atopic dermatitis. The symptoms are made worse by contact with wool, changes in climate, and excessive exposure to soaps, water or oils.

While there exists no firm consensus on the cause of atopic dermatitis, especially in adults, environmental allergens are now thought to play a major role. A substantial body of evidence has been amassed tending to implicate the allergen produced by the house dust mite, *Dermatophagoides pteronyssinus*. This same mite is believed to be responsible for atopic asthma. In particular, there is a close correlation between the production and distribution of *Dermatophagoides* allergen in nature and the epidemiology of the disease in children. However, there is no theoretical reason why other allergenic substances, mite produced or not, should not be the sensitising agent.

The difficulty in determining cause is that there are potentially so many constitutional and environmental factors involved. The universal distribution of dust mites in human dwellings was only recognised in 1966, since when the emphasis was placed on their role in asthma. Their further involvement in general body reactions such as dermatitis is currently attracting attention.

The Allergen: It is believed that if both atopic asthma and atopic dermatitis are linked to *Dermatophagoides*, then the same allergen is involved. The two differing reactions are thought to be due to constitutional factors and to differences in the principal mode of presentation of the allergen. It appears that the allergen, probably a glycoprotein, is freely soluble in water. It is remarkably stable and is able to survive for up to five years without loss of activity. It is also extremely potent.

The mites themselves contain little allergenic material but appear to excrete it. It seems that a sludge of uric acid crystals mixed with other metabolic residues are periodically expelled into the hind gut of the mite, yielding a thick fluid which dries after being smeared on the surface of a suitable substrate. This substrate is usually the detached scales of human skin which form the bulk of house mite dust and also the principal diet of the mites. Thus the allergen is in a relatively coarse particulate form essentially consisting of old skin scales coated with the excretory products of the mites.

In asthma, contact with the allergen occurs through inhalation of aged skin scales. Most of the allergen in a house is on the floor so that activities such as sweeping the floor or beating carpets pose obvious dangers. Human skin in beds is usually comparatively fresh and uncontaminated, so that much of the danger associated with making beds lies with the draughts raised by the process which blow old dust from the floor into the air. Of course, old skin can accumulate in mattresses and pillows but the dangers associated with this have been rather over-estimated. The raising and suspending of house dust has much to do with the electrostatic conditions and is facilitated by humidities at the drier end of the mite's tolerance spectrum.

Dermatitis, however, requires sustained contact between the old scales and the skin. This is best achieved at higher humidities for it seems that there is some hygroscopic quality to the faecal deposits which become stickier in humid conditions. Certainly the scales adhere better to damp rather than dry skin. This may explain why the allergic reaction occurs largely in skin flexures and folds where not only can the scales accumulate but also the water-soluble allergen has the best chance of transferring to the skin.

Old skin and associated mites can accumulate in cots and prams, especially in unhygienic conditions, but even in the cleanest of houses the major accumulation of allergen will be on the floor, and for most infants it will be there that allergic sensitisation begins. Babies spend a considerable amount of time on the floor, especially once crawling, and as a result atopic dermatitis associated with *Dermatophagoides* probably begins at infancy.

There are other mites which can be associated with dermatitis particularly among workers in such occupations as the handling of grain, who may become infected by *Pediculoides ventricosus*. This produces skin lesions a few hours after attack, giving rise to the disease known as grain itch. Likewise, copra itch is caused by the mite *Tyroglyphus longior*. Many other mites are known to infest products and cause skin problems, including *Sarcoptes scabei* which produces scabies. In most instances, effective pest control measures together with protective clothing and good hygiene will prevent recurrence.

Minimising Allergen Production: Ideally contact with mite allergen can best be prevented by avoidance measures focused on house furnishings and cleaning measures and practices. However, by making conditions unfavourable to the mites, not only will they reproduce more slowly but also produce less allergen per mite in any given period of time.

Three factors can be used against them: food supply, temperature and humidity. Clearly routine and regular cleaning will eliminate long-standing accumulations of house dust, their major source of food supply.

Fortunately mites are limited in their tolerance of temperature. Even a modest degree of cooling has a profound effect on their metabolic rate and consequently their output of excretory material. At 20° C their output of excrement is only about one-tenth of that at 25° C. If cooled to 15° C, normal life is no longer possible for them.

Humidity is even more crucial. The mites cannot survive below a 60 per cent level: even below 70 per cent their potential for allergen production is greatly diminished. However, it is the air which immediately surrounds the creatures which counts. If the floor is markedly cooler than the rest of the room, or is damp because of structural faults in the building, then the mites may experience humidities far higher than the general atmosphere of the room. It is for this reason that atopic asthmatics experience temporary relief in mountainous areas: it is not so much that the low humidity found inside high-altitude homes is good for the patient, as that it is bad for the mites. The delay factor masks any immediate beneficial result with atopic dermatitis, but a long-term plan for low-humidity living may warrant consideration.

Acne

Probably the most common skin disorder all over the world, acne is typically seen in adolescents, but also not infrequently occurs in adults. Typical symptoms include blackheads, papules, pustules, cysts, scars and nodules which may appear on the face, chest, neck, shoulders and back.

The cause of acne is unknown, but it can be triggered by factors such as drugs, diet, heat humidity, some cosmetics and certain chemicals, body hormones and emotional disturbances. Certain bacteria in the pilosebaceous follicles may exacerbate the condition and be responsible for infected pustules and the characteristic red, inflamed blotches which dot the faces and backs of many teenagers.

Psoriasis

Psoriasis is a chronic relapsing and remitting skin disease which may appear at any age and affect any part of the body. At any one time, 1-2 per cent of the population are believed to suffer from psoriasis and they account for about 15-20 per cent of all patients seen by skin specialists. Heredity plays a significant part in the aetiology of the condition. If one parent has psoriasis there is a 33 per cent chance of the child developing the condition; if both parents are sufferers, likelihood increases to 50 per cent; and in an identical twin, to 90 per cent. There are six major forms of psoriasis. The most common form, accounting for 95 per cent of all cases, is plaque-like and presents as small, disc-shaped patches on the skin. The patches are scaly, their surfaces rough to the touch, and when rubbed following the removal of the scales, there is a pin-point bleeding — Auspitz's sign.

The Cause of Psoriasis: The cause is unknown, but multiple factors are involved, including heredity. Present opinion also seems to suggest it is an auto-immune disease. This would be explained by arguing that the skin keratinocytes of psoriasis sufferers carry an abnormal surface antigen, which triggers an immune response. Antibodies formed against the abnormal keratinocytes combine forming the resultant immune complex.

Recent studies have also shown that patients with psoriasis have reduced levels of an enzyme which normally converts certain types of hydrocarbons into carcinogenic metabolites. This provides an explanation for the absence of skin malignancy in psoriasis sufferers and the absence of ill-effects when treatment with coal-tar, a known carcinogen, does not increase risk of cancer in psoriatic patients.

Environmental Factors: Environmental factors are necessary for clinical manifestation of psoriasis. They can either stimulate or relieve the condition.

For example, stress can precipitate or aggravate psoriasis; ironically sunlight can both improve it or lead to a deterioration; and similarly, systemic drugs and some local pharmaceutical treatments can cause or exacerbate pre-existing psoriasis.

Chloracne

Though sharing a similar name, chloracne is a completely different skin condition which has become more common lately because of its association with potential adverse reaction to some common drugs.

In reality, chloracne is a form of dermatitis, differing from most other forms in that it is generally caused by occupational exposure. Chloracne has been reported following exposure to chlornaphthalenes, polychlorbiphenyls, polycholorinated dibenzofurans, chlorophenol contaminants and chlorbenzenes. It is part of a syndrome in which there are characteristically acne lesions, usually comedones, but rarely evidence of inflammatory lesions. Contamination often occurs as a result of an explosion or uncontrolled release of the chemical. Perhaps one of the most famous episodes in recent years and which received extensive coverage in the media was the dioxin-induced chloracne in Seveso in northern Italy, which resulted in the evacuation of most of the village for many months.

Acne is the most common feature. The lesions, usually found on the temples on both sides of the face, are not often inflamed. In some severe cases, lesions also occur on the body and may be so severe as to induce other changes such as melanin pigmentation, ophthalmic chloracne, and systemic effects including anorexia, neuropathy, fatigue and impotence. Chloracne may last for years.

Recent studies have failed to identify the causal compound, referred to as the chloracnegen. This appears to be due to a delay between exposure and the time at which skin changes are reported and treatment sought. The same studies also suggest that a chemical such as dioxin produces a long-lasting reaction affecting the sebaceous glands of the skin and epithelial ducts. To date, the most effective therapy appears to be vitamin A, with occasional benefit from benzoxyl peroxide and antibiotic tetracycline or erythromycin, when inflammation is also present.

Unorthodox Approaches to Allergy

Much chronic illness and many skin allergies have been attributed to unrecognised sensitisation to a wide range of substances in foods and in the environment, particularly those resulting from recent human activities. While this may seem a plausible proposition, closer scientific examination often fails to support the imputed link. Today, mainstream allergists base their practice on a sound understanding of immunology and have at their disposal well-validated diagnostic methods and proven forms of treatment.

As a result they view certain controversial theories with scepticism. These include: Rinkel's concept of 'cyclic and fixed' food allergy and of 'masked sensitisation'; and the theories of adaptation, maladaptation and addiction to favoured foods based on the now discredited general adaptation hypothesis of Selye. Particular misunderstandings apply to the nature of adverse food reactions, referred to earlier in this Section. In most circumstances, these responses do not have an immunological basis. Transient non-specific symptoms such as nausea, headache, palpitations, sweating, dizziness and hiccoughs occurring sometimes within seconds of ingestion were attributed to an allergic reaction triggered by food antigens. These symptoms are now recognised as being due either to conditioned food aversions, phobic or anxiety reactions, or pharmacological effects of certain food substances.

Thus, just as the debate continues about the role of the ozone gas depletion in increased incidence of cancer, so too does controversy over the relationship between diet, stress factors, psychology and allergy. Unfortunately, until recently even the most objective observers did not fully appreciate that skin irritations and itching (urticaria), localised swelling and the formations of skin wheals and pustules are frequently due to the release of a non-immunological mediator. As a result, a number of tests were developed to identify possible allergens. These included pulse tests, the 'leukopenic index' and the cytotoxic food test.

In recent years, the same range of non-specific digestive effects and skin symptoms previously ascribed to food allergy are now often said to be the result of a sensitisation to the ubiquitous organism *Candida albicans*. The only evidence put forward to date in support of the concept is the testimonials of apparently satisfied patients. It is possible that the candida syndrome will fall into disrepute together with other theories such as auto-intoxication, intestinal toxaemia and bacterial allergy.

TABLE 4.9: Response of Normal Skin to Sunlight

EXPOSURE PERIOD	RESULTS
12 minutes	perceptible erythema
30 minutes	vivid erythema
60 minutes	painful erythema
120 minutes	blistering erythema

TABLE 4.10: Classification of Photosensitivity Diseases

GENETIC AND METABOLIC
- Xeroderma pigmentosum
- Erythropoietic porphyria
- Erythropoietic protoporphyria
- Porphyria curanea tarda
- Albinism
- Pellagra
- Kwashiorkor
- Hartnup disease

PHOTOTOXIC AND PHOTOIMMUNOLOGIC
- (a) Phototoxic (b) Photoallergic
- Internal — drugs
- External — plants and drugs
- Chloracne

DEGENERATIVE AND NEOPLASTIC
- Sunburn (acute and chronic)
- Squamous cell carcinoma
- Malignant melanoma
- Basal cell epithelioma or carcinoma
- Actinic keratosis

IDIOPATHIC
- Polymorphous light eruption
- Hydroa aestivale
- Hydroa vacciniforme
- Actinic reticuloid
- Actinic prurigo

PHOTAGGRAVATED
- Lupus erythematosus
- Pemphigue follaceus
- Atopic dermatitis
- Bloom's disease
- Dermatomyositis
- Darier's disease
- Acne vulgaris
- Herpes simplex
- Pellagra, and Hartnup disease

TABLE 4.11: Intensity of Radiation
1. distance from the equator
2. season of the year
3. time of day
4. atmospheric conditions
5. distance above sea level

TABLE 4.12: The Cutaneous Signs of Chronic Sun Damage

	Description	Common site(s)
Epidermal benign		
Actinic xerosis	Chronic dry, flaky, itchy skin.	lower legs
Actinic icthyosis	'Fish scale' appearance of skin	anterior aspects lower legs
Solar atrophy	Thin, easily broken skin ulcerates and bruises.	exposed areas, espec. arms, legs, shoulders, upper back
Mottled hypo- and hyper pigmentation	'Lizard skin'.	
Solar lentigines	Large (1 cm) sun freckles; 'age spots'; 'liver spots'; senile freckles.	lips of the young; dorsum of hands and forearms of middle-aged; upper arms and back
Black stellate lentiginosis	Black superficial freckles in people whose other freckles are 'brown sugar' coloured.	forehead, face and neck
Sebaceous gland hyper-plasia	Small 'buttery' 1-2 mm papules.	face, especially cheeks and forehead
Sebaceous adenoma	Aggregation of the above; 3-5 mm 'buttery' papules, discrete border, central dell and occasional punctum.	
The maladie de Favré et Racouchot	Large cystic comedones or black heads.	'crows' feet' areas lateral to the eyes
Guttate Hypo-melanosis	Flat white polygonal 3x2 mm 'rain drops'; no tan.	anterior lower legs (especially females)
Actinic vitiligo	'Burnt out' melanocytes in large white flat areas; no tan.	shoulders and upper back
Stucco keratoses	Pearly white, keratotic papulo-squamous lesions; can be spooned off with fingernails	lower legs

Epidermal pre-cancerous	*Description*	*Common site(s)*
Actinic keratoses	Keratotic lesions feel like sand-paper; great variation in colour, size, tenderness, etc. (from itchy, brown-red, scaly dry papulo-squamous plaques to tender cutaneous horns, may bleed with shaving, or 'smart' with alcohol; many 'light up' in the sun).	all exposed areas
Hutchinson's melanotic freckle (lentigo maligna)	Haphazard border, multi-coloured (often with amela-notic areas) and flat, mainly brown-tan-black in hue; a melanoma *in situ*.	face
Actinic Bowen's disease	Squamous cell carcinoma *in situ*; pink/brown with irregular scales; up to 2x2 cm.	lower legs in females
Epidermal cancerous	*Description*	*Common site(s)*
Basal cell carcinoma (BCC)	Protean in presentation; pearly border with telangiectasia: many variations.	all exposed body surfaces
Squamous cell carcinoma (SCC)	Hard, keratotic compared with BCC; often bleeds, ulcerates.	all exposed body surfaces
Kerato-acanthoma	Arguably malignant; often an ulcer with crater, like SCC	all exposed body surfaces
Malignant melanoma	Ugly lesion, irregular outline & profile. Often multi-coloured, pigment mainly brown-tan-black	more common on exposed skin

	Description	Common site(s)
Combined epidermal and dermal benign Poikiloderma of Civatte (* Erythrosis interfollicularis colle *Peau citrinée)	The 'red neck' syndrome of vivid dermal telangiectasia in areas between follicles with obvious sebaceous gland hyperplasia usually associated with pigment differences on an 'orange peel' look.	sides and 'V' of neck
Acro-kerato elastoidosis	Shrivelling, scaling and ridging of the skin of the dorsum of hands and wrists (one of many chronic UVL syndromes).	backs of hands and wrists
Dermal benign Actinic granuloma	Annular 2-4 cm diameter red depressed lesions with apple jelly nodules on diascopy often mistaken for 'ring worm'.	all exposed sites
Cutis rhomboidalis nuchae	Furrowed straight lines in geometric shapes on back of neck.	back of neck
Telangiectasia	Proliferation of capillaries giving a 'high colour'.	face, especially nose, cheeks and forehead

Source: Australian Prescriber, Vol.9, No.4, 1986, p.85

4.5 The Respiratory System

After the skin, the respiratory system is the part of the body most constantly exposed to the environment and thus most liable to any resultant damage. The respiratory system comprises the nasal passages, including eustachian tube, the trachea and the lungs. Each fulfils a distinct function of its own, and at the same time combines with the others in the defence of the body against intrusion by infective organisms and debris, such as smoke, dirt and foreign chemicals.

The most vulnerable parts of the respiratory system are the lungs. These delicate tissues of the bronchial tree are sensitive to air-borne solids, gases and other suspended matter. Air entering the respiratory passages is filtered by hairs and mucus in the nose. The trachea and larger bronchi are lined by a layer of cells, each bearing a minute hair or cilium which beats constantly, filtering the entering air and removing suspended dust and particulate matter.

Dust, fumes and smoke contain particles which are capable of penetrating the fine tissues of the lung and bronchi. The cilia thereby act as a defence against the intrusion of this material and infective bacteria and viruses which may be carried on it. Glands embedded in the walls of the trachea and bronchi produce a film of sticky mucus which lies on and around the cilia and is constantly carried upwards by their coordinated beating movements. This creates a very effective self-cleansing mechanism and traps the fine particles which escape filtering in the nasal passages.

Dust particles range in size from a diameter of 1 micro-mu (μm) to 150 μm, whereas those in fumes range between 0.12 μm and 1 μm. Industrial fumes often consist of oxides formed from hot or burning metals. Smoke has particles of less than 0.3 μm in diameter. Tobacco smoke is a 'wet' smoke, composed of tiny droplets of tar, the particle size being about 0.25 μm. Normally only particles of less than 10 μm in diameter will remain suspended in air for any time. Those less than 5 μm in diameter are likely to penetrate the lung defences and reach the bronchioles and alveoli.

Alveoli are the minute cavities where gas exchange occurs. They have thin walls lined with capillary blood vessels along which single cells pass thereby facilitating gaseous diffusion. It is the alveoli which are most susceptible to damage and injury from toxic and potentially damaging suspended particles which might evade the respiratory defences by virtue of chemical composition or minute size. Over a prolonged period of time, tobacco leaf chemicals, tar particles and fine smoke inhaled by a passive smoker have the potential to damage these delicate tissues and pass into the blood stream via the alveoli.

In this chapter are discussed respiratory conditions either proved or imputed to be caused by conditions in the environment. These range from natural phenomena, including plant pollens and fungal spores carried by the wind, to pollutants from industrial emissions or mining, and also those which seem to develop as a result of changes in climatic conditions. Such changes can encourage mould growth, increase levels of toxic chemicals and dusts in the atmosphere or alter human behaviour.

Other respiratory health problems arise because of lifestyle changes. Smoking is now well acknowledged as a potential risk to human health and one which affects the whole human body. Certain occupations can also place workers at special risk of future lung problems. The most important of these conditions, their causes and their relationships are discussed in the following pages.

Rhinitis

One of the most common respiratory problems encountered and one which can be traced to an environmental cause, be it biological, chemical or physical, is rhinitis. It is also among those conditions which is highly likely to be allergy based.

Rhinitis is a general term applying to a wide range of inflammatory conditions which affect the upper respiratory passages, principally the nasal passages and the eustachian tubes. The vast majority of instances are allergic in origin and arise as a result of sensitivity of the nasal mucosa to one or more environmental allergens. Symptoms include a constant and irritating runny nose, sneezing, watery eyes and catarrh. These develop from an excess production of mucus by the mucosal cells as they attempt to cope with the allergen.

Animal and vegetable dusts, including mould spores and pollen grains, may be small enough to remain airborne for considerable periods of time and to penetrate the defences of the respiratory tract. Pollen grains trapped in the nasal passages cause local sensitization and irritation, leading to the typical clinical picture of hay fever, but these and other allergens penetrating into the bronchioles may cause symptoms of partial airway obstruction due to contraction of the muscles controlling the diameter of the bronchioles. This again may lead to local obstruction and tissue damage predisposing to secondary infection.

A large number of allergens may be responsible for allergic respiratory problems, including plant pollens, fine plant hairs and spines, fungal spores, animal mites and house dust. The last is caused by the tiny but ubiquitous mite, *Dermatophagoides pteronyssinus*, which resides in bed linen, under beds, in mattresses, and is to be found wherever humans live and sleep. It seems to occur with greater frequency and in highest concentrations in damp, humid environments. Animal danders and feathers are also potential allergens.

Another common form of rhinitis is referred to as hay fever. Strictly, this term describes seasonal forms of rhinitis and is associated with pollens and the flowering by species fertilised by wind-borne pollination. Spring and autumn are times when seasonal rhinitis is particularly prevalent.

Bronchitis

Bronchitis, inflammation of the mucous membranes of the bronchial tubes, is a common cause of death often exacerbated by environmental agents such as dust, fumes, pollen, and chemical atmospheric pollutants. It is characterised by excessive production of mucus by the bronchial glands as a result of stimulation of the body's allergic response mechanisms.

Impairment of the drainage of the bronchi and bronchioles is accompanied by recurrent bouts of chest infection and consequent damage to and fibrosis of lung tissues. In its more advanced stages, alveolar destruction and scarring lead to the development of small cavities (emphysema), and obstruction of blood flow through the capillaries and other small blood vessels in the lungs. Increased resistance to blood flow makes breathing extremely difficult and places an added burden on the heart. This can result in heart failure, particularly when infection is also present, thereby further restricting the lungs' capacity.

Chronic bronchitis is a disease which has long been linked to residence in climates such as that of many areas of the United Kingdom. However, its incidence in other countries with warmer, drier climates has tended to place doubts over early assumptions that places sharing Britain's cold, dank and humid conditions, together with wet drizzly winters and draughts, encouraged chest infections and persistent bronchial congestion. Australia, for example, has a large number of sufferers of acute and chronic bronchitis. With New Zealand, it recorded in 1988 and 1989 the highest number of asthma deaths in the world.

The evidence now suggests that air pollution is an important factor in explaining the distribution of bronchitis. Its potentially damaging effect is demonstrable from an early age. Studies in the United Kingdom confirm that, during the first two years of life, there is a steady gradient of mortality among children living in areas of low pollution, from a relatively low level to higher levels of mortality among those living in heavily polluted areas.

Further studies on sickness rates show substantially higher rates of illness from bronchitis, emphysema and asthma among occupational groups exposed to atmospheric pollution, in the workplace, through residence in heavily polluted towns, suburbs and areas where stagnant air tends to accumulate. Pollution by dust seems to be more important than pollution from fumes. Road dirt, car exhaust fumes, the grime associated with high population densities, and the effects of high-rise buildings on air flow and sunlight combine to make life in a modern city threatening and hazardous for those who are at risk. Cigarette smoking further increases susceptibility to chronic and debilitating respiratory diseases and prolongs recovery from chest infections.

Allergic Lung Diseases
Similar to rhinitis and hay fever is extrinsic allergic alveolitis. In this instance, an abnormal immunological response results in changes to the alveolar tissues of the lungs rather than the bronchi and bronchioles, leading to recurrent bouts of respiratory discomfort and possibly more generalized symptoms such as fever. Over a prolonged length of time, they can lead to progressive disability due to cumulative damage to these delicate lung tissues.

Yet another form of allergic alveolitis associated with acute respiratory symptoms was first encountered by farmers, thereby giving it a colloquial name 'farmer's lung'. In this instance, the symptoms of respiratory discomfort, cough and fever persisting over several days were found to be linked to handling of mouldy hay. An investigation linked the presence of thermophilic actinomycetes which grow well in warm fermenting hay.

In the early twentieth century, the northern Queensland cane-growing belt became the centre for another puzzling but similar condition, bagassosi. If cane fibre from the milling process is allowed to become wet and mouldy, usually as a result of storage outdoors, people who subsequently handle and work with the fibre develop the typical respiratory problems and cough of allergic alveolitis. Here also, the allergen appears to be derived from a thermophilic actinomycete. It is likely that disturbance of the wet hay (or bagasse) disperses the microscopic filamentous bacteria as a fine 'dust' in the air, with the resultant inhalation of the allergen by anyone in the vicinity.

Other comparable conditions include 'pigeon fanciers' lung', where the dust is derived from pigeon feathers and dander, and 'humidifier fever' associated with the use of humidifying air-conditioners in chronically dry climates. There is a superficial resemblance between this latter condition and the early symptoms of Legionnaire's disease. However, Legionnaire's disease is a much graver and potentially lethal disease, caused by a virulent bacterial strain which tends to proliferate in poorly maintained air-conditioner cooling towers. Its course, treatment and prognosis involve a major systemic infection whose outcome depends largely on age, health status and early intervention.

Higher organisms may also produce disease. The grain weevil, *Sitophilus granarius,* may produce asthma in susceptible mill workers. Thus micro-organisms naturally present in the atmosphere and in association with certain species may present a hazard. Filters, sterilization and even airlocks in association with isolation facilities may be necessary where they prove hazardous, for example to workers who may be required to be in constant and prolonged contact with the hazard (Table 4.13).

Asthma

Asthma is one of the most common health problems in developed countries and one about which there is considerable misunderstanding, confusion and mistreatment. It is associated with heavy or difficult breathing arising from a spasm of the involuntary muscle around the small branches of the air-tubes or bronchi. As these muscles contract they squeeze the tubes and impede the flow of air. The effect is partly counteracted by breathing in, because expanding the chest creates a partial vacuum and tends to widen the tubes. However, when the air is expelled the pressure around the bronchi rises and increases the constriction.

The prime symptom of asthma is difficulty in emptying the lungs. A characteristic wheeze or a whistling sound is heard, accompanied by difficulties in emptying the lungs because of bronchospasm, and with resultant 'blow up' lungs. About half of all cases of asthma can be attributed to immunological sensitivity reactions to allergens in the environment which lead to the release of chemical mediators of inflammation and bronchospasm. The range of allergens include plant pollens, fungal spores or house dust.

The condition itself is often difficult to define as asthma sufferers range from people who experience only one or two very mild attacks in a lifetime to others who suffer repeated attacks of a severe nature. Asthma can also appear in childhood with some severity and yet disappear almost completely in adulthood; alternatively it may first appear at an advanced age. These unpredictable aspects of the disease remain unexplained.

The two countries recording the highest annual rate of asthma-related deaths are New Zealand and Australia. In 1988, 826 Australians died from asthma. At the same time, while the number of deaths worldwide from asthma decreases, in Australia deaths continue to increase. This situation applies despite the high quality of health services and the availability of modern medicine and treatments. However, the leading national body, the National Health and Medical Research Council (NHMRC), believes that many of the reported asthma deaths could have been prevented by improved patient education, more accurate diagnosis and early detection of symptoms.

At the same time, many specialists and practitioners treating asthmatics believe that increased atmospheric pollution (particularly in the cities where most Australians live), the seasonal prevalence of pollens from introduced species, and local changes in climate, are equally responsible for the increase in the number of reported cases.

The New South Wales Asthma Foundation in Australia has claimed that asthma is the most common childhood disease, affecting up to 20 per cent of pre-school children, between 10-15 per cent of early school-age children and approximately 10 per cent of high-school children.

The Causes of Asthma: Asthma is most often an allergic reaction which can be provoked by such factors as pollen, dust, dust mites and their faeces, animal hair, some drugs, food preservatives and moulds. Eighty per cent of asthmatics are allergic to the house dust mite which is responsible for other allergic responses such as allergic rhinitis.

A study carried out by the University of Sydney revealed that up to 6984 mites per square metre lived in underblankets and blankets taken from a random number of homes in the metropolitan area. Although avoidance of the particular allergen affecting the asthmatic is recommended, this is almost impossible in cases where the allergen is widespread; for example, dusts, pollens and moulds.

Asthma is sometimes induced by exercise because the amount of air inhaled is increased and is usually colder and always drier than the air already in the lungs. The nose acts as a natural air conditioner and under normal conditions the specialised lining (the mucosa) provides both the heat and the water to condition the incoming air. The more air inhaled during exercise, the more the body has to condition it, meaning more water and heat are lost from the airways.

Some asthmatics react to preservatives such as meta-bisulphite, monosodium glutamate and tartrazine. Even quite small amounts of foods containing these preservatives alone or combined may produce asthma. Aspirin and related pain relievers are common provokers of asthma, with about one-in-five asthmatic adults and children affected.

Asthma: Treatment or Prevention? The best way to minimise asthma is by avoidance of precipitating agents and environments. Where this is difficult or impossible, early correct treatment is best. Conventional medicine offers the most prompt and effective way of limiting the chance of death and rapidly relieving symptoms in extreme cases. Prompt treatment using bronchodilators and bronchial muscle relaxants, assisted by peak flow meters, disodium cromoglycate and cortisone administration where warranted, is proving most effective in reducing deaths. The effectiveness of courses to de-sensitise asthmatics to allergens which provoke asthmatic attacks remains unproved.

The Pneumoconioses
Pneumoconiosis describes a range of lung diseases caused by the inhalation
and accumulation of air-borne particles which the lung is unable to expel,
because of either their size or their nature. Many remain there and do no
harm, though they may 'cloud' chest X-rays where they appear as opacities.
This is referred to as 'benign pneumoconiosis'. Occupational groups concerned with handling calcium-based materials such as limestone, cement,
gypsum, and marble, and certain metals or their oxides including those of
iron, tin and antimony, are most susceptible.

Not surprisingly, coal workers are at potential risk of pneumoconioses.
The dust of coal mining produces fibrosis of the lung tissue if exposure occurs
over a prolonged period of time, a condition commonly referred to as 'coal
miner's lung'. Localized, heavily pigmented foci of dust accumulation and
tissue reaction in or near the bronchioles, together with associated distortion
of adjacent alveoli, can be seen in X-rays. The dust is actually held within
large, phargocytic cells (macrophages) which lie within areas of fibre formation, and it is probably the gradual shortening of these fibres as they mature
which distorts the local alveoli. With continuing exposure to coal dust, the
deposition of fibrous tissue increases, effectively destroying this area of tissue.
The lungs become progressively more impaired and susceptible to intercur-
rent infection.

The extent of the pneumoconiosis is determined only by the frequency,
intensity and duration of exposure of the worker. Exposure to these dusts and
the severity of the resultant pneumoconiosis is governed by the extent of dust
control measures. However, other forms of the condition may not be limited
by the exposure and control measures but, due to the nature of the material
involved, be resultant in local tissue reaction and damage.

Thus, while many inhaled dusts accumulate in the lungs without stimulating a local reaction or causing any recognizable or progressive disease
because of their presence, they and others can serve as nuclei for bacterial
and viral particles. Grime and soot from urban pollution place city-dwellers at
special risk. In contrast, people living in rural areas which are subject to
cleaner air will manage to eventually expel a proportion of the dirty material,
by dilution. Thus, it is not surprising that city dwellers often claim that a stay
in the country is cleansing and invigorating, an excellent tonic against the
polluted city atmosphere and its association with disease.

Asbestosis
Perhaps the most grave of the pneumoconioses is the condition asbestosis, a
progressive form of lung disease caused by the inhalation of fine particles of
asbestos. Asbestos is a fire-resistant substance used as an insulating material

in the construction industry, in ships and in brake linings.

Through inhalation of asbestos particles, the tissues of the lung become fibrous and inflexible. The result is a progressive reduction in the efficiency of the lungs, causing shortness of breath, a predisposition to bronchitis and emphysema. The disease may take anything from ten to twenty or more years to develop and run its course.

Asbestosis may also contribute to or exacerbate cancer in the form of a carcinoma of the bronchus or a mesothelioma of the pleura. The likelihood of asbestosis causing cancer is greatly increased if the victim is a heavy smoker.

Asbestosis cannot be cured, though its symptoms can be relieved. It is a serious and potentially lethal hazard to all who work with the material.

Asbestos: Asbestos is perhaps the most well known occupational health disaster of the twentieth century and has been responsible for death, disablement and disease in every industrialised country.

Asbestos is a broad term for fibrous silicate materials, the most common of which are chrysotile (white), crocidolite (blue), amosite (brown) and anthophyllite. It is responsible for lung cancer and mesothelioma (cancer of the lining of the lung), both forms of cancer affecting the respiratory system. In asbestosis, an associated condition, small, thin fibres of asbestos materials become lodged in the lungs. The harmful effects of asbestos dust are related to fibre size, the smaller the more harmful. There is also good evidence suggesting that asbestos is a cause of stomach cancer and cancer of the bowel.

Both lung cancer and mesothelioma generally take between 25 and 30 years to develop following first exposure, as compared to the 10-20 years for asbestosis. The intensity of exposure to asbestos is also important. Mesothelioma can be induced by relatively short-term exposure to high concentrations, or by smaller, intermittent exposure over a long period.

There is a strong relationship between cancer, asbestos exposure and cigarette smoking. The effect on lung cancer from asbestos and smoking is multiplicative, not merely additive. Despite this, there is a clear and identifiable excess risk of lung cancer amongst non-smokers occupationally exposed to asbestos.

The History of Asbestos Exposure: The dangers of asbestos are not new. As early as 1918, the Prudential Insurance Company of Boston was reluctant to issue policies on asbestos workers justifying refusal on the harmful effects of asbestos dust. In 1930, a report to the British Parliament suggested that inhalation of asbestos dust over a period of years resulted in the development of a serious type of fibrosis of the lungs, and advocated that industry be encouraged to suppress dust.

The first report of cancer of the pleura associated with asbestos exposure was published in 1931. In May 1935, Lynch and Smith published the first report of lung carcinoma in a man with 'asbestos-silicosis' who worked for twenty-one years as an asbestos mill worker (Lee and Selikoff). However, it was to be another twenty years before the insidious nature of the disease was fully characterised independently by Doll and Selikoff. Doll reported a ten-fold increase in the risk of lung cancer among asbestos workers employed for twenty or more years. Selikoff, using union records, found a similar excess risk of lung cancer, several cases of mesothelioma and a higher than expected rate of stomach and bowel cancer.

Perhaps one of the most dramatic examples of recent times comes from Wittenoom, Australia's only blue asbestos (crocidolite) mine in Western Australia, and now the focus for a multitude of claims from the thousands of workers employed over its period of operation. In 1943, the Colonial Sugar Refining Company (CSR) purchased the mine from the original American owners and began twenty-three years of production. During this time, 6000 people worked at the site. Of these over 220 are known to have died from asbestos-related diseases — a figure of less than one-in-28, and one which is likely to reach one-in-25, given the long period between exposure and the manifestation of symptoms.

Despite the early evidence of risk, it has taken an exceptionally long time for authorities, both in Australia and other countries, to introduce regulations to control the hazards of asbestos. Even in Australia, where the problem is so acute, regulations relating to asbestos are not uniform nor enforced in all states and territories. These regulations were, for the most part, introduced only in the mid to late 1970s, twenty-five years at least after sufficient indisputable evidence existed requiring control of the hazard.

Occupational Controls: The occupational standards associated with asbestos have an interesting history. Commencing in 1938, the American Conference of Government Industrial Hygienists was first to adopt a 'safe' standard for asbestos exposure. This standard was based on a simple study of 541 South Carolina asbestos textile workers which contained few who had worked at the plant longer than ten years. This standard remained in place there for thirty years despite the evidence, even before its implementation, of an earlier 1930 UK report that asbestos-related conditions do not manifest until twenty years or more of exposure. Furthermore, the 'safe' US standard was also adopted in the United Kingdom and remained in force for eight years.

A more rigorous UK standard was introduced in the 1969 regulations as a result of recommendations by the British Occupational Hygiene Society (BOHS). These recommendations focused on reducing asbestosis and completely ignored likely cancer effects. Despite this, they were adopted in the US in 1972. The two fibres per millilitre standard of the 1969 UK regulations were revised in 1979 following a three-year investigation by a special UK Advisory Committee on Asbestos. A major finding of the Committee was that the control limit for chrysotile should be 1.0 fibres/ml, 0.5 fibres/ml for amosite, and 0.2 fibres/ml for crocidolite (blue asbestos).

In Australia, there had been a total absence of any prescribed asbestos exposure standard before the mid-1970s. The first standard adopted by the National Health and Medical Research Council (NHMRC) was based on the BOHS recommendations, except that it was doubled to 4 fibres/ml on the basis that Australian workers were, at most, exposed for twenty-five years instead of the fifty years assumed by BOHS early in their estimates. In 1979, the NHMRC issued a statement on health standards associated with the use of asbestos in the construction industry. This statement recommended an average eight-hour asbestos exposure standard of 0.1 fibres/ml for crocidolite, and a 1.0 fibres/ml limit for amosite (recently reduced to 0.1).

Apart from the common problem of non-compliance and the non-enforcement of control measures, recommended exposure standards remain precautions intended to safeguard workers from contracting asbestosis, and cannot be regarded as adequate to provide complete protection against the carcinogenic effects of asbestos. The International Agency for Research on Cancer (IARC) has concluded that it is not possible to assess whether there is a level of exposure in humans below which an increased risk of cancer would not occur. As such, the only complete insurance against the possibility of adverse health effects from asbestosis is its total elimination. While its substitution has increasingly become the most favoured option, the legacy of past and present asbestos exposure will remain for some decades. Whilst estimates of all cancer attributable to asbestos have been put as high as 13-18 per cent, only time can reveal the true enormity of this public health disaster.

Silicosis
Silica is not dissimilar to asbestos in its effects on the lung. Free silica is the principal form which leads to damage. Minerals such as quartz, coesite, cristobile and tridymite are important sources of free silica and thus represent a potential hazard to those handling them. Occupations in which exposure to silica dust is a potential hazard are sandstone and granite quarrying, gold, tin and iron ore mining, and processing industries involving tile making, glass manufacturing, flint crushing, silica milling and silica brick manufacture.

Other easily recognizable forms of silica include the familiar beach sands in which silica is complexed more firmly thereby making it potentially less damaging.

Lung tissues react to the presence of silica by forming localized nodules of tissues around the particles. The nodules tend to enlarge and run together so blotting out areas of lung tissue and impairing lung function. In time, the nodules become infiltrated and fibroused by scar tissue. First signs of increasing disability and difficulties in breathing may take many years to develop and the condition is often well advanced before they become noticeable. At this stage, it may be detected during treatment for chronic bronchitis, pneumonia or tuberculosis. Cigarette smokers are at increased risk of the disease.

The presence of silicates in the rock from which the coal is mined is only considered to play a minor role in the development of lesions in 'coal miner's lung'. These are distinctive and can be distinguished from those found in silicosis.

Metal Fume Fever
Fumes of certain metals — particularly manganese and zinc but also oxides of antimony, arsenic, cadmium, cobalt, copper, iron, lead, magnesium, mercury, nickel and tin — can cause temperature elevations in those who inhale them. This condition is called 'metal fume fever' and is accompanied by a dry throat, chest constriction, fatigue, headache, back pain, nausea and muscle pain.

Principally an occupational problem, it directly affects those working in plants exposed to the air-borne dusts. Long-term lung problems arise in those with respiratory weakness and who are exposed for prolonged periods. However, metal flume related problems associated with similar symptoms have also been reported among people living adjacent to metal-processing plants.

Legionnaires' Disease
In recent years, the condition Nocosomial Legionnaires' Disease has become a worldwide problem. In the summer of 1976, a mysterious epidemic of fatal respiratory disease in Philadelphia led to an intensive scientific enquiry that resulted in the definition of a new family of pathological bacteria, the *Legionellaceae*. Following a Convention of US servicemen (colloquially referred to as legionnaires), 182 attenders contracted a particularly severe form of pneumonia. Twenty-nine men died before the case was traced to a dirty cooling tower, part of the air-conditioning plant on the roof of the Philadelphia Hotel where the Convention had been conducted. *Legionella pneumophila*, a particularly virulent strain, had multiplied to the extent that the 'fresh' re-circulated air from the plant was distributing live bacteria in the hotel.

Though the risk of Legionnaires' disease is slight to the majority of the population, anyone who lives, works or visits air-conditioned buildings may come in contact with its causal species. Theoretically those working in air-conditioned environments, such as hospital nurses, departmental shop assistants, bar and club staff as well as patrons in social clubs and community facilities, are at greater risk by virtue of the extended times they spend in these environments.

Legionellaceae Species: Members of the *Legionellaceae* family were first isolated from clinical specimens in 1943. Unsolved epidemics of acute respiratory disease dating to the 1950s were subsequently attributed to the newly described pathogens. In the intervening years, the *Legionellaceae* have been recognised as important sources of sporadic and epidemic respiratory disease. In Britain, for example, Legionnaires' disease has been claimed to account for about 2 per cent of community acquired pneumonia.

Other forms of atypical pneumonia rank of similar significance. Legionella bacteria are able to multiply in warm water to high concentrations if suitable nutrients, including iron salts and organic matter, are present. In nature they are ubiquitous and are usually found in damp, moist soils and marsh, or tepid pond water. They flourish in temperatures between 22 and 55° C. Only 50 per cent of those known have been confirmed as pathogens to humans. The sources of these infecting bacteria are environmental, the most important being potable water systems. However, the mode of transmission from drinking water is unclear.

Infection is due to the inhalation of infected droplets. However, there is little evidence available as to exactly how the disease is transmitted in any particular outbreak, the evidence tending to be circumstantial. The organism is widespread in nature, being present in water and possibly in soil. Outbreaks have normally been associated with contaminated drinking, washing or cooking water. Cooling towers, showers and similar sources of aerosols are particularly suspect and have been implicated in a number of outbreaks. Hospitals, social clubs, shopping centres, departmental stores and large office buildings are among those likely to be affected.

In most epidemics, air-borne dissemination from cooling towers and evaporative condensers has been shown to be responsible. Here, the Legionella can accumulate and grow together with fungi, slime, rust, scale and corrosive breakdown products. Hence it is necessary that these apparati be regularly cleaned. Conventional biocides are believed to be effective in this instance. Health authorities throughout the world are now framing guidelines and legislation to ensure that all public premises using equipment which may harbour Legionella pathogens are regularly cleaned and inspected.

The presence of Legionella bacterium *per se* in cooling towers is of little medical consequence. While the bacteria remain in the cooling tower water they represent no real threat to the users of the building's air-conditioning system. The risk comes when the bacteria proliferate to reach a dangerous concentration in the water, and the water containing the bacteria is lost from the cooling tower via the spray mist or fall-out. Gravity and the wind carry the bacteria in the fall-out into contact with people.

So, too, if the air-conditioning system's air intake ducts are anywhere near this spray or fall-out, then the bacteria can be introduced into the building, creating a health hazard to those working within or around it. One of the features of the ecology of Legionella which is as yet not known, is how it manages to colonize various water systems. The role of cooling towers in this spread, as well as the spread of the disease, is still to be elucidated. Air-conditioning systems, where aerosols may be generated in humidifiers or washers, may be involved in the spread of the infection. They are also known to be the source of spread of other organisms, and the term 'humidifier fever' has been used to describe symptoms of malaise and fever occurring, particularly, in people returning to work after a break. Such symptoms are commonly worse on Monday mornings following a weekend break and they subsequently subside.

Bacteria, protozoa and fungi have been isolated from the air-conditioning systems in such outbreaks, but identifying the pathogens precisely has proved difficult. The use of steam in the humidifiers, however, has proved sufficient to eliminate these pathogens. Whirlpool baths and spas are a further source of potential infection, although the hazard is reduced if the water is not recirculated but discharged to waste after each use.

The Pathology of Respiratory Infections

The *Legionellaceae* are gram-negative, facultative, intracellular bacteria. In humans, the alveolar macrophage cells are the chief site of growth. In healthy people, host defence mechanisms in the respiratory tract and alveoli effectively remove or contend with invading micro-organisms. However, recurrent sino-pulmonary infections or pneumococcal invasion may mean lower immunoglobulin IgG and IgA titre, and the interaction of alveolar macrophage and lymphocyte is less than normally effective in defence.

Under special circumstances, such as first exposure to a virulent microbe or a large dose of pathogen, contact with pathogenic Legionella species can result in illness. Cell-mediated immunity seems to be the main source of immunological defence, though the role of humoral immunity is not fully understood.

Consequently, individuals with depressed resistance, of advanced age or recovering from prior illness and with a lower level of total immunity, are more likely to be susceptible. In these circumstances, various pathogenic species, such as *L. pneumophila, Mycoplasma pneumoniae* or other members of subgroup 1 of the Legionella species, will invade the respiratory tissues.

The definitive typing of *Legionella pneumophila* depends on monoclonal antibodies and isoenzyme, plasmid and nucleic acid analysis. Biotyping methods have been found to be of little value. The use of monoclonal antibody techniques has permitted comparison of clinical and environmental isolates and allowed the separation of serogroup 1 into subgroups of differing virulence. The subgroup of serogroup 1 called Pontiac is said to be responsible for the majority of sporadic and epidemic Legionella pneumonia in the United Kingdom. The extent to which other members of serogroup 1 and other serogroups of *L. pneumophila* cause disease appears to depend on their prevalence in the environment.

Clinical Symptoms: Whether a person exposed to Legionella bacteria later develops Legionnaire's Disease will depend on a number of factors, such as their resistance to infection, the size of the infected dose and the opportunity for exposure to airborne aerosols which enter the lung. One of the common difficulties in studying the epidemiology of this disease is that some people are particularly prone, for instance, some smokers. Others appear to have resistance. Thus, it is often difficult to identify when initial reports are from isolated cases.

Typical bacterial pneumonia is a rapidly evolving illness that is characterised by an abrupt onset with high fever, rigours, systemic toxicity, shortness of breath and a cough which produces purulent sputum and pleuritic chest pain. Chest signs are evident early and are apparent on X-ray films. A high white-cell count is usual. The term 'primary atypical pneumonia' was first used by the American Armed Forces Commission on Respiratory Diseases to describe an epidemic form of pneumonia that was occurring among the troops. *Mycoplasma pneumoniae* was identified as the cause in this instance and mycoplasmal pneumonia became synonymous with the term 'primary atypical pneumonia'. In recent years, the term has been expanded to include other forms of respiratory infections exhibiting similar clinical symptoms.

Atypical pneumonia is now used to refer to illness due to one of four infective agents: *Mycoplasma pneumoniae, Chlamydia psittaci* (responsible for psittacosis), *Coxiella burneti* (the agent in Q-fever), and Legionella species.

Atypical pneumonia syndromes are characterised by slower development than acute bacterial pneumonia. It is usually three or four days before patients contact their doctor, and at that time they do not appear to be seriously ill. At this stage of the disease, constitutional symptoms of fever, chill, myalgia and headache are prominent. A cough may appear after several days and is initially unproductive with little respiratory distress. Retrosternal chest pain or coughing is common, but pleurisy is absent, so that a disparity often exists between clinical signs in the chest and X-ray appearances. The total white-cell count is usually normal.

Reports of Legionnaire's Disease

Australia is typical of most western developed countries whose climate makes air-conditioner and humidifier use important and provides the conditions under which bacteria pathogens can proliferate in cooling towers. The first Australian case of Legionnaire's disease was confirmed in 1978: in the following years, two further deaths attributed to the disease occurred in Melbourne and another in Adelaide. Australia's largest epidemic occurred in Wollongong in 1987 and was traced to an air-conditioning tower on the building of the Wollongong Shopping Mall. Official figures list ten dead and forty-four people hospitalised with confirmed Legionnaire's disease, one hundred and nineteen suspect cases having been investigated.

November 1988 saw nine cases of Legionnaire's disease recorded in Adelaide. The cause of the outbreak was attributed to garden watering systems, a claim which is regarded as dubious. The reporting of fourteen further cases of the disease in Burnie, Tasmania, in early 1989, so soon after the previous outbreak, attracted the attention of the media. Three deaths occurred. At the same time, Sydney's Prince of Wales Hospital notified the authorities of a death attributed to Legionnaire's disease and Newcastle Hospital confirmed that a 66-year-old man, recovering from a transplant operation, was dangerously ill with atypical pneumonia believed to be due to *L. pneumophila*.

Treatment: Legionnaire's disease is a problem because atypical pneumonias such as those due to *L. pneumophila* do not respond to penicillin, cephalosporin or trimethoprim-sulphamethoxazole therapy, agents commonly used in bacterial pneumonia. The principal agents for therapy of atypical pneumonias, including those due to *L. pneumophila*, are erythromycin and the tetracyclines. Some researchers also suggest that if a patient is exhibiting uncertain pneumonia symptoms making definitive diagnosis difficult and the symptoms fail to respond to beta-lactam antibiotic agents or to trimethoprim-sulphamethoxazole, the atypical pneumonia syndrome agents should be considered. Erythromycin remains the antibiotic of choice.

Passive Smoking

Smoke from lighted tobacco probably constitutes the most widespread indoor pollutant and the one which is potentially the most damaging to human health. Cigarette smoke is the most abundant and as such constitutes the greatest hazard to the health of non-smokers in the vicinity of the smoker. Apart from the health effects, many find the smell of smoke obnoxious, particularly where it lingers in clothing and furniture or fitments. In some sensitive individuals the smoke irritates the eyes and respiratory passages.

Passive smoking may result from exposure to mainstream smoke, which is smoke drawn from the lighted tip through the length of the cigarette and probably into and out of the lungs of the smoker. Alternatively, sidestream smoke may be involved, which is smoke given off from the lighted tip between inhalations.

About 3000 components can be identified in tobacco smoke. These are usually separated into two categories, those in the gas phase and those in the particulate phase. The gas component most likely to produce hazard is carbon monoxide. In the particulate phase, it is nicotine and tar. Low-tar cigarettes which reduce hazard to the smoker, however, do not substantially reduce potential risks to the passive smoker.

Nearly half of the tobacco is burned during inhalation, although this accounts for only about 3 per cent of the total combustion time of the average cigarette. The mainstream smoke produced during inhalation is diluted due to the retention of many of the particulates and much of the gas in the lungs. On exhalation this mainstream smoke tends to be dispersed and further diluted into the surrounding air. In contrast, sidestream smoke disperses slowly and tends to remain in discrete, concentrated clouds for much longer periods and also contains large proportions of some components and smaller proportions of others. For instance, sidestream smoke contains up to 98 times more ammonia and about 33 times less hydrogen cyanide than mainstream smoke.

However, the fact remains that many of the components of tobacco smoke are known or suspected carcinogens. Substances identified include heavy metals, polycyclic aromatic hydrocarbons (PAHs), insecticides and fungicides. Carbon monoxide is usually chosen as an indicator of tobacco smoke pollution. This is present at nearly 4 per cent of the volume of smoke and at these levels is easily detected by portable and relatively unsophisticated equipment. In the blood, it combines with the haemoglobin thereby reducing its oxygen carrying capacity.

The level of carboxyhaemaglobin in the blood of a smoker can be well in excess of the 5 per cent limit recommended by WHO. In chain smokers, this level may approach 20 per cent. In heavily polluted environments such as bistro bars, clubs and offices where smoking is permitted, recent Australian research workers have accumulated evidence that these levels are exceeded by passive smokers. This means that in circumstances where air circulation is poor and smoking is not prohibited, a definite risk to the health of others exists through smoking by members of the group.

It has been shown that coronary heart disease is more prevalent in persons who are heavy smokers and in others who have high carboxyhaemaglobin levels. And in pregnancy, carbon monoxide can cross the placenta and the foetus may have higher levels than the mother. This can lead to reduced birthweight due to an interference with intra-uterine growth. Foetal exposure to carbon monoxide has also been correlated with poor performances by children in mental tests, particularly those involving memory, in locomotor ability and in learning. Passive smoking effects on spouses of smokers have been shown to include a significantly greater risk of lung cancer. The effects of other parameters of ill-health including cardiovascular diseases are currently under study.

TABLE 4.13: Allergic Lung Reactions Associated with Micro-organisms

CONDITION	SOURCE	ORGANISM
Bagassosis	mouldy sugar cane	Thermoactino-mycetes sacchari
Farmers' lung	mouldy hay	T. vulgaris; Micro-polyspora faeni
Malt workers' lung	mouldy barley, malt	Aspergillus clavatus; A. fumigatus
Maple bark lung	mouldy maple bark	Cryptostroma corticale
Mushroom pickers' lung	mouldy, growing compost	T. vulgaris; M. faeni
Paprika splitters' lung	mouldy peppers	Mucor stolinifer
Suberosis	mouldy cork dust	Penicillium frequentans

References and Additional Reading

Australian Standards. 'Air-Handling and Water Control Systems in Buildings: Microbial Control' from *AS 3666*, 1988, pp.4, 12-13.

Brindle, R.J. 'Nocosomial Legionnaire's Disease — Advances in Diagnosis and Typing' from *Journal of Hospital Infection*, 1988, 11/supplement A, pp.196-200.

Christopher, P.J., Noonan, L.M. & Chiew, R. 'Epidemic of Legionnaire's Disease in Wollongong' from *Medical Journal of Australia*, 3 August 1987, 147, pp.127-8.

Edelstein, P.H. 'Nocosomial Legionnaire's Disease: A Global Perspective' from *Journal of Hospital Infection*, 1988, 11/supplement A, pp.182-8.

Edelstein, P.H. & Meyer, R.D. 'Susceptibility of Legionella pneumophila to Twenty Antimicrobial Agents' from *Antimicrob. Agents Chemother.*, 1980; 18:403-8.

Kurtz, J.B. 'Legionella pneumophila' from *Ann.Occup.Hyg.*, 1988, 32/1, pp.59-61.

Lattimer, G.L. & Ormsbee, R.A. 'Clinical Diagnosis' in Lattimer, G.L. ed. *Legionnaire's Disease*, Marcel Dekker, New York, 1981, pp.44-60 (Infectious Diseases and Antimicrobial Agents, volume 1).

Lee, D.H.K. & Selikoff, I. 'Historical Background to the Asbestos Problem' from *Environmental Research*, vol.18, 1979.

Maher, W.E., Para, M.F. & Plouffe, J.F. 'Subtyping of Legionella pneumophila Serogroup 1 Isolates by Monoclonal Antibody and Plasmid Techniques' from *Journal of Clinical Microbiology*, 1987, 25/12, pp.2281-4.

McDade, J.E., Shepard, C.C., Fraser, D.W. *et al.* 'Legionnaire's Disease: Isolation of a Bacterium and Demonstration of Its Role in Other Respiratory Disease' from *N.Engl.J.Med.*, 1977, 297, pp.1197-203

Osler, W. *The Principles and Practice of Medicine*, 8th edn, Appleton, New York, 1917, p.81.

Rasch, J.R. & Mogaobgab, W.J. 'Therapeutic Effect of Erythromycin on Mycoplasma pneumonia' from *Antimicrob. Agents Chemother.*, 1965, 5, pp.693-9.

Reynolds, H.Y. 'Host Defense Impairments That may Lead to Respiratory Infections' from *Clin. Chest Med.*, 1987, 8/3: pp.399-458.

Winn, W.C. Jnr. 'Legionnaire's Disease: Historical Perspective' from *J.Clin.Microbiol.Rev.*, 1988, 1/1, pp.60-81.

Yung, A.P., Newton-John H.F. & Stanley, P.A.: 'Atypical Pneumonia: Recognition and Treatment' from *Medical Journal of Australia*, 3 August 1987, 147, pp.132-6.

4.6 Sensory and Nervous System

The nervous system, comprising the brain and peripheral nerves, is the central controlling organ of the body. Through the connecting network of nerve cells, ganglions, synapses and neurochemicals, the brain fine tunes, co-ordinates and manages messages impinging on the body from the internal as well as external environment. Through the senses, the nerves detect changes in these environments, and convey directions from the brain to manage the changes.

Damage to the nervous system is potentially the most destructive and wide ranging of all, and often the most difficult to address or avoid. It ranges from direct damage to the nerve cells, to effects on neuro-transmitters involved in impulse transmission or in associated structures and processes, whose sensitive control depends on nervous system co-ordinating mechanisms. It can come from inside, through the ingestion and/or accumulation of toxic and foreign chemicals. It can come from outside by direct injury or exposure to similarly toxic chemical agents. For example, the nerve gases, principally members of the cholinesterases when used in modern warfare, can result in rapid and immediate death through interference with peripheral transmission and central system function.

The Nervous System

The basic unit of the nervous system is the nerve cell or neurone. This consists of a cell with a long filament-like structure (the axon), whose function is to carry pulses of excitation along its length from one cell to the next and on to its ultimate effector organ or muscle. Impulses can travel along axons at over 100 cm per second, enabling us to move rapidly in response to stimuli.

Nerve cells are not joined to each other or to the effector cells. When an impulse passes down the axon to the nerve cell, one of a number of specific neuro-transmitter substances is released into the gap between cell body and dendrite fibres of the following cell. These chemicals, acetylcholine or nor-adrenalin, are then broken down by appropriate enzymes thereby completing the impulse-transmission process and permitting the cell bodies to reabsorb and resynthesise new transmitters for subsequent use.

Even though the nature of nerve impulse is essentially the same in all nerve cells, there are distinct sections of the nervous system. Basically the nervous system is divided into the central and peripheral systems. The central system consists of the brain and spinal chord, which co-ordinate nervous activity for the body. The peripheral nerves can be divided into afferent (sensory) nerves which carry impulses to the central nervous system, and efferent (motor) nerves which carry impulses from the central unit to the periphery organs. The efferent peripheral system is further divided into the voluntary and autonomic systems.

The voluntary system innervates striped skeletal muscle and is therefore concerned with movement. The autonomic system controls the 'automatic' functions of the body such as the movement of the intestine, the heart and the lungs. The autonomic nervous system is further sub-divided into sympathetic and parasympathetic systems. In general these two systems work in opposition. For example, stimulation of the parasympathetic nerves to the heart exerts a depressing action on the heart, slowing it down and weakening the force of contraction. Stimulation of the sympathetic nervous system accelerates the heart beat and increases its force of contraction. The actual heart rate is balanced in activity by nervous control mediated by these two systems.

The Senses, The Nerves and The Environment
Environmental chemicals which directly or indirectly affect the nerves range from those in the atmosphere, such as lead and lead oxides which cause derangement of central and peripheral nerve channels, to chemical additives ingested in food consumed, pesticides, herbicides and fertilisers entering the body through skin, the lungs and the food. Many of these substances are first detected by the senses. The ears, eyes, taste and sensations are finely managed by the co-ordinating interplay of sympathetic and parasympathetic pathways of the nervous system.

Sound and Hearing
Noise is a problem for human beings because we have sense organs that are capable of receiving information in sound in various survival-related ways. Animals use sound to stake out territories, to attract the attention of other members of the species, as a warning against intruders, for mating and reproduction, and to locate prey.

The sense of hearing is subserved by three parts of the ear. The outer ear collects and funnels sound-waves to the eardrum which vibrates to the intensity and frequency of the sound. The bones of the middle ear (incus, maleus and staples) then transfer this mechanical motion to the delicate fluid-filled inner ear (cochlea). The inner ear contains a spiral membrane, the different parts of which vibrate in patterns that depend upon the frequency and intensity profile of the incident sound.

Particular groups of hair cells along this basilar membrane become depolarized to a degree that depends on how much the membrane is moving and the stressing of the hairs on the hair cells in their location. The depolarization of these hair cells in turn initiates frequency and intensity-specific patterns of nerve impulses in the associated nerve cells, and these then carry news of noise or other sound along the auditory nerve to the brain.

Noise: The American National Standards Institute defines noise as any 'undesirable' sound. Noise has also been described as unwanted sound or sound without value. Harmful sounds are not always perceived as undesirable and unwanted, and are frequently not classed as noise. Thus, noise is really unwanted sound or sounds of a duration, intensity or quality that causes harm to humans or other living beings in some physiological or psychological manner. Under this definition, noise applies to any loud sound above 90 decibels, regardless of how 'beautiful' it may appear to be.

The Auditory Effects of Noise: The magnitude of noise-induced hearing loss is immense. In the United States, an Environmental Protection Agency survey in 1977 showed that, in their daily lives, about one-in-ten people are exposed to noises of duration and intensity sufficient to cause hearing impairment. In addition, hundreds of thousands of workers between the ages of 50 and 59 could be classed as eligible for workers' compensation because of impaired hearing caused by their jobs.

Further studies have shown that factory workers suffer double the rate of hearing loss of white-collar workers of comparable ages. Also, noticeable hearing loss was evident among 5 per cent of 12-year-old Americans, 14 per cent in 15-year-olds and 20 per cent in 18-year-olds. At the same time, the studies show that people living in environments relatively free of noise do not have hearing loss even in advanced age. Very old Mabaans, African tribal people from southeastern Sudan, have been shown to have about the same hearing acuity as American children.

Specialists confirm that the extent of hearing loss in young people is increasing and hearing acuity among the general population is declining. Most people are unaware of the constant background noise-pollution caused by household electrical appliances, modern equipment, city suburban traffic hum, radios and television. Studies of young people listening to rock music have shown that such music, generated at over 92 decibels throughout the 500 to 8000 Hz range and sustained over one hour or more, can produce a 40 decibel threshold shift in about 10 per cent of the listeners, and somewhere between a 20 per cent and 30 per cent threshold elevation in the remainder. Though most of this threshold shift is usually temporary, repeated exposure can make it permanent. (A threshold shift is an elevation in the lowest threshold of hearing.)

The Nature of Auditory Damage: A number of years ago, researchers found that in guinea pigs exposed to loud noises such as rock music, the hair cells of the inner ear collapsed and shrivelled. These cells are organized in rows along a thin membrane wound in a spiral fashion through the cochlea of the ear. Once destroyed, these hair cells are not replaced. They are responsible for sensing and transmitting the fine incident sounds of human speech in a recognizable form, which results in our ability to communicate with others. Thus, it is these most sensitive communicative speech nuances which signal the first signs of hearing loss.

Hearing Loss: Workers Exposed to Industrial Noise: Noise in the environment, both occupational and non-occupational, is endemic, but noise standards are often vague or non-existent at the community level. However, standards for controlling noise are general in most workplaces in western industrialised countries. Unfortunately, frequently they still fail to recognise that noise damage over a working lifetime is cumulative. For example, the 90 dBA (8-hourly time weighted average) noise standard applying in all Australian states has been estimated to cause 25 dB hearing loss in 28 per cent of those exposed. Unlike age, the other major cause of hearing loss, noise can be prevented or contained once the proportion of loss attributed to each factor is known.

An indication of the extent of occupational noise exposure and its effects on hearing may be inferred from the data of two US surveys, the National Occupational Hazard Survey (NOHS) conducted by the US National Institute for Occupational Safety and Health, and the National Health Interview Survey (NHIS) conducted by the US National Centre for Health Statistics.

The results of self-reported hearing loss among workers aged seventeen years or more surveyed by NHIS were divided into three groups: light exposure, where less than 10 per cent of the workers in these industries were estimated by NOHS to be exposed to industrial noise equal to or greater than 85 dBA; a moderate exposure group, with 10-24 per cent of workers exposed; and a heavy exposure group, with 25 per cent or more of the workers exposed to industrial noise levels greater than 85 dBA. Workers interviewed in the NHIS were further stratified into three age groups: 17-44, 45-54 and 55 or older. Comparison was made of prevalences of hearing loss in workers in the three groups.

The risk of hearing loss attributable to occupational noise exposure was estimated by applying the attributable risk for each industry group, categorised according to noise exposure, weighted according to the proportion of total hearing loss cases occurring in that group, to give an overall summary estimate of attributable risk. This led to an estimate of 20 per cent cases of self-reported hearing loss in the adult male population being attributable to occupational noise exposure. Furthermore, prevalence of self-reported hearing loss in males increases with both age and prolonged exposure to industrial noise.

The data show no effect of occupational noise exposure in females. They also show that the prevalence of hearing loss is less for females than for males in all categories, including the industries with least probability of noise exposure. In other words, females have a lower prevalence of self-reported hearing loss from non-occupational causes (age, disease, background noise) as well as from occupational noise. Explanations for the low prevalence of hearing loss from industrial noise in females may include reduced exposure in women, since more work part-time or are allocated to less noisy areas, even in noisy industries.

It might be argued that using self-reports as a basis for diagnosis of hearing loss lacks objectivity, and in fact the US hearing loss data suggest that the indicated prevalence of hearing loss is low compared with findings of another US survey and the National Health and Nutrition Examination Survey based on audiometric studies. On the other hand, it might be argued that subjective awareness of hearing loss is a more important effect than a defect detected by audiometry. Nonetheless, the findings confirm that hearing loss disabilities have been seriously underestimated by industries and health authorities: studies about the effect of differing levels of background noise and of loud music exceeding recommended maximum levels, as in discotheques, appear warranted.

Other Physiological Effects of Noise: Anyone who has experienced sound near the threshold of pain would not be surprised to learn that sound in the range of 120-150 dB can affect the respiratory system and balance to the extent of dizziness, disorientation, nausea and vomiting. Even at the level of 70 dB, sounds can have measurable physiological effects, which may not result in immediate impairment, but do emphasise the magnitude and variety of impact noise has on human beings.

Since nerve pathways permit both ears to communicate with the brain, other body functions can reflect receipt of noise at a subconscious level. Loud noises and explosive sounds can incite the sympathetic nervous response referred to as the 'flight or fight' reaction. Overall, loud noises cause an increased production of most hormones of the pituitary gland; among the most important of these is the adrenocorticotrophic hormone (ACTH). ACTH then stimulates the adrenal gland, which secretes hormones which sensitize the body to adrenalin, increase blood sugar levels and suppress immune responses. The result is a heightened sense of awareness, a constant background level of tension above that normally experienced.

Sleep is one of those important physiological functions which is affected by noise. Sleep deprivation occurs with noise interference and heightened levels of noise in the environment. A continual background noise will make sleep more difficult, and is possibly mediated by the raised hormonal levels referred to above.

Studies done in France suggest that noise may cause as much as 70 per cent of the neuroses found in major cities like Paris. A study reported in *The Lancet* (Abey-Wickrama *et al*, 1969) showed that for certain categories of mental illness, admissions to psychiatric hospitals in London occurred at significantly higher levels for people living in areas of maximum noise levels near Heathrow Airport. The authors suggested that while noise may not actually be the cause of mental illness, it might aggravate conditions leading to the appearance of symptoms.

Extremes of Temperature
Where extremes of temperature are due to weather patterns, heat waves and seasonal extremes of cold, effects vary depending on age, health and people's capacity to guard against them. Excess heat in particular serves as an occupational health problem. In confined environments such as factories, where performance schedules are tight, the presence or absence of air-conditioning, number of persons per metre of space and the requirements of the job tend to make conditions the more aggravating. However, environmental heat and cold are physically hazardous only under certain conditions.

Environmental heat is likely to prove dangerous only when the temperature exceeds the limit where an individual can maintain a deep body temperature at or below 38° C. In increasing order of severity, exposure to extreme heat can cause prickly heat (miliaria), heat cramps, heat exhaustion and heat stroke (hyperpyrexia). Cardiovascular collapse, heart failure and dehydration are most likely in the elderly or the very young.

Extremes of cold have both localised and systemic consequences. Local effects include chilblains and frostbite to the toes, fingers, ears, nose and cheeks. Acute exposure results in whole body chilling or hypothermia. Mountain climbers, Arctic explorers, sailors in open boats on the high seas, and divers are all at particular risk. At special risk are sailors swept overboard and forced to wait hours for rescue in cold water, where the combined effects of shock and cold make them especially prone to hypothermia.

In industry, cold environments may be hazardous for those working an eight-hour shift in temperatures below 10° C, especially in wet and windy conditions. The provision of adequate warm clothing and protection such as gloves, head and face protectors, is now routine for industries where work activities are centred on cold rooms, refrigeration units and plants.

Vibration

Exposure to vibration is not a major cause of ill-health in the general population though it may be found in certain industries. Low levels of vibration are known to affect comfort and reduce work efficiency. Prolonged exposure to intense vibration can cause severe physiological dysfunction and sensory numbing. Chronic exposure to vibration affects either the whole body or a local area, usually the hands or arms.

Exposure to local vibration, particularly in cold working conditions, leads to an injury known as 'industrial white finger' with symptoms resembling Raynaud's disease (blue extremities, sensitivity to the cold — very common — due to lack of circulation; nerve damage). Vibration disease caused by whole body vibration, however, is characterised by gross central nervous system changes and is identified with dystonic changes to blood vessels and a polyneuritis syndrome of the lower extremities. Diencephalic disorders and disorientation may also result from prolonged exposure.

Drivers of such vehicles as semi-trailers, tractors and cranes, and workers in sustained body contact with vibrating machinery like weaving looms, are prone to whole body vibration disease. A wide range of hand-held power tools and such control mechanisms as steering wheels can be sources of hazardous local vibration, particularly if poorly adjusted and fitted.

References and Additional Reading

Abey-Wickrama, I., A'Brook, M.F., Gattoni, F.E. and Herridge, C.F. 'Mental-Hospital Admissions and Aircraft Noise' from *Lancet*, 1969, vol.7633, pp.1275-7.

New Scientist, 29 July 1989 and 18 November 1989.

Rosen, S., Buergman, M., Plestor, D., El-Mofti, A. & Hamad-Falti, M. 'Presbycusis Study of a Relatively Noise-free Population in the Sudan' from *Ann. Otol. Rhinol. and Laryngol.*, 1962, vol.71, pp.727-43.

The Sydney Morning Herald, 23 November 1989, 'A Consumer's Guide to the Environment' (8-page supplement).

The Weekend Australian, 25-26 November 1989, p25.

4.7 The Unexplained and Unexplainable

The final chapter concerning specific environmental effects on human health deals with some of medicine's most difficult and baffling conditions. Many of these are classed as psychological or complex conditions involving a whole range of interacting processes. Others are designated psychosomatic illnesses, a convenient category covering behaviours and sub-clinical states which resist explanation in conventional medical terms.

Hence the chapter will deal with the 'difficult to diagnose' whose symptoms may be blamed on the weather, changes in temperature, wind direction, humidity or even the season. Some explanations are offered and theories proposed.

The Psychosomatic Illnesses
Perhaps it is a reaction to the pre-eminence and pre-dominant authority granted science that we see, as the twenty-first century approaches, psychosomatic illnesses being resurrected from the medical closet. Conventionally trained, respected medical practitioners, specialists, psychologists and other health professionals are beginning to take a second look at some of the ideas of the alternative practitioners and faith healers on whom they once poured scorn. And they are beginning to re-examine the variety of symptoms reported to them by patients they formerly dismissed as hypochondriacs and cranks.

The symptoms of these conditions were usually difficult to describe, impossible to measure and intermittent in frequency. Being vague, non-specific but persistent over time, they were difficult to fit into any precise disease category. They ranged from emotional to psychological, physiological and dietary problems. They made sufferers feel sick or 'less than well'. Many found that they could no longer work with efficiency, lacked concentration and became depressed.

The alternative practitioners had long argued that the symptoms described frequently served as the first clinical signs of illnesses which could be diagnosed with greater certainty only some time later. Often this proved to be after the condition had worsened and was more resistant to treatment.

Other therapists considered the patients were suffering from sub-clinical ill-health which either did, or did not warrant treatment, depending on the skills and techniques provided or available to the practitioner. The range of signs and symptoms reported by the sufferers and the treatment approaches used by those who attempted to 'manage' or 'cure' them were drawn principally from naturopathy and homeopathy, areas of medicine dismissed by orthodox medical doctors.

With the increased attention and recognition given 'alternative' therapies in the last twenty years has come a greater willingness to accept that many of the symptoms and conditions classed as psychosomatic may indeed be 'real' diseases, rather than the figments of over-active imaginations. Depression during winter, irritability in summer and even pains in the joints when dry westerly winds are blowing: is some rational explanation possible? There are other conditions which medical doctors have found equally difficult to accept or classify. For example, people seem to differ in their ability to recover from illnesses, and their recovery rates do not seem to accord with former apparent health status. Recovery after viral and other illnesses is often delayed. Why is this so?

Climate and Human Health
The weather has long been claimed to have an effect on health. Heat stress, asthma attacks, irritability, depression and aggressive behaviour can be explained by changes in the weather. Health is even claimed to be a function of modern lifestyle and work conditions. The 'sick building' syndrome is no longer considered to be a fiction of an overactive imagination.

Changes in the seasons are most often blamed for changes in health status. In affluent western countries, complaints about depression, irritability and weight gain accompany talk about winter. Chest infections, coughs, aches and pains abound, particularly among the elderly housebound by rheumatic joints and susceptible to draughts. With spring comes a return of better moods. Energy levels are high, weight loss is common and jokes abound. Minor ills are forgotten except among allergy sufferers, bothered by runny noses and smarting eyes.

Summer induces a relaxed, languid state tempered by the occasional gastric upset caused by over-indulgence or eating heat-affected food. Heat rashes, sunstroke and skin problems seem to predominate. Autumn brings the first rash of colds and sneezes as colder weather appears before heavier clothing is brought into service.

Seasonal affective disorder (SAD) is all these behavioural symptoms and moods rolled into one. It is the name given to a condition which is basically psychological in nature, and appears to be related to the moods and physiological health experienced by certain susceptible individuals in different seasons of the year.

Seasonal Affective Disorder (SAD)
The condition was first reported in a serious and reputable scientific journal in the early 1980s, by an American doctor, Alfred Lewy. Lewy described a patient who experienced dramatic decreases in energy, mood and functioning during winter, accompanied by over-eating and over-sleeping. These symptoms were worst in mid-winter; in mid-summer the opposite effects could be observed, climaxing in an almost manic cheerfulness.

Dr Lewy treated his patient by attempting to lengthen his day artificially, using a series of bright sunlamps. Effectively this extended his photoperiod and Lewy theorised that this was caused by suppression of melatonin, the hormone secreted by the pineal gland. The presence of melatonin in the blood suppresses physiological function and prepares the body for sleep. The bright sunlight of the lamps suppressed melatonin activity, thereby 'adjusting' his body cycle.

At the time, it was thought that SAD probably affected only 5 per cent of the population. By 1985, that figure had been substantially revised to nearer 25 per cent as many Americans reported being affected by seasonal mood and body changes. Of the claimed sufferers, 85 per cent are women and their symptoms are far from consistent in their nature. Most experience cravings for carbohydrates, gain weight, feel irritable and begin to sleep longer. They lose interest in sex, performance at work declines and concentration skills deteriorate. Many sufferers first complain in their mid-twenties and experience 'anticipatory anxiety' even before the end of summer. For the majority, complaints become worse with age.

Other researchers, writing in the *American Journal of Psychiatry*, describe a similar complaint which is characterised by milder symptoms than those of Lewy's original subjects. People in this category — including a large number of those judged to have SAD — could be experiencing a Hibernation Response. This phenomenon is managed effectively by counselling and behaviour change strategies. Its scientific basis is equally tenuous.

There are real difficulties in obtaining the information which would link the psychological and physiological signs of SAD and Hibernation Response with quantifiable changes in weather and seasonal variations. The emergence of the new field of biometeorology offers some hope that scientists and medical experts will now seriously address the problems.

Local Weather Changes and Health
Among the problems to be resolved by biometeorologists will be the 'old wives' tales', such as those linking dry winds with irritability and marriage break-ups, and the imminent arrival of rain and cold with aches, pains, stiffness and rheumaticky joints.

Cold fronts are correlated with increased heart attacks: some researchers claim there is almost a doubling in their incidence. Extremes of hot weather bring increased stress to cardiovascular systems, particularly among the elderly. In Greece during the extended periods of very hot summer weather of 1988 and 1989, admissions to several hospitals reached peak levels. Elderly people experienced cardiac distress symptoms and congestive heart failure. Hyperthermia, headaches and dehydration were common among adults and the very young.

Post-operative complications often coincide with changes in the weather. Sixty per cent of complications occur with cold fronts, 30 per cent with the arrival of a warm front. Asthma and glaucoma attacks are frequently triggered by passing cold fronts. Cases of angina pectoris follow a marked seasonal pattern, peaking in autumn and winter. Finally, a study of 1.6 million patients with circulatory ailments showed a peak in January and February.

The Effects of Wind: Wind, too, may be troublesome. Folk tales abound describing the miseries of the foehn in Austria, France's mistral, Israel's sharav or the Santa Ana of California. Central Europeans describe a range of symptoms which they refer to, without embarrassment, as 'foehn psychosis'. These include physical weakness, irritability, headache, depression, anxiety, insomnia, apathy, nausea, nightmares and a tendency to quarrel.

The effects of the wind have been blamed on an over-abundance of positive ions. Negative ions, on the other hand, seem to have a calming or healing effect on humans and animals. Researchers believe that serotonin or 5-hydroxytryptamine is the key factor linking ions and human comfort. Serotonin is a brain neuro-transmitter and mood-altering substance essential to transmission of impulses along sympathetic pathways. It is also believed to play a major but unappreciated role in body function at the cellular level. For example, serotonin mediates aspects of red blood cell aggregation and adhesion, is involved in inflammatory responses and has a role in infection control.

A high positive ion concentration appears to depress serotonin levels. Bioclimatologists working at Hebrew University in Jerusalem have found that the unpleasant effects of depressed serotonin levels are felt as much as twelve hours before the sharav arrives. At the same time, the normal atmospheric ion ratio of 5 positive to 4 negative ions climbs to 132 positive to 4 negative.

Temperature and Mortality: A number of recent American research reports have concentrated on assessing the possible effect of increased temperatures and global warming on human health as well as on the environment. For example, the Center for Climatic Research at the University of Delaware in

Newark has been studying the impact of weather and climate on human mortality. Its work surveyed weather patterns and mortality data from 48 selected US cities and found that differences could be identified based on season and region. Northern cities such as Pittsburgh and Detroit, which did not often suffer from excessive hot weather, recorded higher climate-mortality rates during unusually hot summers than was evident in cities and regions where consistently hot and/or steamy conditions were routine. The elderly were particularly badly affected and experienced as much as a 50 per cent increase in summer mortality. Furthermore, the earlier in the summer the heat waves arrive, the higher the death rates attributable to weather.

Psychologists have also concluded that hot temperatures were clearly correlated with aggressive, violent behaviour. Reporting in the *Psychological Bulletin* of 1989, Dr Craig Anderson of the University of Missouri, Columbia, described an exhaustive survey of two centuries of scientific data on temperature-aggression connections. He catalogued incidence of homicide, assault, rape, genocide, riots and beatings with high temperatures (in excess of 30° C) in nearly endless permutations of variables, clinical controls and geographic regions. He concluded that the hot, humid months of July and Autumn are characterised by more murder, rape and assault in the northern hemisphere.

Anderson theorises that temperature effects are two-fold and impact first on the individual, and second on others with whom that individual reacts. Unbearably hot weather apparently stimulates the hypothalamus gland, which is located in the brain and is the locus of the human thermoregulatory system. This gland regulates body temperature. It also plays a role in testosterone, cortisol and serotonin control, hormones which are important in aggressive behaviour.

Sick Building Syndrome
Related to the weather and coming within the category of non-specific environmental illnesses is a condition labelled 'sick building syndrome', reported in several countries over a number of years, and recently with increased frequency. As the name implies, the disease originates within the building or its services. Occupants in such cases are prone to periods of illness, although usually only to the extent that a small proportion of them are affected at any one time. Symptoms may vary from patient to patient but the WHO reported that common symptoms include:
   * eye, nose and throat infection
   * sensations of dry mucous membranes and skin
   * erythema
   * mental fatigue
   * headache

* high frequency of airway infections and cough
* hoarseness, wheezing
* itching and unspecific hypersensitivity
* nausea and dizziness

In many cases the symptoms are so vague that they are likely to be dismissed as common colds or influenza, or to be misconstrued as psychosomatic. However, their frequency in association with buildings has given some substance to the idea of 'sick' buildings, particularly with respect to high-rise buildings which are essentially utilitarian and lacking in visual appeal. In Denmark between 15 and 30 per cent of persons surveyed reported symptoms of the type listed above relating to certain types of buildings, whilst in Sweden the figures approached 40 per cent. In some cases, the illness amongst occupants would appear to result from pollutants evolved from new installations, such as paints or other synthetic materials. In others, the 'sickness' appeared more long-term and a number of common features could be identified, such as a ventilation system involving some form of air conditioning, a relatively warm internal air temperature, or a large proportion of surfaces covered by materials such as carpets. In most cases, it is not possible to identify a particular cause for the effects reported by occupants, which might range from dusts, to gases, vapours and micro-organisms. Formaldehyde, nitrogen oxides, ozones, and carbon monoxide have all been identified in different buildings.

The possibility also exists that positive ions associated with a dry atmosphere or the flickering of fluorescent lights might be responsible, particularly under distinctive weather conditions. Negative ion generators have been used to improve conditions, and in many instances have proved successful. Nonetheless, scientific evidence to support their claimed benefits remains elusive.

Air Pollution and Feeling Good

According to many reports, people do not function very well on smoggy days — they suffer impaired efficiency. For some people at least, pollution brings a decrease in the internal feeling of wellbeing. US studies have shown more absenteeism and higher accident and suicide rates during unfavourable weather and on overcast and smoggy days. They have been supported by evidence of personality changes, increased levels of irritability and accentuated traits such as forgetfulness. Furthermore, these effects are most marked on the first day of a pollution episode. The California Department of Public Health reports that as the level of photochemical oxidants rises so does the motor vehicle and personal accident rate.

One of the pollutant substances likely to be responsible is carbon monoxide, the chief environmental problem of which is that it combines with blood haemoglobin and competes for oxygen. In an atmosphere with more than 10 ppm carbon monoxide even in the presence of normal amounts of oxygen, the blood can be rendered unable to carry sufficient oxygen to the tissues. If as little as 5 per cent of the haemoglobin is carrying carbon monoxide, the oxygen-carrying capacity of the blood is reduced to the extent that certain kinds of performance tests reveal physiological impairment. With an affinity for haemoglobin 200 times more than that of oxygen, even extremely low levels of carbon monoxide are believed to cause impairment. Symptoms similar to those described earlier are probably due to early carbon monoxide effects.

Exposure to 1000 ppm of carbon monoxide for less than an hour is fatal. A level of 3-4 per cent carboxyhaemoglobin has been associated with reduced work efficiency; decreased vigilance, altered perception and decreased manual dexterity occur at a level of 5-7 per cent or greater (US Environmental Protection Agency, 1984).

Hydrogen sulphide is another major asphyxiant sometimes encountered in high concentrations around sewers and in various occupational settings. This toxic gas impairs tissue oxygenation indirectly by paralyzing the breathing control centre in the brain. Although hydrogen sulphide's 'rotten egg' odour gives it away at low concentrations, at higher concentrations it rapidly paralyzes the sense of smell and prevents any warning. Hydrogen sulphide apparently affects neurons by decreasing their ability to conduct impulses. People chronically exposed to low levels of this gas through residence or work, report persistent lethargy and feeling unwell.

Lead monoxide from petrol fumes, ozone from photochemical oxidation and nitrous oxides have all been listed by the US Environmental Protection Agency as major low-level contaminants likely to affect the well-being of people, particularly those living in cities.

Psychophysiological Effects of Crowding and Environmental Ugliness
Crowding and ugliness tend to have effects on humans which are similar to those coming from weather and changes in the nature of the atmosphere. Crowded environments resulting from over-population, over-building and concentrated high-rise settlements are oppressive and ugly. They represent a unique product of human activity, a form of visual and environmental pollution which is interpreted differently by those people involved.

Beauty and ugliness are in the eye of the beholder: one person's crowd is another's party. Hence crowding and its effects are highly subjective: they depend as much on the circumstances as the intent and ideas of the individual. As an individual's mood changes so does his/her interpretation and response to the same situation.

The Effects of Crowding in Animals: In order to investigate whether crowding people together in confined spaces for short and long periods of time has physiological effects, a number of biologists decided to study crowding effects in animals. The first significant change observed was a reduction in animal reproductive capacity, caused sometimes by lower ovulation rates, and sometimes more directly by deterioration of organs such as the liver, spleen and kidney and in adrenal gland activity.

It is also suggested that the reproductive effects are consequences of a crowding-induced stress syndrome. The implication is that while the type of stress involved most often simply serves as a stimulus to members of the same species to move apart and distribute themselves in an optimal way, it can have gross pathological effects if for some reason the redistribution does not occur.

It has also been suggested that the crowding-induced stress syndrome is derived from the same biological roots as territoriality and pecking order. Wynne-Edwards was one of the first to hypothesize that territoriality and social hierarchy in animals is an important form of biological control. The activities that subserve territoriality and pecking order and even the singing of birds may all serve to alert members of the same, and perhaps even related, species to 'presence', influencing behaviour in a way that ultimately limits population density. Perhaps in species like *Homo sapiens* the same kind of intraspecific stress, in some vestigial if not functional form, continues to provide a source of biological stress.

A crowding-induced stress syndrome has been reported in many species of animals including rabbits, deer and lemmings, and even in certain other more exotic species housed in zoos. This shock disease has been described as a kind of hyper-insulinism; that is, its effects resemble an over-secretion of insulin by the pancreas. Changes occur in the blood vessels, the fatty content of the liver, the size of the adrenal glands, and the heart.

Crowding Effects on Human Beings: A limited amount of the data from animal studies has relevance to human experiences. Observations of human beings under chronic stress conditions indicate that the human pituitary-adrenal complex responds in a very similar way to that of mice, rats and deer. Much has been written about cities as agents of stress, deviant behaviour and various forms of mental illness, but establishing real connections has proved difficult. Perceptions and psychological feelings of 'being oppressed', lacking 'territory', and being deprived of privacy are common complaints from people living in dense crowded cities, camps and restricted areas.

The reason for lack of detailed studies on the effects of crowding may have much to do with establishing the distinction between crowding and what people interpret as opportunities for meaningful interaction with others. Van Potter defines an optimum environment as one that delivers an optimum amount of stress. People apparently need a certain amount of stress or stimulation that requires some kind of reaction. Other researchers, however, claim that many kinds of disease, particularly mental illness, occur with highest frequency in the central parts of cities. There may be many reasons for this; overcrowding appears to be at least one of them.

In a study by Faris and Dunham, schizophrenia, depression, senility, psychoses and other kinds of mental disorders were found to be most frequent in the central parts of cities such as Chicago, Illinois, Rhode Island and Providence. Furthermore, similar results were true in both relatively poor and relatively well-to-do parts of the cities. In another study undertaken in Nottingham, England, nearly 70 per cent of mental hospital patients were found to live within four kilometres of the city.

Whether the city produces mental illness, or those with mental illness tend to concentrate in the inner city, needs to be examined. In a study of residents of midtown Manhattan, only 20 per cent of the people interviewed were found to be relatively free of significant symptoms of mental illness, although only about 3 per cent of the remainder were really incapacitated by their mental disabilities. This raises questions about the definition of normal.

Noise, atmospheric pollution and poor housing, along with the social disorganization characteristic of cities, contribute to a complicated picture. Calhoun (1962) blames crowded environments for high levels of crime, drug abuse and other forms of social pathology. Murder, suicide, rape, assault, domestic violence and marital disharmony are highly correlated with urban environments. The anonymity of the city, its impersonal character and frequent change of neighbourhood and/or job provide an environment which other researchers associate with various forms of deviant behaviour. People act irresponsibly, vandalise amenities and form gangs of delinquents through the need to 'belong', or feel they play a role — even that of the criminal — in the society.

High-Rise Buildings: Ineichen (1986) has conducted a review of health problems associated with high-rise living. In his paper he attempts to correlate symptoms of depression, psychological stress and neurosis to residence in these structures in regions throughout the United Kingdom. Many of the studies cited poorly designed residential flats in which little planning forethought had been given to layout, design and the needs of future occupants for light, heating, cooking, laundry, privacy, and facilities for children to play.

Problems were exacerbated by poor maintenance, continual lift failure, faulty rubbish disposal systems, graffiti, litter and vandalism. Difficulties in monitoring semi-public space around high-rise flats contributed not only to crime, but to fears about personal safety and difficulties for mothers trying to supervise children at play.

Also projected as relevant were the isolation of inhabitants in high-rise buildings, where they were often denied the opportunity to interrelate socially as a result of building design; the comparative difficulty which high-rise dwellers have in modifying and influencing the physical environment around their homes; and the isolation of residents from the natural world of trees, shrubs and birds.

The Psychological Need for Privacy: A number of studies identify privacy as a basic human need akin to the territorial imperative, but one which is acknowledged only when it is lost. Privacy in the home satisfies the needs of the individual for self reflection, intimate conversations, study, etc. When internal design or overcrowding frustrate these needs, the result appears to be withdrawal, depression or illness. Considerations of design and layout are relevant to privacy, and not simply overcrowding.

Gabe and Williams (1986) found such a significant association between low and high internal density and psychological distress among women, that they proposed the inclusion of internal density considerations in state guidelines for architects. Worry about communicable diseases to children, another public health problem, smoking, drug abuse and use of tranquillisers were also raised as concerns among women living in urban high-rise.

Marital difficulties and money worries were aggravated by the feelings of isolation and perceived lack of friends and neighbours who might be able to share confidences in private. Fears of gossip together with inter-family tension and rivalry in the high-pressure environment reduced opportunities for friendship and cooperation.

However, the final chilling note on the residential environment and crowding is sounded by Keeley (1972). Keeley reported that studies of mice born and reared in normal uncrowded conditions, mice born crowded and raised crowded, and mice born crowded and raised uncrowded, showed that offspring of the last two groups manifested reduced activity and a decrease in responsiveness to various stimuli when compared with the mice born and reared uncrowded.

It appeared as if they had been 'lulled' into apathy. In addition, mice from the last two groups were more aggressive with other members of the group and had lower rates of survival. Keeley accounted for his results by suggesting the existence of some sort of stressing factor which may, in fact, even pass through the placenta to the embryo. If extrapolated to human communities, these and other findings raise serious questions for future town planners as they create the social conditions for future urban metropoli.

Chronic Fatigue Syndrome, Chronic Mono-nucleosis or
Myalgic Encephalitis (ME)

Recently accorded disease status is Chronic Fatigue Syndrome, once commonly referred to as Yuppie Flu, a complex viral disease that can have debilitating effects which are persistent and difficult to treat, especially by conventional methods. The symptoms include:

* lethargy, weakness and extreme fatigue making even the most simple tasks arduous
* persistent and chronic allergies
* severe muscle and joint pains or weakness
* headaches or feelings of cranial pressure
* poor co-ordination, vertigo and circulation
* stomach and bowel problems, most particularly nausea, bloated feelings, constipation and poor digestion
* inability to handle stress or recover from emotional trauma and personal loss.

These symptoms are found to a varying extent in individuals though the most common and extreme remains that of profound and persistent tiredness. Similarly, the degree of physical incapacity suffered by ME victims varies. Sufferers may feel nearly normal and suddenly suffer an acute onset of symptoms.

Patient histories show that victims suffer degenerative illness of a physical, dietary, mental, emotional and environmental nature due to multiple stress factors, which slowly reach the stage where extreme stress or a viral attack results in the appearance of ME symptoms. Consequently, naturopaths will tend to claim that at this stage the adrenal gland has been exhausted and reacts by forcing the body to rest and recover.

Under these circumstances, ME could be termed a post-viral illness. Such an explanation would coincide with the onset of symptoms immediately following a severe bout of flu or some other serious viral illness. Because antibiotics are ineffective against viral conditions, the body is forced to marshal its own immune defences. As these illnesses become increasingly virulent, so greater demands are placed on the body's immunological mechanisms.

Although hard-working, overly ambitious men and women appear to be the prime targets for chronic viral conditions such as ME, it is encountered in people of all ages and social backgrounds. Yet, despite its rapidly increasing incidence in developed countries, it is nothing new. As early as 1880 a Dr

George Beard recognized Chronic Fatigue, naming it 'neurasthenia' since he assumed it to be a neurosis accompanied by weakness and fatigue.

More recent outbreaks of the illness were reported in Los Angeles in 1934 and in New York in 1950. An outbreak in Iceland in 1948 affected 6.7 per cent of the population of Akureyi, generating the name Icelandic Disease. It also appears to have been responsible for a mysterious outbreak of viral illness which affected more than 200 staff members, mainly nurses, employed at London's Royal Free Hospital in 1955.

It is not easy to explain why there has been such reluctance to acknowledge the existence of a chronic fatigue-type syndrome and to describe the range of common symptoms. One explanation must certainly lie in the holistic and wide-ranging nature of the symptoms encountered. Another may lie in mis-diagnosis because of the medical profession's tendency to dismiss vague, non-specific symptoms of the type which are reported by ME sufferers. Nonetheless, the recent increase in its incidence means that researchers are now redirecting their interests and have begun to search for solutions.

Claims and Explanations: To date medical scientists can make only one claim with any semblance of confidence: Chronic Fatigue Syndrome is a real disease which has specific though wide-ranging symptoms. Explanations, however, tend to come more from the unconventional domain. There has been a suggestion that it could be related to exposure to pathogens such as Ross River Fever virus, *Coxsachie bacillus*, Epstein Barr virus (EBV) and Human B Lymphotrophic virus (HBLV).

The very latest information comes from the *International Clinical Nutrition Review* (October 1988), which states that the red blood cells of ME sufferers in relapse undergo changes in shape. The deformation includes spherocytosis stomatocytes and some unusual forms best described as a dimpled spherocyte. Membrane surfaces on the most aberrant forms are rigid and have a rough granular appearance.

In the United Kingdom, the ME Action Group, a consumer body agitating for funds for research, classes the illness as a holistic condition. According to its records, practitioners who use natural methods obtain the highest rates of success in treatment. They use a combination of eliminative and supportive therapies, ranging from cranial osteopathy to diet, detoxification, counselling and stress management. These techniques are based on the philosophy that all healing ultimately comes from within, and that the role of the practitioner is to stimulate the body's own inherent healing mechanism while at the same time providing psychological and moral support, guidance and knowledge.

Although treatment varies from one practitioner to the next, depending on their orientation and skills, the fundamentals of a natural approach are generally three-fold. First, the body is cleansed of stored wastes which interfere with normal cell function and stifle life energies. Immunity is then strengthened through diet and lifestyle therapies. Finally, the endocrine system is rebalanced so that health can be maintained and future disease threats managed with success.

The basic concept underlying this approach is that viral conditions develop most easily in a biologically toxic and exhausted body, ravaged by poor diet, stress, lacking fitness in body and mind, in which organs systems function at less than optimal efficiency. This idea is not accepted by conventional medicine though individual doctors claim success in treating ME sufferers by detoxifying the body.

Before World War I a British surgeon, Dr Arbuthnot Lane, describing a condition whose symptoms resemble closely those of Chronic Fatigue Syndrome, believed that the condition resulted from a massive build-up of waste products or toxicity in the body, particularly in the intestines. His observations stemmed from personal experiences of intestinal organ damage which so ravaged the body that they required surgery. He argued that diet particularly, together with irrigation of the intestines and altering patients' lifestyles, would permit the body to cleanse itself from inside. So seriously did his own profession regard his observations that Lane's hypothesis was accorded almost 400 pages in the *Proceedings of the Royal Society of Medicine* for 1913. Despite this, they were largely forgotten soon after and have only recently come to light.

Detoxification: Dr Patrick M. Donovan, director of the detoxification programme at the Clymer Clinic in Pennsylvania, has incorporated Lane's ideas into new treatment methods following research into the etiology and development of disease symptoms and the accompanying changes to blood cell morphology. These new methods begin by reducing the patient's exposure to environmental toxins such as drugs, food additives and foods which may induce an allergic response, as well as 'flushing out' long-term wastes, yeasts, heavy metals such as lead or mercury and undesirable bacteria. Elimination and cleansing mechanisms associated with the liver, bowel, kidneys and lymphatic system are stimulated by botanic medicine (natural herbs), colonic irrigation and exercise.

The patient's diet is then reformulated. Sugar, coffee, fried and salty foods, alcohol and highly processed items are eliminated, and a wholefoods diet appropriate to the patient's lifestyle is developed. These techniques are consistent with the philosophies of practitioners who maintain that it is the twentieth-century lifestyle — exposing individuals to dangerous environmental chemicals, excess stress, smoking, under-nutrition, over-nutrition, etc. — that suppresses immune functions and results in the accumulation of toxic wastes.

Potentiating the Immune System: Strengthening the immune system and restoring the integrity of its function lies at the crux of the natural approach to chronic viral conditions and is responsible for ensuring the final third stage of the treatment process. Each programme, diet and regime developed for an individual is designed to meet the specific needs of that individual. Strengthening the body's immune functions through detoxification and subsequent rebalancing of organ function permits body healing and develops resistance against future ills.

Non-Specific Psychological Ills

There is no doubt that the experiences of natural and humanly engineered environmental disasters suffered by individuals and whole communities affect the future lives of survivors. Memories of earthquakes which entomb for days and even weeks, and of chemical warfare and modern war, so profoundly affect survivors that they pass on changed attitudes, beliefs and values to their offspring.

In the same way it now appears that the attitudes and values of the children of parents who experienced environmental catastrophes such as nuclear explosions are affected by their fears as well as limited by any injuries or resultant genetic physical abnormalities. Have we failed to realise the long-term psychological and physiological effects such events can bring?

Explanations: Does The Weather Affect Human Health?

So, does the environment, specifically natural weather changes, influence human health? Can we blame weather extremes for heart attacks, accidents and sudden deaths, even for day-to-day changes in minor aspects of mood or well-being? Weather is determined in part by events beyond earth and in the solar system as much as in and around the planet. The eleven-year cycle of sunspot activity, the transit of comets and the entry of stray asteroids all affect the weather as much as natural volcanic activity, earthquakes, factory emissions, massive oil-field fires and nuclear explosions.

The relationships between the sun, the weather and human functioning are complex. Scientists report that solar radiation is amongst the most important climatic elements and is the primary source of atmospheric energy. The sun is subject to periodic energy discharges called sun-spots, which come in predictable cycles with peaks and troughs and overlapping mini- and maxi-cycles. Solar fluctuations have a major impact on the earth's magnetosphere, a kind of magnetic transfer zone about 5000 km above the surface. Changes there may directly influence weather much lower in the vaporous zone between ground level and ten kilometres out.

The earth is normally bathed in solar winds, streams of hydrogen ions which emanate from the sun. Sunspot activity enhances the solar wind, resulting in magnetic storms in the earth's atmosphere. The storms particularly affect the magnetic north and south poles of the planet and can produce the aurora borealis, the fabulous northern lights. Magnetic storms markedly decrease the strength of earth's electromagnetic forces, affecting short-term weather and longer-term climatic conditions. Sudden magnetic field fluctuations which are preceded by solar eruptions by about two days, are actually predictive of major atmospheric storms such as blizzards.

Magnetics, Electromagnetic Forces and Health

Magnetics completes the connection between sun, weather and health. This connection is mediated by electromagnetic fields which physical scientists now measure and include in calculations to explain events in the solar system and on earth. A number of forward-thinking scientists and healers believe that human life depends on the flow of energy from the extremities to all parts of the body.

Many medical scientists tend to dismiss arguments suggesting that electromagnetic forces energise the body, and will dismiss proponents of such arguments as unproven, philosophical and unmeasurable. Unlike modern scientists, Indian Ayurvedic practitioners and doctors of traditional Chinese and Unani medicine have long believed that living beings are energised by powerful, unexplained forces which flow through all body organs and systems. Health and well-being, they argue, depend on maintaining balance and smooth flow through energy channels or meridians. These meridians connect and integrate the operations of the physical body and the mind so that the two work in harmony. Ill-health is a result of disturbances to the smooth flow of energy through poor lifestyle practices, the accumulation of toxins and wastes, and the presence of disease.

Until the last twenty years, western medical doctors have dismissed arguments about life energies and unexplained forces as adequate explanations for treatments by Indian, Chinese and Arabian systems of traditional medicine. However, in attempting to understand how acupuncture 'works', research by scientists and western doctors has tended to confirm a role for some type of electromagnetic forces. This role seems to be related to impulse transmission through the nervous system.

Evidence also comes from other sources. A Russian scientist, Dubrov, claims to have found that short-term variation of the geomagnetic field may directly affect the central nervous system and may trigger cardiovascular dysfunction. Eclampsia, glaucoma and epilepsy may be triggered by changes in local magnetic forces which occur during thunderstorms and periods of sunspot activity. He also details increases in road traffic accidents, blood pressure changes, small changes in leucocyte counts and sub-cellular chromosomal changes. Dubrov further postulated that the acupuncture points which are used to stimulate and tone meridian flow in the body, are ion exchange sites for the electrical and magnetic activities of the earth's electromagnetic field and the body's own biomagnetic field.

The New Field of Biometeorology

Biometeorology is a new multidisciplinary field of study that has emerged in recent years to study the effects of atmospheric phenomena on all life. However, while biometeorology successfully identifies and studies connections between atmospheric effects and specific biological responses, it has had less success in explaining observed phenomena and developing theories which can be tested.

Among the more rational theories are those of Thomas Landscheidt, working at the Schroder Institute for Research in Cycles of Solar Activity in Belle Cote, Nova Scotia. Landscheidt began his investigations over twenty years ago in Bremen, at that time in West Germany, and his predictive work on solar activity is well regarded by scientists.

Landscheidt suggests that the earth, the atmosphere and human life represent one whole and complete system in which each component interacts with the others to produce a final result. For example, the weather experienced locally, on a day-to-day basis, represents one transient and short-term phase of climate within that hemisphere at a particular period of the solar year. In terms of size, even minor events in the solar system, or, on a smaller scale, in our own planetary system centred on the sun, seem massive to earth, a very small part of these larger systems. Even infinitesimally small changes at that level can be responsible for major changes in climate and significant alterations in the local weather we experience. Sunspots and solar flares are examples.

The eruption of a medium-sized solar flare releases power equivalent to 200 million hydrogen bombs. The massive energy discharge destabilises the earth's electromagnetic field, producing magnetic storms and violent weather. The atmospheric instability spreads and produces local effects with dust storms, tornadoes and willy-willies (small tornadoes of dust).

But Landscheidt does not limit his discussions to explanations of weather and the solar system. He quotes a series of energetic magnetic storms which hit earth in 1789 and 1968, blaming them, respectively, for the global instabilities which resulted in the French Revolution and the American political upheavals that followed Kennedy's assassination. He claims that his findings confirm that socio-historical events as well as human health are intimately connected (Landscheidt, 1989).

At the same time, it is possible to argue from the opposite perspective, that of earth. A series of major fires in the Arab oilfields in Kuwait may produce a pall of smoke which spreads around the globe, concentrated at the latitude of Kuwait. It has been predicted that this pall of smoke would darken the earth and bring substantial changes to climate. Just in that latitude alone, temperatures may be lowered by as much as 10° C, resulting in massive crop failures with social and economic effects. However, for earth as a whole and for the sun and the solar system, such an environmental catastrophe is minuscule and is subsumed within the larger whole.

In such a scenario, humans will experience effects on their health. In reality therefore, this means that human health is subject both to large-scale solar climatic effects and local changes in weather.

Behaviour and the Weather: Social psychologists estimate that as many as 20 per cent of the population are weather sensitive. Most biometeorologists will grant a correlation specifically between heat and violent behaviour and a more general link between some aspects of weather and human functioning. However, the multiplicity of variables involved makes theories and predictions impossible.

Biometeorologists try to explain the confluence of weather variables by talking about weather matrices. Each weather matrix involves absolute measures of temperature, humidity, sunshine hours, wind speeds, precipitation and geomagnetic activity, together with the rate at which changes in each occur. The experts maintain that weather matrix influences over a six-month period of time will account for about 20-30 per cent of daily mood changes.

Others complain that weather effects are individual because everyone responds differently to the same meteorological stimuli. They suggest that the links may be neurophysiological, particularly where changes in mood are involved. Psychologists interested in a link hypothesise that personality type or cognitive style is related to the temporal lobe sensitivity in the limbic system, the part of the brain thought to control behaviour, emotions, smell and other sensory perceptions.

Osteoarthritis and Exercise

Finally, to deal with the problem of those aching joints in winter, stiffness when it rains or when the weather promises to change for the worse. Most sufferers of rheumatic and arthritic conditions will tell you that they feel stiff, and that mobility and grip strength are reduced in cold weather, particularly during winter. A few will also tell you with all confidence that rain is due the next day, because their knees or hip are 'playing up'. Chances are the forecast will prove correct.

The majority of scientists dismiss these remarks as 'old wives' tales'. A few take them a good deal more seriously and are now trying to establish the reality of any links between joint pains and the onset of periods of rain, dry wind and changes in temperature.

Trauma

To date, the results have proved inconclusive but some facts are becoming clear. One is the relationship between trauma and osteoarthritis. Trauma is believed to play a significant role in the genesis and perpetuation of osteoarthritis. It is often the result of accident or occupation and the environment is frequently blamed.

In general populations, the prevalence of osteoarthritis increases with age. Radiological evidence of osteoarthritis has been shown in 86 per cent of women and 78 per cent of men over the age of 65 years. Clinically, the disease is characterised by joint pain, tenderness, limitations of movement, crepitus and variable degrees of local inflammation without systemic effects.

Cartilage is considered the primary target of osteoarthritis; however, many of the joint structures are vulnerable to mechanical trauma. Injuries to these structures, even in the absence of direct damage to the cartilage, may play a role in genesis of osteoarthritis.

Examples such as congenital dysplasia of the hip, angular deformities of the knee and a wide variety of post-injury states are well known to be predictors in early onset of osteoarthritis. While the first two are basically hereditary or due to birth-related problems, the last is not. Increased incidence of osteoarthritis has been found to occur in the finger-joints of cotton mill operators and in the knees of individuals with chronic obesity, especially females in domestic work.

At the same time, little proof can be found of a relationship between the lack of the disease in the hips and knees of physical education teachers or the ankles and knees of parachutists, individuals who might be expected to be sufferers. In fact, although anecdotal evidence of the association of osteoarthritis with trauma is quite strong, very little firm data exists to show that chronic occupational trauma or even acute injuries lead to osteoarthritis (Maguire, 1989).

It is important to note that stresses acting on the joints are necessary for normal maintenance of their health. When stress is reduced through lack of exercise or enforced immobility, significant changes in the cartilage occur, presumably mediated by the cells. Evidence suggests that prolonged periods spent standing in awkward positions, repetitious arm and shoulder-joint work as in the operation of machines, and kneeling and lifting as required in a number of manual occupations, are associated with higher levels of osteoarthritis, muscle strain, and work-related bone and joint injuries. Confirmation of a definitive link is hampered by the pre-existing high levels of osteoarthritic and joint problems in the normal population.

References and Additional Reading

Calhoun, J.B. 'Population Density and Social Pathology' from *Scientific American*, 1962, vol.206, pp.139-48.

Gabe, J. & Williams, P. 'Women, Housing and Mental Health' from a Paper presented to the Conference, *Unhealthy Housing — A Diagnosis*, Warwick University, 14-16 December 1986.

Ineichen, B. 'Mental Health and High-Rise Living' from a paper presented at the Conference, *Unhealthy Housing — A Diagnosis*, Legal Research Institute & Inst. Environmental Health Officers, Warwick University, 14-16 December 1986.

Keeley, K. 'Prenatal Influence on Behaviour in Offspring of Crowded Mice' from *Science*, 1972, vol.135, pp.44-5.

Landscheidt, T. *Sun, Earth, Man*, Urania Trust, 1989.

Maguire, K. 'Osteoarthritis and Exercise' from *Patient Management*, vol.13, no.5, May 1989, pp.103-13.

Potter, V.R. *Bioethics: Bridge to the Future*, Englewood Cliffs, New Jersey, Prentice-Hall, 1971, and cited in Porteus, J.D. *Environment and Behaviour*, Addison-Wesley Publishing Co., Reading, Mass. 1977.

Whybrow, P. & Bahr, R. *The Hibernation Response*, Random House-Morrow, 1988.

5.0 THREATS AND HAZARDS

We have now reached the stage when we begin to seek explanations for some of the effects not so far explained by scientific theories of disease causation. In this Section entitled Threats and Hazards, we begin to look at the more complex issues about human activities and the changes they have wrought in the environment.

Over aeons of time the survival of the ecosystem, comprising planet earth and all living systems in it, has depended on the continued supply of solar energy, its capture by green plants and, through the process of photosynthesis, its synthesis into the energy building blocks essential for life. This unique system and its finely interrelated parts are now under threat.

Human Activities

It is undeniable that the principal threat to planet earth and the living species resident there comes from these organisms themselves, because of the way they have usurped the role of nature and upset the delicate balance on which all life depends. In the course of evolution and the history of the earth, *Homo sapiens*, the human species, emerged. In an extraordinarily rapid period of time it has moved to a position of dominance over all other species. For most of this time the human species lived in harmony with the environment on which its survival depended. In time, nature could recycle the wastes produced. The natural balance was maintained and nature continued to provide adequate water, food and territory.

Unfortunately, humans proved to be remarkably fecund, particularly when they learned how to make use of nature. They began to expand in numbers and to intrude into all domains of the planet, laying claim to all land suitable for habitation. They developed new ways to harness nature to their own desires. Initially, these changes were relatively small on a global scale and the ecosystem adjusted to restore the natural balance. But with the twentieth century, the changes escalated in scope and speed of innovation, until today the whole ecosystem is under threat. And there is no guarantee that, even in time, it will find mechanisms to cope. Ironically, the species responsible now finds that its own activities which have proved so damaging to earth, also threaten its own survival.

This Section examines the changes in the environment which pose the most serious risks to human health. The first chapter considers the species itself, and the extraordinary increase in numbers through natural processes and the results of improved public health. Changes to the atmosphere and to the quantity and quality of the water supply are addressed in following chapters. The role of chemicals and their inappropriate use, albeit for the 'right' reasons, are discussed, emphasising their beneficial as well as adverse effects on human health. The final chapters attempt to deal with the more complex issues and interventions associated with human activities. Again the emphasis is on their implications to health.

The Ultimate Hazard?
Human beings are only one of the natural species to evolve on earth. In early times, they lived in harmony with nature and other species for practical reasons. Excessive removal of trees, over-hunting of game, killing for pleasure or wasting water meant that food resources were depleted and no longer adequate to sustain them. Humans roamed in bands, and numbers were kept in check by natural means. Age, misadventure, accident, childbirth and infection were the main causes of death. Even to the end of the Pleistocene epoch, humans appear to have had no major impact on the environment.

However, by this stage they were beginning to have an effect on selected species which served as food. For example, early Pleistocene humans are believed to have improved hunting techniques by the use of fire. Advancing lines of hunters following grass-fires easily drove prey into traps or swamps where they would be killed, a method so successful it is believed to have been a major contributing factor to the extinction of as much as 40 per cent of the game species in parts of Africa (Martin, 1967).

Some anthropologists believe that as humanity advanced through North America and eventually into Central and South America, many of these species, some already endangered by environmental and habitat changes, were rendered extinct long before selection could produce more wary varieties.

The Change to a Settled Life
As humankind began to develop the beginnings of today's agricultural patterns, humans began to have more chronic impacts on the environment. Some of this impact was damaging but some might be considered benign.

Scholars as early as 347BC described deforestation and overgrazing around the Mediterranean Sea leading to the drying up of springs and the erosion of soil. Both ancient and modern scholars attribute the collapse of the Babylonian Empire and other civilisations of the Middle East to such ecological disasters. There were also environmentalists in those early days. Around 40BC, Virgil recommended that the crops be rotated, that legumes be planted with other crops, that land be left fallow in alternate years and that soil be regenerated with manure, ashes and the like. Pliny wrote of human activities and their potential to alter local microclimate by practices such as diverting the course of rivers and draining of lakes. He reported that grapes and other food crops were destroyed by frosts once the temperature moderating effects of certain bodies of water were removed (Cole, 1966).

Early agricultural humans were known to use fire to clear land for crops and to create pasture. Undoubtedly this resulted in the destruction of forests and produced temporary upsets on the balance of nature in and around the Mediterranean and other parts of Europe dating from the Middle Ages. In Europe, great destruction of forests began around that time as land was cleared for farming, and as more and more lumber was used for ship building.

As the effects of humans became more widespread, the impact extended to some benign but nevertheless major changes in flora and fauna. Cows and horses were moved to new areas, birds migrated following changes in the planting and distribution of seeds and grain crops by humans. Although wheat first arose in parts of the Middle East, it has been scattered throughout the globe. Today some 600 million hectares of land are covered by wheat for a substantial proportion of the year, supplanting thousands of species of grasses, trees and shrubs.

Human horticulturalists have often acted as agents of natural selection as they learned to identify good seeds, to cross-pollinate and favour those seeds which yielded disease-resistant, strong and productive varieties. Selective breeding techniques have become increasingly sophisticated to the extent that today there is concern at the loss of simple, natural species, the gene stock.

Human Effects on Other Species

Homo sapiens has also served as a passive factor in the evolution and development of other species. Many organisms have learned to exploit humans and the niches humankind has created. The rise of humanity was paralleled by an increase in success for a number of disease organisms that wreaked havoc on human populations throughout most of their history. Some of these diseases were transmitted by rodents that had adapted very well to human-made habitats. A number of organisms have adapted to live in harmony with humans in the habitats created by them, and without much direct effect on humans. These include certain weeds, songbirds, field-mice, squirrels and other organisms like dogs that find some of the human-made niches very suitable, or at least not incompatible with their biological success.

Thus humankind has had a great impact, both positive and negative, on nature since its introduction into natural systems. Extrapolation from the past and the events of the present seem to indicate that this pattern will persist. However, the power of humans over nature has altered and it is now possible for humans to create major long-term and irreparable change on the environment, greater than any in the past.

Poverty: A Phenomenon Unique to The Species

Humans have also effectively jeopardised their own survival and undermined the quality of their own lives, as well as that of the environment. Poverty is one such case. Poverty is defined as a relative and/or absolute lack of basic productive and economic resources. In the developing world, to talk of poverty implies an absolute lack of food, clothing, consumer goods and shelter. In the affluent, developed world, where few people go hungry, it is more a relative sense of 'being without' or 'feeling deprived'.

In the developed as well as the developing world, poverty can therefore be as much a product as a cause of a degraded environment. A heavily populated Inner London suburb with high unemployment, poor schools, buildings defaced by vandals and graffiti, and health services over-burdened, is likely to record a level of health among its residents below that of the rest of the population and possibly equal to that of a developing country.

In other words, poverty is essentially a social problem. It is caused more by human actions, politics, struggles between rich and poor, and population pressures on scarce resources as much as chance, inheritance and country of birth. The solutions to human health, environmental health and poverty are closely interwoven and interdependent. Understanding them is equally difficult and complex.

One of the best examples probably lies in the developing world. Over-population, hunger, scarcity and exploitation of nature's finite resources have set up a depressing cycle of poverty with its own agenda. To progress and achieve a better standard of life, that agenda requires an abandonment of traditional ways, makes further inroads into limited resources and calls for yet more sacrifice at the individual level.

A Depressing Cycle: Poverty, Food Aid and Dependency

In the developing world, the poor are driven by their need for food to over-crop the land, to carry excessive numbers of grazing animals for the available pasture and natural water, to strip all shrubs and trees for firewood, and to use cattle dung for fuel rather than as a natural fertiliser for crops. When rains fail and seasons prove harsh, crops fail to grow, valuable seed is planted to compensate and when that, in turn, again yields poorly, the spectre of hunger looms.

It has been estimated that in 1990, 10,000 children per day died through endemic disease and poor health associated with malnutrition and the accompanying food deficiencies. About one-quarter of sub-Saharan Africa's population of more than 100 million faces chronic food shortages. Africa now imports more than 40 per cent of its food. In the 1950s, it was self-sufficient.

When the crops failed for three then four consecutive years in Niger and Malawi in the early 1970s, locust and grasshopper plagues followed, grazing animals died, and people moved south placing pressures on marginal resources also lacking adequate rain. Behind those moving south, the land became windswept, eroded and useless. Famine and further poverty resulted. When the rains finally came, much of the land was so badly degraded it could no longer be reclaimed.

Warfare and strife further complicate the problem in Africa. Large-scale national and local warfare such as that occurring in Ethiopia and Eritrea, the Sudan and between rival tribal groups in individual countries such as Burundi, Zaire and the Central African Republic, means that soldiers raid and devastate local farms. Denied basic food for themselves, the farmers frequently lose incentive to replant. Continual conflicts create additional uncertainties and further limit food production. Stores become depleted even without drought, and yields tend to decline.

As a result farmers abandon their lands, swelling the ranks of refugees and leaving vast tracts of formerly productive farmlands untended and liable to erosion and further deterioration. They themselves become habitual clients for food aid, suffer declining standards of health and life, and lose the independence and self-esteem associated with their former lives.

Famine and Chronic Food Shortage: Looking for a Cause

Famines have always occurred in Africa but they are becoming more common. Continual endemic warfare aggravates food shortages and means that famine tends to recur. Quite simply, Africa is not growing enough food to meet the needs of an expanding population. Crippling international debts have drained the meagre resources of the poorest countries, and as a result of colonialism much of the best land is also devoted to cash crops for export.

Low prices for agricultural commodities have increased the poverty suffered by rural peoples, especially those in the newly developing countries of Africa. Vocal urban populations have demanded such unrealistically low food prices that farmers have had no incentive to grow more food than required to meet their own immediate needs. The result has been increased dependence on food imports and a lack of stocks to meet short- and medium-term shortages, particularly when crops or rains fail.

Over the last thirty years food shortages have hit all but five sub-Saharan countries. More than two million Africans starved to death in 1985. Another four million currently face starvation in Ethiopia as a result of drought and war. While South Africa's history of destabilisation has brought hunger to

Mozambique and Angola, the shortage of food is spread throughout the continent. Its cause is a lack of adequate food production.

The problems of food supply are social and political in nature. In most African countries state marketing boards buy cash crops from farmers at unreasonably low prices. Farmers in the main coffee, cocoa and cotton exporting countries earned only about half the export value of their produce in 1984. The remainder went into state coffers to meet the demand for foreign exchange for development.

Yet even farmers producing crops for export are well off compared with those supplying food for domestic consumption. The inevitable result is that cash crops continue to be grown on land which could be, and formerly was, used for growing food. When a drought hit the Sahel in 1983, food had to be imported.

However, in the same year five Sahelian countries — Burkina Faso, Mali, Niger, Senegal and Chad — harvested a record 154 million tonnes of cotton fibre for export. Food aid and cheap food imports conspire to maintain low prices so that farmers have little incentive to grow more than their own immediate needs. Drought depletes available reserves and famine results.

Solutions are not easy to negotiate. The only way to ensure that more food is grown is to guarantee farmers higher prices and sufficient support in case of crop failure. Raising food prices in the cities provokes dissatisfaction in the urban population which is politically more powerful.

The problem stems from colonial times. The imperialists levied taxes to be paid in cash, which forced Africans to grow crops like cotton and cocoa that were needed by the distant factories of the imperial parent countries. Meanwhile the colonial economy fostered in Africans a drive towards a consumer economy. African colonies provided a new market for goods like clothing, shoes and bicycles manufactured by those same distant factories.

Independence did little to alter this colonial pattern. With the coming of independence, many fledgling Asian and African governments set themselves the task of trying to catch up with a West that had been industrialised for 150 years. In many newly independent and emerging countries from the Commonwealth of Independent States (CIS) to India, China and now Africa, industry was identified as warranting preferential government assistance. Agriculture has been relegated to a minor role, despite the fact that in many of these countries it has provided, and continues to provide, up to 80 per cent of that country's export income. Political pressures forced them to seek to satisfy the urban hunger for consumer goods, rather than concern themselves with possible future food shortages.

On the advice of foreign experts, many newly independent countries started up industries which relied on imports of essential components from the West — thus keeping money flowing from the poorer to the wealthier nations. Many examples of the resultant problems can be identified. Abidjan, the capital of Ivory Coast, concentrated on assembling luxury cars for export; Cameroon, Ghana and Nigeria manufactured beer, tobacco, textiles, flour, footwear and cement. Kenya produced in bulk, low-value commodities such as soft drinks and canned foods; Zimbabwe, Zambia and Zaire made light engineering goods linked to their export mining industries. The result was a 15 per cent annual growth in the industries of sub-Saharan Africa between 1965 and 1973.

Though the majority of foreign exchange continued to be earned from exports of primary commodities such as coffee and copper, by 1973 manufacturing industry accounted for over 15 per cent of the gross domestic product in eighteen African countries. However, by 1980 commodity prices had fallen sharply and the oil price rises meant that there was little hard currency available for purchasing essential imports. Heavy borrowing from western governments and banks bridged the gap but initiated the massive debts now borne by these countries. With its oil exports, Nigeria alone could afford to maintain industrial growth.

In the course of the 1980s, Africa's debt burden increased faster than elsewhere in the Third World. In sub-Saharan Africa, the total debt rocketed from $6 billion in 1970 to $34 billion in 1988. Between 1986 and 1987 alone, debt-servicing obligations drained one billion dollars from Africa to the International Monetary Fund (IMF).

Structural adjustment programmes devised by the IMF and World Bank effect an economic squeeze. The educated urban middle-class population are capable of making the necessary adjustments. However, for the poor rural peoples, poor commodity prices and cuts in public spending from which they benefit mean that scarce resources are depleted, unemployment rises and the extent of poverty increases.

Faced with the massive debt crises of the 1990s and increasing numbers of poor, the World Bank now admits doubts about the efficacy of its own policies for tackling economic declines. The cancellation by France of all aid debt owed by 35 sub-Saharan countries alone offers some hope for these African countries. In Burkina Faso, a self-help group called the Naam movement has been working since 1967 to make villagers responsible for their own development. Their starting point is a respect for traditional peasant knowledge allied to technical expertise. Naam groups have built gardens, planted village woodlots and built village shops and mills. They use any available low-cost tools and materials and are conscientious in their care for the environment.

Potential Solutions?

Given the return of peace — an important ingredient to solving the crises over food — Africa must prioritize food production. There is a need for governments to encourage peasant farming and provide the necessary local incentives to achieve these goals. Small successes are already apparent.

When Zimbabwe achieved independence in 1980, among the first decisions of the new government was the adoption of a policy intended to increase food production. While aiming to maintain productivity among the 5000 largely white commercial farmers the government sought to increase the amount of food being produced by the new nation's 850,000 peasant families. Under the former government, communal farmers had never delivered more than 80,000 tonnes to the Grain Marketing Board in any one year. Since that time, the amount has increased ten-fold. As a result, Zimbabwe is self-sufficient in its basic food needs to the extent that commercial farmers can be free to produce food solely for export.

As Africa enters the 1990s, Zimbabwe's agricultural successes are attracting increased attention. Today, the peasant farmers grow not only food items but are responsible for 55 per cent of the nation's cotton, 90 per cent of sunflower production and 30 to 40 per cent of groundnuts. Overall agricultural exports account for 45 per cent of Zimbabwe's total exports, sustain half its industries and employ the largest proportion of the country's workforce.

Achieving these successes has involved a mix of policies and programmes. Under the present government, land not used productively by white farmers under the former regime of Ian Smith is being purchased and peasant farmers settled there. One-third of the 165,000 families originally targeted for resettlement have already been re-located to these new lands. With a population growth of 4 per cent per annum, however, resettlement will only prove a stop-gap measure.

Pricing policy is an additional tool available to governments to encourage farmers to produce food. While food subsidies to urban communities were progressively cut between 1982 and 1984, the price of maize, Zimbabwe's staple food, more than doubled over the same period. The price subsequently remained flat for three years because of a huge stockpile and low prices on international markets. Prices have since been readjusted to around US$25 per tonne, enabling farmers to make a decent profit.

Increasing the credit facilities to farmers has also encouraged enterprise among peasant farmers. Rather than insisting on collateral to secure loans, funds were lent on the basis of viability of a particular programme. As a result, lending to peasant farmers rose from a 1979/80 level of Z$250,000 to Z$30 million annually in ten years.

Other changes have also assisted the farmers; for instance, the changing status of women has meant that they too are eligible for loans. Improved education standards enable many more peasant farmers to understand the principles of crop rotation, fertilising and marketing. They are better able to read about and use fertilisers and pesticides with safety.

Farmers have also been helped by an improved marketing infrastructure. Elsewhere in Africa, a major impediment to commercial farming has been the deterioration of roads and railways due to budget cuts. In Zimbabwe, more secondary and feeder roads have been constructed to reach rural areas. The number of Grain Marketing Board depots has been increased from 43 to 66, with new ones located in peasant farming districts. New crops and more inter-cropping practices have been adopted. For example, in mountainous areas, winter wheat is being cropped between coffee bushes; on the plains, farmers growing beans rotate them with maize. Irrigation schemes have been commenced in some areas and are planned for others.

Ahead
So much for poverty and the role of humans in destroying their own health as well as that of their environment. Other major systems affected by humankind and other changes wrought are the substance of the remaining chapters of this Section.

References and Additional Reading

Cole, L.C. 'Man's Ecosystem' in *Bioscience*, 1966, vol.16, no.4, p.243.
Martin, P.S. 'Pleistocene Overkill' from *Natural History*, 1967, vol.76, no.4, pp.32-8
Scientific America, September, 1989.
US Department of Agriculture. *Annual Report of National Statistics, 1988-9.*

5.1 The Population Problem

The issue of population and human numbers is probably *the* major issue for species survival and integrity. The same inherent tendencies and qualities which aided human development now serve as the threats. Population controls, limits to growth and arguments about the quality of life are particularly difficult problems to resolve for a variety of biological, political, moral and religious reasons — the conscious exercise of deliberate measures to limit fertility runs contrary to nature. It is also deeply offensive to peoples in many societies and creeds throughout the world. However, population control and the management of existing resources to meet the needs of all those alive today are matters that must be solved. Furthermore the present problems first appeared less than 500 years ago.

It took only 200 years for the total world population to double itself after 1650; after 1850, only another 100 to double yet again. This phenomenal increase in numbers by a living species lies at the root of the world's most pressing economic problems. It means that limited resources are being forced to support a continually increasing number of people who are constantly using up and not replacing earth's natural resources of minerals, plants and animal species. These people have acquired survival strategies against epidemic disease, famine and natural catastrophe.

The Early Findings on Human Numbers
Before the nineteenth century, reliable estimates of population are hard to come by. The first reasonable guess about human numbers is attributed to Bodio and dates from the time of the Emperor Augustus (first century AD). It suggests a population of 54 million. Other estimates differ, including those given by Mulhall who argues that the total population of Europe could not have exceeded 50 million before the fifteenth century.

Nonetheless, by the turn of the last century it was generally agreed that the world's population approximated 1.6 billion. This estimate was based on actual population tallies in countries like Britain, France and America, together with informed guesses from experts and overseas diplomats in India, China and the less settled regions of the world (Table 5.1).

The Rate of Population Growth
The rate of population growth can be judged most reliably from the average annual percentage increases between recent censuses. Certain points emerge

from these tables. The population of Asia, even excluding the CIS (formerly USSR), is over half the world total. Taking the period 1920-60, the population of Europe rose by only 23 per cent, and that of the CIS by 35 per cent, whereas increases in the population of Asia exceeded 50 per cent and those of America and Oceania even more.

In recent years, reliability and accuracy of population estimates have improved substantially, thanks to increased use of formal census-taking methods. They show, for example, that low annual increases in Belgium of 0.30 per cent, England and Wales (0.46%), France (0.66%) and Italy (0.68%) compare with much higher rates of growth in Ceylon (2.84%), Australia (2.46%), China (2.41%) and Brazil (2.39%) (Table 5.2).

It is apparent that in those areas of the world regarded as under-developed or economically depressed, substantial growth of population has occurred in recent years. It is easy to misinterpret these figures by assuming that they represent a higher fertility rate in countries which tend to be regarded as Third World nations. The obvious exception to the above is Australia. Census data confirm that the rate of natural population increase corresponds to that of other developed industrialised countries such as Europe. Growth in total population in excess of replacement levels is due to implementation of formal immigration programmes, recruiting new Australians from countries of the British Isles and Europe, and recently from Asia.

The first official register of marriages, births and deaths comes from Geneva and was begun in 1549, when the average length of life was only twenty years and the average marriage produced more than five children. By the end of the eighteenth century, life expectancy was greater than thirty-two years and the average family comprised three children. By about 1840, the marriage produced only 2.75 children and life expectancy had increased to forty-five years.

The History of Population Theory

As the conservative and restrictive thinking of the Middle Ages succumbed to a renaissance in ideas and philosophical debate, Francis Bacon, the critical conscience of Henry VII's England, began to ponder the fate of the human race. From his prison in the Tower of London in 1597 he published a series of twelve essays, one of which addressed the issue of human populations. It was note-worthy for its shrewd, pithy but unpopular observations which included suggestions on the relationship between the growth in human numbers, disease, war and pestilence, all of these common concerns of the age.

The next man to seriously ponder the dynamics of the growth and decline in human populations was Malthus. In his essay on 'The Principle of Population as It Affects the Future Improvement of Society', published in 1798, he argued that population increase occurs in a geometrical ratio while subsistence increases only in an arithmetic ratio. Consequently, unless there exist checks on population, there would soon be more people living than there was food to feed them.

The ideas of Bacon and Malthus attracted wide-spread criticisms. Both men proposed that individuals should restrain their procreative urges, and governments legislate to prevent human numbers outpacing their capacity for self-sufficiency. Yet neither of these intellectuals specifically referred to the role of nature and its annual cycle, on which depend all human existence and economies.

Warnings and Explanations: The population debate ebbed and declined in the early years of the twentieth century. The Club of Rome Report, *Limits to Growth*, not only revived the debate in intellectual circles but, thanks to television and radio and a better educated public, thrust discussion on human numbers, population increase and the use of earth's finite resources into the living rooms of every man and woman. Following as it did on Rachel Carson's *Silent Spring* and E. Schumacher's *Small Is Beautiful*, the Report served notice that the modern conservation movement and the era of Zero Population Growth had arrived.

There is no single or simple explanation for population explosions except where migration into a country is exceptionally heavy. The principal reason is the marked decrease in the death rate. Factors contributing to the decline in death rate which is being witnessed in one country after another reflect those which occurred first in Britain and the developed European nations. Tables 5.3, 5.4 and 5.5 provide figures for the United Kingdom and United States.

The Death Rate and Population Figures
As is evident, when the death rate declines there tends to be a lag before a comparable reduction occurs in the birth rate. As a result, the under-developed parts of the world are currently experiencing population growths in excess of 2.5 per cent per annum, compared to the developed countries where growth rate has declined to less than one per cent. To a large extent, this is explained by reluctance to change traditional preferences for large families and to practise fertility control.

Contributions to the decline in death rate are also the result of advances in medical science, including antisepsis, vaccination and the interdisciplinary science of epidemiology. Tracing the pattern of occurrence of a disease to the origin of its outbreak and the identification of a specific causal organism, greatly improved the management and prognosis of deadly infectious diseases such as cholera, diphtheria and plague.

A gradual increase in food supplies and the nutritional standards of the population have improved the ability to cope with minor and major disease. A well-nourished mother is able to give birth to, feed and rear more healthy children. With more children reaching adulthood, themselves in a better nourished state than their forefathers, they are capable of producing more and healthier children for a longer time than ever before, and living into old age where their offspring can support them.

Living Conditions and Populations

Until comparatively recent times, people in Britain and Europe lived under appallingly unhealthy conditions. They were ignorant of basic hygiene, lacked piped water, efficient sewage disposal and an understanding of the relationship between sanitation, infection and disease. Town dwellers often drew water from local streams polluted by human, agricultural and animal wastes. They worked long hours in over-crowded, dingy and badly lit factories. They lived in poorly constructed buildings where families shared ill-ventilated rooms in which they ate, slept and performed personal ablutions, often jammed six persons and more to a small room. Poor nutrition, an inadequate diet and poor health served to further encourage the rapid spread of disease. Epidemics of infectious disease such as cholera, typhoid and diphtheria raged through the town poor.

Today, similar conditions are associated with 'under-developed' countries such as India, Brazil and Bangladesh, and the population curves for India, China and Venezuela resemble those of eighteenth-and nineteenth-century Britain. For example, in the slums of Karachi, densely packed shanty dwellings lack access to supplies of pure running water: there is inadequate drainage, and open sewers carry human, animal and household debris.

It is salutary to remember that almost identical conditions were widespread in Britain and other, now 'advanced and developed' western nations only a few decades ago. The flattening of the population curve for Britain early in the twentieth century was caused by a decrease in birth rates. This offset a continual decline in death rates which has accompanied the social, economic and medical advances referred to in Table 5.6.

The Role of Migration

Any analysis of the world distribution of population must take into account the factor of migration. Migration is the term used to apply to the movement of people within definite geographic boundaries (internal migration) and from one nation, continent or region of the world to another with the intention of permanent settlement (emigration). Large-scale movements of peoples across both land and sea have been commonplace throughout the entire history of mankind.

In the last 4000 years, European migration has been responsible for sudden and rapid population increases in America, Canada and Australia. Most significant was the enforced migration of tens of millions of African Negroes to the Americas during the notorious Slave Trade of the eighteenth century; and again the population of the United States grew from 5.3 million in 1800 to 76 million in 1900. In addition there have been sizeable movements of Asians within Asia. Nearly 2 million Japanese have migrated to adjacent mainland China and its coastal islands; some 10 million Indians and 20 million Chinese have settled throughout South-East Asia.

In the past, migration served to accommodate local land shortages, provide opportunities for commerce and a new home for refugees and others displaced by warfare, religious conflict or natural catastrophe. More recent and large-scale movements of people have resulted from politics. Examples include the separation of East and West Pakistan from mainland India following Partition in the late 1940s, and the 1989 emigration of Eastern bloc Communist peoples to seek work and a new life in Germany, France and other parts of western Europe.

Population planners are now asking whether a global policy of migration might serve to relieve population pressures in those parts of the world where living standards are below par, and available land, food and work cannot support their present populations. Most generally agree that the answer is no.

There are other reasons to reject transmigration or enforced repatriation as a means of resolving population crises. Firstly, the number of people involved is too great. In India alone it would be necessary to tranship some 14 million people annually merely to stabilise the population. A slightly greater 15 million people is yearly added to China's 1.9 billion living off a land mass of which only 7 per cent is arable and productive. The shipment, rehousing and resettlement of such very large numbers of people would entail virtually insoluble administrative problems.

Secondly, there are no large regions of potentially prosperous farmland to which millions of subsistence farmers can be moved. Most regions of the world favourable to agriculture have already been settled, the effective limits of ready cultivation having been reached in all continents.

There are also serious political barriers to the free international movement of people. Recent attempts at enforced migration include the forceful re-location by Indonesia of a million people to Timor and West New Guinea, regions of low population, from Java, a small island with a population approaching 64 million people. Many of those shipped from Java have since sought illegal re-entry or lost their lives attempting a perilous return by poorly constructed and unseaworthy craft. British attempts to repatriate back to mainland Vietnam the thousands of economic refugees who have thronged into Hong Kong from North Vietnam, have met with resistance from the people and criticism from many nations, themselves unwilling to accept the refugees.

Most states severely restrict immigration, permitting entry only to those migrants who can readily be employed and in possession of those skills in demand. In this way, their entry does not generate resentment among existing residents who might otherwise protest against opportunities and jobs denied them. In addition, many communities both white and coloured impose regulations to restrict the entry of those of different colour or ethnic background. The White Australia Policy limited entry of Chinese and Asian peoples during the nineteenth century; the Japanese restrict settlement by foreigners, and white South Africa has pursued an exclusivist apartheid policy for almost all of the twentieth century.

Finally, the costs of emigration are high and would-be migrants are usually poor and often suffer diseases of deprivation. This means that they impose a heavy burden on any accepting country.

Yet, even with these reservations, there is general agreement that parts of countries such as India, Bangladesh and Brazil are grossly overpopulated and suffer from famine, disease and poverty. At the same time, there exist many regions of Canada, for example, which are underpopulated and in need of development. The country also possesses the potential to support many more people than the present 36 million. Australia also appears vastly underpopulated but does not offer the same potentially productive land or resources for ready exploitation. Significantly, both these countries have chosen to increase greatly the density of the population, a policy which does not necessarily entail overpopulation (Table 5.7).

Modern Migrations

The 1980s saw the emergence of the largest and most recent migrations of peoples for over two centuries, in two distinct regions. The first began in the late 1970s as conflicts in Asia decreased and repressive regimes came to power in Vietnam, Cambodia and Burma. A large movement of refugees followed, continuing into the 1990s. Political, economic and social refugees seek asylum by moving south and southwestward, to the southern Pacific and the Americas.

The second large wave began in 1989 with the decline in power, and ultimately the demise, of Communist governments in Eastern Europe. Thousands of people swept westwards into countries seen as more advanced, developed and socially permissive than their own. Ethnic Hungarians were forced to flee their homes in Roumania and Czechoslovakia, while East

Germans sought homes and jobs in the West. With the prospect of a united Germany becoming a reality in the 1990s, Germans living in Roumania flooded into eastern Germany in search of a better life. Peoples throughout the area are on the move.

Concerns About Migration

The Australian Migration Dilemma: Quality v Quantity

Particularly to the peoples of overpopulated Asia, Australia appears a vast expanse of unsettled country with a small population confined to the thin strips of land bordering the Pacific and Indian Oceans. Such assessments fail to acknowledge a lack of adequate water, a preponderance of poor quality land ill-suited to farming or grazing, great distances and isolation.

Since the 1930s, prominent Australians have recognised that the country's climate, rainfall and soils mean that it can never comfortably sustain more than 30 million people. Suggestions that Australia could accommodate a population of 50 million, as voiced by some leading politicians, would only be achieved by unacceptable reductions in current standards of living for all Australians.

Three issues continue to be crucial to on-going debate in Australia over continued support for a migration programme, its nature and its balance, particularly in the context of scarce water resources and high levels of unemployment:

* that Australia's present population of 17 million is 'rapidly degrading the environment and is therefore not sustainable in the long run' (Australians Against Further Immigration (AAFI) Submission to the Fitzgerald Inquiry 1987). To seek to double the population by the year 2030 through net immigration of 170,000 yearly as urged by some, or to 26 million as achieved with an intake of 126,000 annually, would pose an unacceptable burden on the degraded environment.

* that the present population is rapidly ageing and immigrants are needed to augment the workforce, provide for national defence and inject new skills and technologies into the workplace.

* that Australia's recent migrants have come predominantly from Asia, an area socially, culturally and ethnically at odds with that of the present population. Evidence of the formation of ethnic ghettoes, an increase in the level of crime and violence and an alarming growth in community disharmony are said to be threatening the fragile social fabric of the nation.

Data also indicate that in the last thirty years, the rural workforce in Australia continued to decline in numbers despite a doubling in volume of agricultural production and export. While Australia has a relatively minor pollution problem compared with Europe, North America and Asia, the fossil fuels of Australia are low in sulphur content. Some 73 per cent of the population live in cities of 100,000 or more, and major pollutants frequently approach or exceed threshold levels.

The question of the carrying capacity of the environment and its ability to sustain a large population centres on the quality of Australia's land. Of the available 6 per cent arable land, the area under crop is roughly equal to only that of two US States, Iowa and Illinois combined, added to which is a now well defined land use problem.

Australia's arid inland and Canada's permafrost reduce the potentially arable land of two similarly sized countries to about 10 per cent of the total area. In Australia, the actual area of arable land in use is about 6 per cent, roughly the same percentage as relatively tiny New Zealand but less fertile. In 1988, Australia was able to feed about 40 million and clothe about 200 million through exports. It has been claimed that by consuming these exports and forgoing all luxury food imports, Australia could support perhaps 50 million, but in doing so it would need to accept Indonesian food levels and Third World standards of living.

The 1978 national study of Australian soil degradation estimated that 66 per cent of crop land had already been degraded through erosion and rising salt levels. It is now further degraded. France produces over 40 million tonnes of cereals a year compared with Australia's 15 to 20 million tonnes.

Fossil fuel energy is pollutant. Australia is energy-rich but Australians are profligate energy users. Some 16 million Australians consume the same amount of energy as 8 million Americans, 19 million Russians, 69 million Mexicans and 490 million Indians (OECD Report, 1989).

Another limit to growth in Australia is the supply of water. While water-rich on a per capita basis, the population spread is misplaced relative to the source. Far northern Australia has 2 per cent of the population, 4 per cent of the potentially arable land and 52 per cent of annual mean surface run-off. Southern Australia has 82 per cent of the population, 65 per cent of potentially arable land but only 27 per cent of annual mean run-off. Any large-scale movement of water from north to south is practically and economically unrealistic.

Additional problems arise because increased urbanisation places additional demands for physical and social infrastructure and services often beyond the capacities of local communities.

The Ethnic and Cultural Mix: Nevertheless, and despite the qualms, in the early 1990s Australia maintains an active migration programme and is currently receiving around 140,000 new settlers annually. In official terms, immigrants enter under one of three categories — refugees, business and skilled migrants, and family reunion schemes.

Recent controversies centre on the substantial number of Asians entering under all sections of the programme, and about those entering illegally or staying beyond the expired times of visas. It is claimed that Asians far exceed those of any other ethnic group and that these peoples are culturally and socially incompatible with Australians. Furthermore they tend to settle in ethnic ghettoes, do not learn English, and bring their own traditional values, conflicts and crime to the country of settlement, adopting unacceptable standards of business practice and increasing the level of violence and disharmony in the general community.

TABLE 5.1: Population (in millions)

	1920	1930	1940	1950
Europe	328	355	380	393
Asia	967	1073	1213	1368
USSR	158	176	196	202
Africa	140	155	172	198
America				
north	117	135	146	168
middle	30	34	41	51
south	61	75	90	111
Oceania	8.8	10.4	11.3	13
TOTALS	1810	2013	2249	2503

TABLE 5.2: Average Annual Rates of Growth
Years 1950-80 (percentage)

	1950	1955	1960	1965	1970	1975
World	1.8	2.0	2.0	2.0	1.9	1.9
Developed	1.3	1.3	1.2	1.0	0.9	0.7
Third World	2.1	2.3	2.3	2.4	2.4	2.3
Africa	2.1	2.3	2.5	2.6	2.7	2.8
Asia	2.0	2.2	2.2	2.3	2.3	2.2
Latin America	2.6	2.8	2.8	2.7	2.6	2.8
North America	1.8	1.8	1.5	1.1	0.9	0.8
Europe and USSR	1.1	1.1	1.1	0.8	0.8	0.6
Oceania	2.6	2.7	2.6	2.5	2.4	
Australia and NZ	2.3	2.2	2.1	1.8	1.7	1.0

Source: World Population 1980, *US Dept. of Commerce, 1981.*

TABLE 5.3: Birth, Death and Infant Mortality Rates
(United Kingdom)

	1851	1901	1931	1961	1983
Birth rate (per 1000)	34.3	28.5	15.8	17.6	12.7
Death rate (per 100)	22.0	16.9	12.3	11.9	11.7
Infant Mortality (per 1000 live births)	153	151	66	21	10.8

TABLE 5.4: Life Expectancy Years
(United Kingdom)

	1838-54	1901-10	1930-32	1950-52	1981-3
Males					
birth	39.9	48.5	58.7	66.4	71.1
age 21	38.8	42.2	46.0	48.7	52.5
Females					
birth	41.9	52.4	62.9	71.5	77.1
age 21	39.6	44.9	49.0	53.2	58.2

Sources: Registrar General's Annual Reports for Statistical Reviews for England & Wales; UN Demographic Yearbook 1983; Registrar-General's Decennial Supplement, English Life Tables (1931-1981).

TABLE 5.5: Demographic Summary
(United States)

	1900	1970
Population	76 million	205 million
Life expectancy	47 years	70 years
Median age	23 years	28 years
Births per 1000	32	18
Deaths per 1000	17	9
Immigrants per 1000	8	2
Annual growth	1.75 million	2.25 million
Growth rate	2.3 per cent	1.1 per cent

Source: p.84, Eisenbud M. Environment, Technology and Health, New York University Press, 1978.

**TABLE 5.6: Factors Contributing to Decline In
Death Rates in United Kingdom**

* Social legislation

1848	establishment of Boards of Health to provide mains drainage, pure water supply, street cleaning facilities.
1850	cesspools abolished in the City of London.
1868	Artisans & Labourers' Dwellings Act, empowered local authorities to condemn property unfit for habitation.
1875	Public Health Act established local sanitary districts and provided for appointment of inspectors, medical officers and health surveyors.
1907	Act required notification within 48 hours of birth; attempts to establish efficient midwifery training centres
1911	National Insurance scheme established to protect against sickness costs.
1921	Local Councils made responsible for tuberculosis control.
1946	National Health Service Act provided free medical services for all residents of England and Wales.

* Advances in medical science

1796	vaccination against smallpox.
1800-50	improvements in surgery, midwifery, military hygiene; 154 new hospitals and dispensaries established, including most of the famous London teaching hospitals.

* Gradual increase in food supplies due to improvements in agricultural techniques and practices: e.g. crop rotation, manuring; new implements, e.g. cultivators, seed drills.

* Rise in nutritional standards: e.g. cheap imports of grain to Britain from New World after 1870.

* Social changes: e.g. availability of cheap washable cotton clothing after 1750 made personal cleanliness easier and reduced infection.

**TABLE 5.7: Immigration and Emigration
(numbers per year)**

	Immigration	Emigration
Australia	107,171	92,342
Canada	128,618	41,746
New Zealand	40,705	34,147
UK	202,000	185,000
USA	544,000	n.a.*

* US statistics fail to record numbers of emigrants

Source: Table, p.10, Converse, P. & J. How We Compare, Five Mile Press, Melbourne, 1986

References and Additional Reading

Demeny, P. 'The Populations of Underdeveloped Countries' from *Scientific America*, 1974, vol.231, p.149.

Chadwick, E. 'Report on the Sanitary Condition of the Labouring Population of Great Britain' (1842) edited with introduction by M.W. Flinn, Edinburgh University Press, Edinburgh, 1965.

Greep, P.O. 'Population Growth, The Environment and Fertility Control' from *International Journal of Environmental Studies*, 1974, vol.7, pp.51-5.

5.2 Populations: The Human and Natural Environment

An Issue of Quality
The quality of the lives people lead is about individuals having adequate personal space, food and water for fulfilling more than just the basic survival needs. It is about opportunities for humans to achieve more; and about the 'little bit' that makes life worth living to them and to others. Increases in human numbers intensify competition at all levels of society. In the most fundamental sense, there is competition for food, shelter, work, and sex. In a more sophisticated sense, increased numbers are about urbanisation, poverty, resources exploitation, disharmony and conflict.

To a large extent, an increase in population density in a city, town or country means that a given quota of land is being required to support a growing number of people. An increase in the number of people living within a defined area or space results when fertility rates are high and/or rising, or where substantial migration has occurred.

The Capacity of the Land
The effect of unbridled population growth on land pressure in the under-developed countries can be gleaned from an estimate that, in 1975, the population density averaged forty persons per square kilometre. This is more than twice that of the more developed nations. During the fifteen-year period between 1960 and 1975, it was necessary to accommodate more than twelve additional people per square kilometre.

Many parts of Europe, particularly southern Italy and Greece, have substantial populations who derive their livelihood from semi-subsistence peasant farming and from industries based on agricultural production. Olives, grapes for wine, and cereal grains are favoured crops. In many respects, their lives resemble those of rural peasantry and hill tribes-people of south and eastern Asia and of those living on the rich, alluvial plains of India and China. Ignorance, superstition and poor health standards are shared by both. There is also the same relentless pressure of increased numbers on limited land resources. Acceptance rather than willingness to address problems, fear of attempting new methods, futility and fatalism are common.

In both Europe and monsoonal Asia, clusters of dense population have developed where there is fertile land, where opportunities exist for irrigation and a temperate climate aids rapid crop growth. Unique types of intensive agriculture have resulted, including the terraced rice fields of Indonesia and Bali, the irrigated fertile huertas or 'gardens' behind Valencia, Spain, and the seasonal cultivation of river flats of the Brahmaputra, the Ganges, the Mekong and the Hwang Ho rivers.

In many under-developed countries of the world, such as India, Venezuela and China, the problems are especially critical as an already large and increasing population is attempting to live from land which has been devoted to agriculture for thousands of years. Smaller holdings, more intensive cultivation and a declining income per individual tend to result.

Poverty, Land and The Role of Population
These pressures are not new: the cultural response enforced existing preferences for large families. Village peoples have reacted, not by limiting the size of their families so that there will be only as many mouths to feed as their land will support, but rather by adding children to the workforce to guarantee that, if members are lost or die, there will still be others to work and hold the family farm. The goal is to continue to hold the land in the family. Unfortunately the result is a depressing cycle of rural poverty, particularly in circumstances where improved health standards lower rates of infant death and infectious disease.

Alternatively, families unable to feed all their members may encourage younger sons to seek non-traditional work in nearby urban centres and in town factories. The result is a simultaneous increase in population densities in towns, and the growth of cities where unemployment is characteristically high, particularly among the unskilled former farm workers. Slums and squalid living conditions result as available housing is quickly exhausted. In these circumstances, increased poverty, poor health standards and inadequate nutrition are common (Tables 5.8, 5.9).

Today, the populations of many under-developed countries are becoming urbanized at a rate of over 4 per cent per annum so that their towns are doubling in size every fifteen years. In Venezuela, for example, the capital city of Caracas grew from 359,000 in 1941 to 3,550,000 in 1985, and other Venezuelan cities either equal or exceed this rate of growth. At the same time, the Venezuelan farming population is also expanding rapidly at about 11 per cent every decade, whilst the amount of cultivated land is actually decreasing. The result is that the Venezuelans are finding it progressively more difficult to feed not only their rapidly growing urban population, but also the increasing numbers of people living in the countryside (170 people per square kilometre of cultivatable land).

Generally speaking, those areas where population densities are greatest are located in parts of the world where geographical conditions favour intensive cultivation or the development of modern industries, or both. Almost without exception, the vast arid and semi-arid regions of the world contain few people. It is unrealistic to suggest, for example, that China's 1.9 billion people should be more uniformly distributed throughout a country where only 7 per cent of the land is arable, and where vast deserts, high plateaux and rugged ranges preclude most permanent settlement.

Despite large-scale irrigation schemes the deserts remain largely drought-stricken and forbidding wildernesses. Exceptions include the gold-mines of the West Australian desert. At Kalgoorlie, for example, gold was discovered in 1892, and in spite of high temperatures and lack of water a township of 23,000 people rapidly came into being. Even in 1990, water is still being brought 564 kilometres to Kalgoorlie from the Mundaring Weir on the Helena River.

A similar situation pertains in Mount Isa in central north Queensland where copper, lead and zinc are mined. In these and other similar settlements, the dry heat and the quantity, quality and accessibility of adequate water are not conducive to good health. The hot dusty conditions create wind storms, and the open-cut mining favoured at Mount Isa results in clouds of dirt and coloured dust which settle on and in buildings, clothes and the lungs. As a result, the incidence of respiratory problems is well above the national average.

Tropical rain forests, also conspicuous as regions of few inhabitants, are generally associated with high humidity and rainfall, soils of low quality, and unhealthy, smelly swamps infested with bacteria, insects and parasites such as those responsible for malaria, sleeping sickness and dysentery. All attempts to encourage increased human settlement in such areas have failed, and have been costly in terms of deaths and ill-health suffered by settlers.

The Tendency Towards Urbanisation
Consistent with increased growth in human numbers has been the greater movement of populations from rural areas into towns, and the consequent rise of densely populated cities. If the city states of Singapore and Hong Kong are excluded, international comparisons of the level of urbanisation suggest that in the period up to 1980, over 80 per cent of Australia's population lived in towns in excess of 20,000 people, compared with 55 per cent in the United States, 60 per cent in Canada and a world proportion of about 35 per cent.

Despite its area, Australia is unique among these nations, with a low average rainfall per square kilometre together with a vast expanse of land which is either non-productive or of marginal use for agricultural purposes. This means that the majority of the population inhabits the coastal areas where there is adequate rainwater for drinking, and where there has traditionally been ample land for living and sustenance.

As a result, the population of 16 million people is concentrated in small bands and coastal strips along the south-east coast of the continent, leaving the inner centre and the north deserted or carrying sparse numbers. However, Australia remains unique in that these people are concentrated in the cities and hinterland of Sydney, Melbourne and Brisbane.

Along the north-east coast of North America (another region, for example, where population densities are comparable to those of Australia), between Chicago and New York and in the adjoining St Lawrence Valley in Canada, the distribution of population is reasonably uniform and tends to correlate closely with modern industry, power resources, good transport routes and port facilities.

In many respects, therefore, this thickly settled region of North America resembles the coalfields and port conurbations of Europe. The result is a large number of towns with populations greater than 20,000 scattered throughout the region and dependent on the production associated with locally produced petroleum, natural gas and hydro-electricity. These industries and the consequent health and lifestyles of residents in towns dependent on them are closely interdependent. Any changes or improvements to resolve associated health problems need to bear this dependence in mind.

Using the 20,000 population threshold, England and Wales (72%), Belgium, Netherlands and Japan closely follow upon the Australian preference for city life. However, in all these countries the land is used essentially for capitalist enterprise and as part of internal, national and international trading systems. Some 84-93 per cent the nations' workforces still continue to earn their livings from non-rural occupations. The growth of very large urban settlements similar to those of Western Europe, North America and Japan has become one of the characteristics of human society of the mid-twentieth century. Until the 1930s the world's largest cities were confined to economically advanced regions.

Population Pressures and Urbanisation — Some Results
What happens when population pressures and decreasing availability of land force rural dwellers to seek a better life in the cities? What happens in shanty towns and slums which have grown up as increased numbers have failed to find accommodation and jobs, or natural disasters such as flood or cyclone have further aggravated population pressures in large towns and cities? Exploitation, graft, corruption and the continuing cycle of poverty (Tables 5.10, 5.11).

In the harsh dry summer of northern India, lives and scarce possessions are lost and poverty accentuated when regular fires burn down upwards of 15,000 huts of slum dwellers. These constructions of plastic bags stretched over rusty bicycles or umbrella spokes, equipped with raffia mats, stolen bricks and a fireplace for cooking, burn like tinder. With the mid-year sun, the jhuggies (very poor unemployed slum dwellers in Indian cities) are the first victims of the heat in these sprawling slums reaching sometimes 30 kilometres beyond a city's limits. The number of jhuggies or bustees is estimated to be around 1.5 million. Official figures suggest that they form 34 per cent of India's urban populations. Their main preoccupation, apart from scavenging the rubbish tips for food, is to avoid the attentions of Mafia-like exploiters.

In India, people are allowed to work once they have an address. Having an address also enables people to acquire precious ration cards, which can be redeemed at state-owned shops. Indian children can then enjoy access to school education and their parents become eligible to vote. Unfortunately, access to these benefits is possible only to the slum-dweller who can effectively employ graft, flattery or trickery.

Each slum has its own pradhan, an elected person who does the talking with government officials. Corruption of the pradhan is part of survival. It is he who distributes the ration cards whenever there are any available. Although slum dwellers are supposed to have precedence when it comes to allocating the few new homes that get built, most of these end up in the hands of the wealthy middle-class, many of whose members were former pradhans.

Increased population, greater movement to the cities, deteriorating conditions and exhaustion of the environment promise ill for the future and doom for many. India and its people are but one example of a situation affecting all parts of the Third World.

Food Supply and Human Numbers
Public attention was focused for the first time on the problems of food supply arising from a rapidly increasing population in the late eighteenth century. In his famous *Essay on the Principle of Population*, Malthus states that 'the power of population is infinitely greater than the power in the earth to provide a subsistence for man'.

This hypothesis implies that mankind is faced with the dismal prospect of never being able to solve the problem of famine. Malthus is arguing that geometric increases in population cannot be maintained by mere increases in food supplies occurring through arithmetic progression. In other words, assuming that, to begin with, food supply could be doubled after twenty years, this means that within a century the supply of food would increase six times: meanwhile the population would have multiplied thirty-two times. The amount of food available per head of population would fall until numbers were checked by 'natural' catastrophes such as war and famine, plus epidemics and infectious diseases to which the nutritionally deficient population lack any resistance.

The Malthusian thesis caused great concern in Britain but was soon found inadequate. At the time Britain was experiencing large increases in productivity, technological improvements in industry, transportation and agriculture. These made it possible for the country to absorb the subsequent rise in population from 10 million in 1801 to 37 million in 1901. The importation of cheap foodstuffs and raw materials for industry assisted and continued to balance a decline in the death rate which accompanied the birth of modern medical science. Furthermore, the rate of population increase in the United Kingdom slackened considerably in the twentieth century.

Nonetheless, the Malthusian arguments do have relevance even today, and in particular for situations in the Third World. In many developing countries in Africa and Asia, there is very limited capacity available to boost agricultural and industrial production to a level which will support the high fertility rates, increased infant survival, improvements in medical care, and better management of 'natural' emergencies. Resources, international commercial pressures and the political power of many of the developed nations rather tend to widen the gaps between the two political blocks.

There are those countries which have managed to accommodate population increases in past decades and are now in a negative population growth cycle, and those which have as yet been unable to reach this hiatus. Unfortunately the power and commercial leverage lies with the already developed wealthy countries, which control to their own advantage the prices for basic commodities and raw materials required for their factories and economies.

Despite the fact that, after 1950, food output per head rose in all major regions, in some countries it has declined. Since 1960, this has been particularly marked in Africa. Continued tribal fighting in Ethiopia, Somalia and the Ogaden regions of East Africa have laid waste to already marginal lands, forced herdsmen and agricultural workers to take up arms, and left the land idle and unplanted, crops and beasts untended. Warfare and internal dissent have been responsible also for falls in agricultural production of basic food stuffs and exports in Cambodia, Afghanistan, Vietnam and China.

Hunger, Undernutrition and Malnutrition

Today, with a world population approaching 7 billion, many people do not have enough food to satisfy even their minimal energy needs. Most notably in Brazil, the Honduras, Bangladesh, India and Ethiopia, medical surveys conducted under the auspices of the World Health Organization, UNESCO and the United Nations find that nearly 20 per cent of the world's children show clear signs of protein deficiency and malnutrition.

Any discussion of the adequacy of food supplies should distinguish between famines (which cause mass starvation) and the more subtle effect of chronic malnutrition. The most frequent cause of famine in modern times is widespread failure of crops due to drought or flooding. Until the late 1980s, many population experts believed that human numbers would outrun the growth of world food supplies and that the proportion suffering from malnutrition would become greater. Emphasis upon the famines in India during 1966-68, in the African Sahel in 1972-74 and again in 1981-83 substantiated these claims.

Nonetheless, as Table 5.12 indicates, world food output more than doubled between 1952-54 and again between 1978-80: the output per capita increased by 24 per cent in the same period. Between 1950 and 1980, the total area of land devoted to arable farming increased in both developed and under-developed countries. Average crop yields and the area committed to major food crops also increased, with the result that total world output rose by 125 per cent. In contrast, world population for the period was boosted by 85 per cent. While food production in Third World countries grew at a faster rate than in the First World, so too did its population.

In 1986, the available evidence suggested that about 13 per cent (453 million) of the developing world live in countries which have experienced a continuous decline in food output per capita since 1961, and 18 per cent (628 million) in countries where there has been a decline in food output per capita since 1969-71. Using this data, conservative estimates suggested that by the mid 1980s about 800 million people, or one-quarter of those living in the Third World, would be starving or underfed.

Quality of the Diet
A balanced diet that supplies all the constituents needed to meet a person's individual energy requirements and the raw materials for body building and replacement must include carbohydrates, proteins, fats, vitamins and minerals. Carbohydrates tend to be the main source of energy in under-developed countries where they can supply 90 per cent of the energy requirements, compared to a level of as low as 40 per cent in some industrialised countries. In agricultural societies, these carbohydrates come from dependence on starchy foods such as potatoes, cassava and plantains. Industrialised populations obtain their carbohydrates from simple sugars and fats.

The materials which build and repair the body come principally from proteins. The quantity of protein in the diet is a good measure of the economic health of a country. Fish, beef, milk products and soy beans are foods with the highest protein content. Soy beans are the most important source of vegetable protein and, along with fish, play a major role in supplying the needs of many Asian populations (Tables 5.13, 5.14).

The amount of fat in the diet varies from country to country and has been linked with the level of economic development. In poor countries, the total amount of energy derived from fat alone is about 15 per cent, but in more prosperous countries may rise to 45 per cent.

Studies of the dietary habits of New Guinea tribes-people indicate an average daily calorie intake for adult males of 2600 calories. Measurements on Ethiopian peasants and Fulani tribesmen in Niger suggest an average intake of only 1800 calories per day. These variations and the adequacy of the diets are partially dependent on individual body weight, regional temperature and the amount of agricultural work undertaken.

If the minimum calorie requirement of an Asian population is set between 1600 and 2000 per person per day, the fact still remains that the world already carries more than 2 billion permanently hungry people. Furthermore, any increase in these numbers is unlikely to be met by a comparable increase in the quantity and quality of available food.

In many regions of the Third World, the inadequacy of the diet is especially marked at the end of the agricultural year when stocks of food are becoming short. A failure of the rains and subsequent crop germination often means that remaining old grain is used for food. Since this is not replenished by a harvest, not only do the people have no food to tide them over until next spring but they will have used any remaining seed grain being held as a reserve for new plantings. In such marginal subsistence conditions, famine becomes inevitable. While waiting for germinated crops to reach maturity and yield grain for food, near-famine conditions may apply. This means that at the busiest and most demanding time of year, farm labourers are not working efficiently because of hunger and the lethargy which this generates.

The quality of foodstuffs is as important as the quantity. Equally important as total energy of food is the presence of adequate protein in the diet. A minimum daily protein intake has been set by different authorities between a low value of 30 grammes per day (London School of Hygiene and Tropical Medicine, 1986) and a value closer to 70 grammes (US Department of Health, 1985). While requirements for fat are significantly dependent on climate and level of activity, a minimal intake of essential fatty acids is crucial to health. Intake of fats also tends to be low among many Third World people. Infants and young children have a special need for fats as a store of energy and a reserve during the difficult period of weaning from breast milk to a normal mixed diet.

A particular problem arises with the diets of tropical peoples. Their diet tends to lack important vitamins and minerals such as phosphorus, calcium, iodine, sodium and iron, normally absorbed as salts which growing plants obtain from the soils in which they grow. Heavy rains in these regions result in leaching of these minerals from the soil, so that the rate of growth and nutrient content of vegetables is variable and inadequate.

Predicted changes in the climate of the tropics suggest increases in the severity of cyclones and hurricane activity, resulting in stronger winds, greater sea swells and heavier rainfall. Increased damage to homes and property, crop failure and further greater loss of soils would press tropical peoples further towards marginality.

Food taboos, religious practice, local tastes and tradition restrict the range of food cultivated and consumed. Hindus depend on a vegetarian diet which is low in proteins. Orthodox Jews will not eat pork; the Sinhalese associate milk with many diseases, and some inhabitants of Grenada believe that milk gives children worms. The Masai herdsmen of East Africa never eat meat and milk on the same day for fear that by infringing this religious taboo their cattle would become diseased. Solomon Islanders forbid their women from fishing and some fish are forbidden them as food. Some African tribes specify that menstruating women should eat no meat; strict religious taboos specify that warriors going into battle eat only specially prepared and selected cuts of meat. In all these instances, improvements in diet and the standards of nutrition depend largely on education. These are difficult enough for uneducated people to understand, even without the additional changes in lifestyle required of them to prepare for anticipated climate changes.

Arable Land and Food Supplies
There are about 32 billion acres of land in the world, of which about 8 billion are suitable for cultivation and a further 8 billion for grazing but not crop production. It has been calculated that of the 8 billion acres that were potentially available in the early 1970s, about 3.5 billion only were under cultivation and more than 4 billion used for grazing. In Asia, about 83 per cent of the arable land was under cultivation compared with 11 per cent in South America.

Agricultural and industrial innovations in the nineteenth century have greatly enhanced the productive yields from land and improved the speed and efficiency with which crops are grown and harvested. They have had profound effects on the social structure of all food-producing countries. In the United States at the start of the nineteenth century, 95 per cent of the population lived on farms where they were self-sufficient in food and produced an excess for the remaining 5 per cent and for export. By the mid 1970s, this number had declined to a mere 9 per cent able to raise sufficient food to feed the remaining 91 per cent and to supply an export market.

Thus, while only 50 per cent of the arable land of the world is farmed and in some countries by only a small number of people, productivity has been greatly enhanced. Fertilizers which supply nutrients to the soil have been among the most important aids to production.

Prior to the nineteenth century, soil nitrogen essential for the healthy growth of plants could only be obtained by composting (adding animal and vegetable wastes to the soil), and by crop rotation techniques. Phosphorous, another essential soil nutrient, was supplied from minerals that were mined then applied to the soil in a soluble form to facilitate their incorporation. In the nineteenth century, crop improvements depended on increased use of mineral and natural fertilisers such as potassium sulphate from Nauru and Chilean nitrate deposits (guano). With time came a better understanding of plant nutrition, modern techniques of soil tillage and water conservation and the use of herbicides and pesticides.

Pesticides have been particularly important both to agriculture and to disease control since the first quarter of this century. Early pesticides were based on discouraging pests by the use of companion plants such as garlic and marigolds, and the use of deterrent plant materials such as pyrethrum. More recently, synthetic plant chemicals have been formulated into over 80,000 commercial products which can kill or maim a wide range of animals, insects and micro-organisms. They are used for agricultural production, animal health, public health and forestry. Methods for protecting agricultural products from damage by pests are a necessary part of the strategy for producing crops, livestock and timber products because pests are destructive at all stages of plant growth and yield. They also destroy food when it is in storage and during transportation.

There are at least 10,000 species of destructive insects in the United States, and plant diseases can be caused by as many as 160 bacterial, 280 viral and at least 8000 fungal species. During the years 1963 and 1964, insects and rodents destroyed grain in India to the extent of 13 million tons, which would have been sufficient to supply a daily loaf of bread to 77 million families for one year. In nineteen corn-producing states in America, the grain destroyed by birds in one year could have fed more than 300,000 pigs.

One of the earliest modern methods of protecting crops against plant diseases was to select disease-resistent varieties, a remarkably successful technique enabling farmers to cultivate wheat, corn, soybeans and rice which are resistant to fungi, insects and other infestations which were formerly destructive. Nonetheless the chemical methods of control have proved the most successful and have greatly increased land productivity in many parts of the world since their large-scale use after World War II.

There are some excellent examples of the modern use of similar techniques. 'Miracle' strains of rice were optimistically introduced to poor Asian peasants of the Philippines in the 1970s, together with promises that self-sufficiency in rice would be achieved in a short time, and with an excess for export. The expected 'Green Revolution' brought an immediate increase in yields of rice from the rapid-growing, disease-resistant new seed. For a short time, it looked as though the expectations of biologists had been realised. However, the new strains proved to be costly to farm, requiring substantial amounts of expensive fertilisers. Neither were farmers able to gather and store them each year in the traditional manner to serve as seed for the next season's crop. The progeny of the hybridised rice proved slow to germinate and did not retain the qualities of the original parent. Finally, the heavy use of fertilisers on the irrigated crops encouraged the growth of water weeds which choked the streams and proved a menace to river transport.

It is now some forty years later, and communities and their governments are becoming aware of the risks and dangers which fertilisers pose to the health of humans and the environment. Furthermore, although fertilisers, weedicides and new hybrid cereal seeds have boosted production, they have failed to accommodate the increasing numbers of people. Their use may not always be appropriate as they may result in new problems of their own.

Population Dynamics
A number of theories have been advanced to try to explain changes in populations which are evident despite factors such as migration, urbanisation and a trend towards denser settlement patterns. It is frequently suggested that demographic growth takes place in a number of distinct stages, referred to as the population cycle. This concept derives from the European experiences as described in the United Kingdom. It has four stages, each marked by a high but fluctuating birth rate, followed by an early boom in numbers. This stage evolves into one marked by a high birth rate and a declining death rate, leading finally to a decreasing birth rate.

Today's demographers attempt to place non-western countries into one or other stages of this European-derived cycle. This is not always appropriate. Many countries such as Sri Lanka are presently marked by sharply declining death rates with equally startling rates of natural increase. Improvements in nutrition and health-care further boost birth rates and limit death rates. Examples include Algeria (3.3%), Malawi (6.3%), Kenya (4.1%), Mexico (3.9%), Ecuador (3.5%), Iraq (3.5%), Pakistan (3.0%).

Decreases in birth rates with resulting moderation in population growth now extend over most of Europe, Anglo-America, temperate South America, Japan, Australasia and the Commonwealth of Independent States (formerly USSR). Explanations are not hard to find. There is now no perceived need for couples to plan for many children. In former times, children were needed to help in the fields and to act as agricultural or factory workers. The low survival rate to adulthood, the high death rate due to infection and the harsh conditions of daily life meant that families needed to be large to ensure that some members survived. Similarly, in countries where warfare was endemic, couples tended to have many children to replace sons and daughters killed in fighting. In recent times, the high birth rate in Iraq has been attributed to the effects of warfare.

In western developed nations, there are equally compelling reasons to explain population trends and cycles. Many European couples deliberately limit family size, explaining their decisions by reasons such as the high costs of feeding, educating and providing for children in a modern consumer-oriented society. They also aspire to the acquisition of many of the consumer goods and benefits offered by the modern society, and which were not available to any but the privileged aristocracy of earlier generations.

Significant decreases in birth rates became apparent in many less developed countries during the 1970s. Even where birth rates remain high, such as in Sri Lanka, there is evidence of potential for decline. Between 1970 and 1985, the rate of total world population increase slackened to 1.8 per cent per annum, compared with an average of 2 per cent in the 1960s (Lowry, 1986).

Under-nutrition occurs when protein intake is low or minimal despite a diet containing an adequate supply of food energy. Normally this occurs where carbohydrates form the major food item in the diet. This situation is evident in under-developed areas where the people depend on crops of sweet potato (Papua New Guinea), taro and manioc (East Equatorial Africa), millet (Niger and North Africa), or rice (South and East Asia). Malnutrition arises where total food intake as well as the amounts of basic proteins, carbohydrates and fats are below levels considered to be sufficient to maintain health and an adequate supply of energy.

TABLE 5.8: Housing and Population (1986 data)			
	no. of dwellings	rooms per dwelling	persons per room
Australia	5,100,000	5.0	0.6
Canada	8,200,000	5.7	0.5
Israel	1,130,000	2.6	1.2
Japan	34,000,000	4.6	0.7
New Zealand	1,260,000	5.2	0.6
USA	88,207,000	5.3	0.5
Zambia	1,195,000	2.7	1.8

TABLE 5.9: Dwelling Standards (1986 data) % households			
	piped water %	*flush toilets* %	*electricity* %
Australia	99	99	99
Canada	99	99	99
Israel	98	71	98
Japan	93	46	98
New Zealand	98	98	99
USA	99	99	99
Zambia	63	31	77

TABLE 5.10: Urbanisation, Effects and Implications	
Effect	*Implication*
* local weather change due more heat rising from given area, pooling of air, emissions from industry, altered local topography (high rise)	* increased incidence of people suffering extremes of heat and cold, especially very young and elderly
* increase in urban unemployment with entry of rural people seeking work	* decline in wage levels, more competition for jobs, with resultant decline health standards
* problems of food supply, storage and distribution to demands of large cities from distant agricultural centres	* increase costs of basic foods; subsequent decline in quality of nutritional standards
* increase in household size	* less individual living space, privacy; increased nos per room, crowding
* more multi-storey urban high rise	* physical and social (see Table 5.11)
* increase wastes	* accumulation of non-degradable plastics, etc.
* increase in traffic and associated infrastructure	* increase in pollutant emissions (e.g. carbon-based fuels, lead): more space taken for roads, parking facilities
* local water shortages, supply & quality problems	* decrease in water quality and resultant decline in standards of personal hygiene
* increase in demands for energy by industry and individual consumers	* difficulties in meeting demands by environmentally sensitive systems; therefore increased dependence on nuclear energy
* increased demand for public health services & social welfare	* lack of adequate resources, funds and trained staff to meet increased demands

TABLE 5.11: High-Rise Settlement

* *Physical and Environmental Effects*

1. tendency to unsightly, uniform low-cost flats and small apartments lacking views, privacy

2. extremes of heat and cold in buildings, due to difficulties in controlling internal temperature of large buildings to suit individual needs, location and exposure of individual units

3. creation of wind tunnels between buildings, and other problems associated with local changes to weather (increase in heat rising from buildings, localised pooling of air with pools hot and cold air creating eddies)

4. waste and garbage disposal problems

5. monotonous and unimaginative landscapes, absence of trees, parks and nature reserves as all available space used for housing residents

* *Social and Health Effects*

1. social isolation, loneliness and absence of social support, resulting in anxiety, depression and increase in suicide rates, particularly among women and the aged

2. communal, age and ethnic tensions, and violence to people and common property

3. increased incidence of respiratory problems associated with living in polluted city environment

TABLE 5.12: Population and Food Output 1952/6-1981

Region	Food output % change		Population increase % change		Food output per head % change	
	1952-6	1969-71	1952-6	1969-71	1952-6	1969-71
W. Europe	46	20	14	6	29	14
E. Europe & USSR	71	15	21	9	41	5
N. America	32	35	26	12	5	20
Oceania	68	31	38	16	22	13
Developed world	48	21	19	10	24	11
Latin America	57	46	53	33	3	11
Near East	61	41	49	36	9	4
Far East	56	42	41	27	11	12
Africa	41	23	44	37	2	10
Asian Centrally Planned	—	41	—	20	—	17
Developing World	54	40	44	27	7	10
World	51	29	34	20	12	6

TABLE 5.13: Nutrition in Under-developed and Developed Regions

Item	Economically under-developed regions*	Economically developed regions**
Calories (per person per day)	2210	3340
Total protein (grammes per person per day)	63	90
Animal protein (grms per person per day)	10	44
Population (UN estimates for 1985)	3487 million	c.1100 million

Source: UN Demographic Yearbook
* Far East, Near East, Africa, Latin America
** Europe, USSR (CIS), Anglo-America, Argentina, Uruguay, Australasia

TABLE 5.14: Consumption of Animal Protein

	per person per day	
New Zealand	74.8 g	
Uruguay	61.9	
United Kingdom	53.4	good
Austria	47.5	diet
Israel	36.3	
Chile	29.2	
Spain	23.4	moderate
Mexico	23.4	to poor
Japan	16.9	diet
Egypt	12.2	
Guatemala	8.5	very
Pakistan	7.7	poor
India	5.9	diet

TABLE 5.15: Health Problems Associated With Large Populations

PART A

1. Advanced industrial societies (USA, Japan, Western Europe)

* increased urban settlement resulting in growth of slums and shanty towns (e.g. Mexico, Djakarta) with open drains, lacking piped water and sewerage facilities

INCREASE IN INFECTIOUS AND WATER-BORNE
EPIDEMIC DISEASES

* increase in unemployment and under-employment, resulting in more competition for jobs, lower wages. Decrease in family incomes, forcing women and mothers into workforce, bringing decline in nutrition and diet

INCREASED DEPENDENCE ON PUBLIC SUBSIDISED
HEALTH CARE
INCREASED USE SOCIAL WELFARE SUPPORT NETWORKS
DECLINE IN QUALITY OF CHILD CARE
DECLINE IN NUTRITION AND DIET
INCREASE IN STRESS-RELATED ILLNESS AND DEPRESSION

* increase in youth unemployment

INCREASE IN YOUTH SUICIDE RATE
INCREASE IN INJURIES ASSOCIATED WITH VIOLENCE
AND DRUG ABUSE
INCREASE IN ALCOHOL ABUSE

* increase in mature age unemployment

INCREASE IN ALCOHOL ABUSE
INCREASE IN STRESS-RELATED ILLNESS AND DEPRESSION

* increased population density, leading locally to increase in community tension and small group social support mechanisms

INCREASED DEMAND, MENTAL HEALTH COUNSELLING

PART B

2. *Under-developed Societies (e.g. China, Sri Lanka, Zambia)*

* increase in production demands made on countryside, resulting in more intensive cropping, failure to rotate or fallow land, etc., bringing a decline in quality of crops and in productivity of land

DECLINE IN HEALTH STANDARDS AND NUTRITION OF RURAL PEOPLE: GREATER SUSCEPTIBILITY TO DISEASE

* increased urban settlement resulting in growth of slums and shanty towns (e.g. Mexico, Djakarta) with open drains, lacking piped water and sewerage facilities

INCREASE IN INFECTIOUS AND WATER-BORNE EPIDEMIC DISEASE

* increase in unemployment and under-employment, resulting in more competition for jobs, lower wages. Decrease in family incomes, forcing women and mothers into workforce, bringing decline in nutrition and diet

DECLINE IN NUTRITIONAL STANDARDS, MALNUTRITION
DECLINE IN INFANT AND MATERNAL HEALTH

* increased population density, leading locally to increase in community tension and small group social support mechanisms

References and Additional Reading

Berg, A. *The Nutrition Factor*, The Brookings Institute, Washington DC, 1973.

Ennis, W.B., Dowler, W.M. and Klassen, W. 'Crop Protection to Increase Food Supplies' from *Science*, 1975, vol.188, pp.593-6.

Farallones Institute. *The Integral Urban House: Self-Reliant Living in the City*, San Francisco, 1979.

Lowry, J.H. *World Population and Food Supply*, 3rd edn, Edward Arnold, Victoria, 1986.

Morgan, E. *Falling Apart: The Rise and Decline of Urban Civilization*, London, 1978.

Revelle, R. 'The Resources Available for Agriculture' from *Scientific American*, 1976, vol.235 (3), pp.164-78.

5.3 Atmospheric Pollution

The layer of air which surrounds the earth is called the atmosphere. It is essential for most plant and animal life: it also retains much of the heat of the earth's surface. The upper layers of the atmosphere protect earth from an avalanche of glaring heat and biologically damaging ultraviolet radiation which reaches earth from space.

Until about the middle of the eighteenth century, the air we breathe was thought to be a simple elementary substance of which all other gases were modifications. As a result, oxygen was first spoken of as 'deflogisticated air', nitrogen as 'flogisticated air' and carbon dioxide as 'fixed air'.

Air is now known to be composed of a mixture of gases: 78% nitrogen, approximately 21% oxygen and smaller quantities of carbon dioxide (0.03%), water vapour (varying between nil and 2%), with the remaining 1% composed of argon, hydrogen, helium, neon, ammonia and others. Table 5.16 summarises the constituent elements of dry air.

The Nature of Air Pollution

Air is considered polluted when it contains substances which are harmful or likely to be harmful to humans and the environment. Air pollutants may be solids, liquids or gases.

Solid air pollutants have particle sizes measured in micrometers and remain suspended in air. Large particles which settle from the atmosphere tend to range in size from 20 to 300 micrometers. These particles mostly consist of fly ash (tiny particles of impurities from coal), soot, refractory materials and water-soluble salts (usually originating from salt water). Smoke, for example, contains many solid particles which either prevent penetration of sunlight or deflect and scatter rays which strike them. On a global scale, this effect can mean shading of the earth, a reduction in temperatures and the onset of an Ice Age.

Small dust particles less than 10 micrometers and sometimes as small as 0.1 micrometers remain suspended in the air indefinitely. They settle only when coming into contact with solid surfaces, where they tend to clump together or serve as a nucleus for raindrops and fog droplets. These small particles also tend to settle out in the damp, warm air of the lungs, and may irritate the delicate respiratory surfaces of asthmatics and sufferers of chronic respiratory diseases.

However, it is the liquid and gaseous wastes in air which probably present the most serious environmental problems in the short-term. These wastes influence life on the planet and can be modified by direct action. They are produced and added to the atmosphere every day, where they may combine to form the smog so typical of many industrial cities (see page 221). Other gases, such as freon (widely used as refrigerants and in aerosols), drift into the upper levels of the atmosphere and react with molecules there.

Each day, pollutants are produced by human work and by industry which are called primary pollutants, and they are added directly into the air. They include carbon dioxide, carbon monoxide, sulphur dioxide and some of the oxides of nitrogen. These pollutants become energized by sunlight and may interact with each other and with other gases in the air to produce new compounds known as secondary pollutants, which include ozone, nitrogen dioxide, aldehyde and complex organic molecules.

Agricultural practices also lead to emissions of several gases. The burning of forests and savanna grasses in tropical and subtropical regions to create pastures and cropland, yields additional large amounts of carbon monoxide, methane and nitrogen oxides. Moreover, soil exposed after forests are cleared emits nitrous oxide, as do nitrogen-rich fertilisers. The breeding of domestic animals is another major source of methane, from anaerobic digestion in the gastric tracts of cattle and cud-chewing animals. Rice paddies are also responsible for increased atmospheric methane.

Air pollution is usually more obvious than land or water pollution. It is easily visible, spreads quickly with the wind and can be detected by smell, taste or contact with eyes and skin. Thus, it is hardly surprising that primary and secondary pollutants can affect the health of human beings.

For example, air heavy with smog is taken into the lungs where particulate matter can be deposited: chemicals such as aldehydes and sulphuric acid irritate delicate eye tissue and affect sensitive skins. In addition, among these chemicals are substances like ozone which has a dual action. In the upper layers of the atmosphere it acts to protect life by capturing damaging radiation; at ground level and as produced by reactions from industrial chloro-fluorocarbon wastes, it is toxic.

Laws Against Pollution of the Air
The first law to control air pollution was passed in England in 1273. In common with the Royal Proclamation of 1306, it prohibited the burning of coal in London because of the large amounts of soot and unpleasant odours produced by the low quality coal then used. Much more coal was burned during the subsequent Industrial Revolution and this caused the introduction of stricter controls on smoke emission. In 1847, the English Town Improvements Act contained a section dealing with factory smoke, and smoke

abatement was again considered in the Public Health Act of 1875. The 1956 British Clean Air Act finally sought to contain smoke emissions and to enforce the use of smokeless fuels in domestic heating. Its enactment followed a worsening of autumn smog in Britain's major industrial cities and the resultant effects on human life and health. This coincided with increased levels of policing compliance and stricter penalties for infringement of laws and regulations.

Today, the WHO and specific country health authorities set limits and recommended maximum levels for atmospheric pollutants, usually expressed in terms of parts per million (ppm) or parts per hundred million (pphm). Deposition rates which measure the amount of dirt and grit that falls from the atmosphere are measured in grams per square metre per month. Many countries now also monitor pollution on a daily basis and publish daily readings as public health warnings.

Sulphur dioxide rarely constitutes as much as 50 parts per billion of the atmosphere even where its emissions are highest, and yet it contributes to the aesthetic nuisance of decreasing visibility, acid deposition, and the corrosion of stone and metal. The nitrogen oxides, similarly scarce, are important in the formation of both acid deposition and what is called photochemical smog, a product of solar-driven chemical reactions in the atmosphere. The chlorofluorocarbons (CFCs), which as a group account for just one part per billion of the atmosphere, are the agents primarily responsible for depleting the stratospheric ozone layer. In addition, rising levels of CFCs together with methane, nitrous oxide and carbon dioxide comprise the so-called 'greenhouse gases' believed to be the main cause of increased temperatures of the atmosphere.

The hydroxyl radical (OH), a highly reactive molecular fragment, also influences atmospheric activity even though it is much scarcer than other gases, with a concentration of less than 0.00001 part per billion. Hydroxyl, however, contributes to the cleansing action of the atmosphere, although its abundance is likely to diminish in the future.

Certainly, some fluctuations in the concentrations of atmospheric constituents can derive from variations in rates of emission from natural sources. Volcanoes, for example, can release sulphur and chlorine-containing gases into the troposphere (the lower 10 to 15 kilometres of the atmosphere) and the stratosphere (extending roughly from 10 to 50 kilometres above the surface). These emissions apart, it is now clear that the activities of human beings account for most of the rapid changes witnessed during the past 200 years, activities such as the burning of the fossil fuels, coal and petroleum for energy, industrial factory emissions and agricultural practices such as biomass burning and massive deforestation. Chemists, meteorologists, solar and space physicists, biologists and ecologists are presently trying to ascertain the reasons behind the changes and to predict future conditions for planet earth and its inhabitants.

Europe's Crisis

In the 1970s in Western Europe, evidence of a growing wasteland of dying forests, smog-bound cities and sick workers began to claim the attention of politicians. The realities and grave warnings of the conservationists Carson and Schumacher looked like materialising. And people began to realise that precautions involved more than just control of noise, smog and coal-based discharges; they also involved invisible gas emissions like sulphur dioxide, nitrous oxide, hydrogen sulphide and other invisible gases.

The crisis was fuelled when large numbers of trees, particularly those of the Black Forest, were found to be dropping leaves prematurely and dying. People were complaining of sore eyes, nausea and tiredness. The numbers diagnosed with respiratory problems such as asthma and bronchitis began to climb, and more people became chronic sufferers. Medical doctors, health professionals and social scientists lent their professional expertise to the swelling voices of concerned people. Explanations were called for and studies to identify causes and offer solutions were demanded.

Smog

The word 'smog' was coined in 1905 by the President of the British National Smoke Abatement Society, to describe the pea-soup fogs of London. First official use occurred in 1926 when the US Weather Bureau in Indianapolis began to include comment on smog in local weather reports.

Smog occurs when smoke provides fine particles which act as nuclei for the condensation of fog, hence the word (a combination of smoke and fog). It is particularly liable to form in areas where temperature inversion develops. These conditions arise on calm, clear nights when the earth cools rapidly as heat radiates away from the surface, which becomes cooler than the air higher up. Being denser, the cold air near the ground is denser than the air above, so that it hugs the ground and pollutants accumulate in it. The temperature inversion is broken up by wind or by sunshine which warms the air near the ground. The inversion will last as long as the windless, overcast days persist, sometimes for as long as a week, depending on climate and topography.

The most severe London-type smog on record occurred in London in December 1952. In a five-day period, 4000 people died of symptoms associated with air pollution. The deaths were attributed to sulphur dioxide, but its concentration of 4mg/cubic metre (1.3 ppm) was insufficient to cause death in all but the most infirm. Recent accounts blame the deaths on hydrocarbons (petrol vapours and refinery emissions) and on solid particles suspended in the air.

Photochemical Smog: In and around cities photochemical smog is a common feature of modern life. It is particularly prevalent in industrial cities sited in areas with a uniform topography characterised by flat alluvial floodlands or plains, and where natural air circulation is minimal. The term 'photochemical' refers to the undesirable mixture of gases formed in the lower troposphere when solar radiation acts on anthropogenic emissions, especially those containing nitrogen oxides and hydrocarbons from vehicle exhausts. The result is a mixture of reactive gases that can be destructive to many forms of life.

The city of Los Angeles has become associated with this type of smog. First identified in the 1940s, the secondary pollutants produced irritate the eyes and mucous membranes, cause damage to plant surface tissues and have a characteristic odour. Efforts to control ash and sulphur dioxide emissions rarely succeed in controlling photochemical smog, which is also experienced by Sydney, Rome and Athens, cities located in temperate latitudes and lying close to a large expanse of ocean or sea water.

Ozone is a major product of photochemical reactions and is itself the main cause of smog-induced eye irritation, impaired lung function, crop and tree damage. The severity of the smog is therefore generally assessed on the basis of ground-level ozone concentrations.

Investigators first began measuring ozone levels in the late nineteenth century, initially at ground level, then, in the 1930s and 40s, in the atmosphere using increasingly sophisticated airborne devices. Early measurements in Europe suggest that the 'natural' level of ozone close to the ground was around 10 parts per billion. Today the typical ground-level concentrations in western Europe are up to four times higher. Readings of more than ten times higher than the natural level are often recorded in western Europe, California, eastern United States and Australia.

Photochemical smog has also begun to make an appearance in Asia and the tropics. In the latter instance the increase has been dramatic, and is greatest when felling and clearing of timber lands is followed by burning of the undergrowth and of low-level bush and savanna grasslands. These practices release immense quantities of the precursors to smog. The strong solar radiation in these latitudes means that firing is followed by a rapid sequence of photochemical reactions. The result is that ozone levels can climb to five times in excess of normal. Increases in population in these areas together with the rapid pace of industrial development, the typical floodplain focus for city growth and the absence of effective government controls, mean that significant pollution of the atmosphere will soon become a feature of many Asian and South American cities.

Nitrous Oxide

Nitrous oxides in the atmosphere are believed to contribute to three major components of atmospheric pollution: smog, decreased visibility at ground level and acid deposition on buildings (through combination with water droplets in the atmosphere). It is uncertain the extent of their contribution to stratospheric ozone depletion. Nitrogen oxides destroy ozone but they can also interfere with this cycle. For example, nitrogen dioxide molecules can remove chlorine monoxide from circulation by combining with it to form chlorine nitrate. Since they are not among the greenhouse gases, they make no direct contribution to atmospheric warming.

Together with water molecules, the nitrogen oxide gases freeze to form particles, making up what are known as polar stratospheric clouds. The cloud particles facilitate the chemical reactions that release chlorine from compounds that do not themselves react with ozone, such as hydrochloric acid and chlorine nitrate.

There is some dispute over the effects of exposure to nitrogen dioxide emissions into the atmosphere. Clinical tests suggest that sensitive people such as children, the elderly and asthmatics experience respiratory problems when exposed to nitrogen dioxide for brief periods, even at low levels. Tests on animals exposed to higher levels form the basis of concern among some researchers that long-term human exposure could result in irreversible lung disease, a weakening of the immune system and possibly cancer.

Nitrogen dioxide emission occurs from unflued indoor gas heaters. The WHO has set a goal that limits the amount of nitrogen dioxide emissions outdoors to 18 per cent. Australia's Air Quality Committee of the National Health and Medical Research Council (NHMRC) sets a recommended outdoor level of 0.16 ppm and opts for a goal of 0.1 ppm. Industry argues for a higher level of 0.3 ppm, claiming that a lower limit would cost the Australian industry several hundred million dollars in the first three years of implementation.

In Britain and Europe, the severity of winter has long mandated the use of efficient large-scale heating for schools and offices. However, in some other countries such as Australia, concerns have been raised about the levels of nitrogen dioxide emissions from the 80,000 unflued gas heaters used in NSW Government schools. Tests conducted in 1989 indicate that heaters in many schools are defective. For example, a reading of 2.9 ppm (almost 19 times the recommended outdoor level) was recorded at Georges Hall Public School in an unoccupied room and without ventilation.[1] One Lithgow Public School reading was 1.28 ppm; another for Orange High School was 1.01 ppm. Alternatives to the costly replacement of heaters include much needed repairs to existing heaters, gradual replacement as old heaters fail, and attention to the need for adequate cross-ventilation.[2]

Sulphur Dioxide

Sulphur dioxide plays a major role in air pollution. Vast amounts are released into the atmosphere, both naturally and through the activities of society. The natural sulphur dioxide comes from volcanic and biological sources and is widespread over the earth. The sulphur dioxide generated by human works comes mainly from the combustion of petroleum materials (7%), smelting (4%) and the burning of coal (24%). By far the largest contributions come from bacteria (45%) and sea spray (20%).

The composition of burned coal ranges between 0.3 and 3.0 per cent. It is not an efficient process and has a 'conversion rate' of 80 per cent, and only 20 per cent remains as ash. About half of all the sulphur dioxide responsible for air pollution comes from fossil fuel combustion: the remainder is blamed on metal ore smelter industries. Until recently, this sulphur dioxide was released into the atmosphere. Sulphur dioxide has an accelerating effect on the corrosion of exposed metal surfaces on buildings and equipment. The resulting fumes sometimes killed surrounding vegetation.

Conversion to sulphur trioxide and sulphuric acid occurs via combination with atmospheric oxygen under the action of sunlight. Being readily soluble in water, sulphuric acid returns to earth as rain and is commonly referred to as 'acid rain' because of its corrosive effects on buildings, statues, marble and other carbonate rocks.

Damage to plants from sulphur dioxide occurs at concentrations as low as 1.5 mg per cubic metre, while in people 30 mg per cubic metre causes throat irritation. Eye irritation and coughing occur at 60 mg per cubic metre and concentrations of $10mg/m^3$ will have a noticeable effect on the respiratory passages.

Acid Rain

Acid rain, also termed acid snow, fog or dew, develops mainly as a by-product of atmospheric interactions involving nitrous oxide gases and sulphur dioxide. Through various reactions such as combination with the hydroxyl radical, these gases can be converted within days into nitric and sulphuric acids, both of which are readily dissolved in water. When the acidified droplets fall to the earth's surface, they constitute acid rain.

Because water droplets are removed from the atmosphere rapidly, acid rain is a regional or continental rather than a global phenomenon. In contrast, the atmospheric lifetimes of several other trace gases, including methane, carbon dioxide, the CFCs and nitrous oxide, are much longer and so the gases spread rather evenly throughout the atmosphere, causing global effects.

Since the beginning of the Industrial Revolution in the mid-eighteenth century, the acidity of precipitation (as measured by the concentration of hydrogen ions) has increased in many places. For example, it has roughly quadrupled in the north-eastern US since 1900, paralleling increased emissions of sulphur dioxide and the nitrous oxide gases. Similar increases have been found elsewhere in the industrialized parts of the world. Acid rain has also been detected in the virtually unindustrialised tropics, where it stems mainly from the release of nitrous oxide gases and hydrocarbons by biomass burning.

Wet deposition is not the only way sulphur and nitric acids in the troposphere find their way to the earth's surface. The acids may also be deposited 'dry' as gases or as constituents of microscopic particles. Indeed, a growing body of evidence indicates that dry deposition can cause the same environmental problems as the wet form.

Acid deposition clearly places severe stress on many ecosystems. Although the specific interactions of such deposition with lake fauna, soils and different vegetation types are still incompletely understood, acid deposition is known to have strongly increased the acidity of lakes in Scandinavia, the north-eastern US and south-eastern Canada, leading to a reduction in their size and the diversity of fish populations. Such deposition also appears to play a major role in the forest destruction occurring in Europe (see above).

There is little doubt that acids deposited from the troposphere also contribute to the corrosion of outdoor equipment, buildings and works of art, particularly in urban areas. These cost hundreds of billions of dollars for repairs and replacement. Particles containing sulphate ions have other effects as well. By scattering light so efficiently they decrease visibility; but by also influencing cloud albedo (reflective power, whiteness), particle size may have further implications for climate change. Particles also prove a source for skin irritations and irritate delicate lung tissue in sensitive individuals.

Ozone

Ozone poses a unique problem. Ground level ozone poses a risk through its contribution to smog, particularly that produced by solar radiation (photochemical smog). A decrease in ozone near the ground would therefore benefit polluted regions. However, ozone plays a major beneficial role in the upper atmosphere where it is effective in limiting the penetration of damaging ultraviolet radiation. Any decrease in stratospheric ozone is disturbing because it permits more damaging ultraviolet radiation to reach earth's surface, where it is responsible for eye damage, increased risk of skin cancer and cataracts, and of tissue and chromosomal damage to plants and phytoplankton.

In the upper atmosphere, ozone is concentrated in the stratosphere approximately 30 kilometres above earth's surface. Stratospheric ozone forms as a result of dissociation of atmospheric oxygen molecules under the action of ultraviolet radiation and subsequent combination of these highly reactive oxygen radicals with other oxygen molecules.

$$\text{(i) } O_2 \xrightarrow{hv} O_0 + O_0 \qquad \text{(ii) } O_0 + O_2 \rightarrow O_3$$

Ozone is an important atmospheric oxidizer or cleanser. It acts to transform gases into water-soluble products, thereby facilitating their subsequent removal through precipitation. It also participates in the formation of another important oxidizer, the hydroxyl radical. This forms after ultraviolet light dissociates ozone, releasing a highly energetic and hence highly reactive oxygen radical that then combines with a water molecule.

$$\text{(i) } O_3 \xrightarrow{hv} O_0 + O_2 \qquad \text{(ii) } O_0 + H_2O \rightarrow 2OH$$

The extent of stratospheric ozone depletion has been most dramatic over Antarctica, where an ozone 'hole' — a region of increasingly severe ozone loss — has appeared each southern spring since about 1975. In the past decade, the springtime ozone levels over Antarctica have diminished by about 50 per cent. A more global assessment of stratospheric ozone is presently underway, but, in the past twenty years, depletions of 2 to 10 per cent have apparently begun to occur during winter and early spring in the middle to high latitudes of the northern hemisphere, with greatest declines in the higher latitudes.

The Role of Chlorofluorocarbons (CFCs)
The chlorofluorocarbons were first developed as a replacement for ammonia in refrigerators. These agents, particularly CFC-11 (CFCl3) and CFC-12 (CF2 CL2), are the major agents responsible for ozone depletion. The emissions of these chemicals have grown rapidly since their introduction on a widespread basis during the 1950s for refrigeration, as aerosol propellants, solvents and blowing agents. The CFCs are highly prized because they are highly stable, virtually non-toxic, non-inflammable, non-corrosive and poor conductors of heat. However, the same inert properties which once made them so valuable to industry, mean that they persist unchanged in the atmosphere for long periods of time. It takes twenty to twenty-five years for the CFCs to reach the stratosphere where they are broken down.

It has been calculated that a single chlorine atom can destroy up to 100,000 ozone molecules before being inactivated. The process of CFC breakdown begins with the combination of a chlorine molecule with an ozone (O_3), with the formation of chlorine monoxide and molecular oxygen.

$$Cl + O_3 \rightarrow ClO + O_2$$

The chlorine monoxide then reacts with an oxygen radical (formed by photodissociation of another ozone molecule) and liberates the chlorine, which is free to initiate the cycle again:

$$ClO + O_0 \rightarrow Cl + O_2$$

Particularly in Antarctica, frigid temperatures hasten the chlorine-catalyic cycles by removing nitrogen oxides which, paradoxically, possess the potential to minimise the chlorine-catalysed ozone depletion.

Thus a 'chain' reaction is initiated which tends to use up more ozone molecules. As a result one CFC molecule can be responsible for the loss of thousands of molecules of ozone. This means that with greatly increased amounts of CFCs reaching the stratosphere in the next few decades, the integrity of the thin ozone mantle, so essential to life on earth in its present form, is placed seriously at risk. Even if chlorofluorocarbon emissions ceased today, chemical reactions causing the destruction of stratospheric ozone would continue for at least a century.

Concerns about the effect of CFCs on ozone were first raised in the early 1970s, and in the United States resulted in a ban in 1978 on nearly all uses of CFCs in aerosol propellants. However, by the end of the decade it had been concluded that while CFCs certainly depleted the ozone layer, the effects were likely to be less serious than had originally been thought. Ironically, this easing of concern meant that the companies which had been working on developing benign CFC substitutes wound up the research programmes which have now, with much urgency, been resumed.

Timetables for Change

In 1985, British scientists discovered a 'hole' opening up in the ozone layer over the Antarctic. At the end of that year, the problem of ozone layer depletion and its effects was formally recognised as serious when a treaty known as the Vienna Convention established the principle that signatory nations would take measures to protect the ozone layer. In 1987, the United States, the European Community and twenty-three other countries, including Australia, signed the Montreal Protocol which set a programme to reduce global use of a list of 'controlled substances' — five CFCs and three of their close relatives, the halons, with high ozone depleting potential. The target set was a staged 50 per cent reduction in consumption of these CFCs by 1990, and a freeze on halon consumption at 1986 levels by 1992.

Following scientific findings that the threat to the ozone layer is even graver than first recognised, the Montreal Protocol was updated in 1989 by the Helsinki Declaration. Though not binding, this committed the eighty signatory nations to pursue a phase-out of the five CFCs by the year 2000, and to controlling and reducing other ozone-depleting substances. Proposals by developing countries for an international fund financed by the developed countries to help them buy CFC-free technology were rejected by the US, UK and France, who argued that existing aid channels were adequate.

By June 1990, a pre-determined date for review of the Declaration, signatory countries were able to confirm their continued commitment to a phase-out of CFCs, but refused to speed up the process despite attempts by Australia, in particular. On the contrary, a disillusioned Australian Youth Delegation, returning to Sydney in early July, reported that the major CFC producers had convinced the Conference nations to refuse any attempts to hasten elimination of CFCs, claiming that the industry was already outpacing government schedules and did not warrant further pressure.

The Culprits

The five CFCs targeted for immediate attention and their estimated ozone depletion potentials or ODPs (derived from an index system which gives CFC-11 a value of one), are CFC-11 and CFC-12 (ODP 1.0), CFC-113 and CFC-114 (ODP 0.8) and CFC-115 (ODP 0.4).

Used mainly in fire-fighting applications, halons or bromofluorocarbons (BFCs) have much greater ozone-depleting potential than CFCs — the three halons controlled as a result of the Montreal Protocol have estimated ODPs of 3.0, 6.0 and 10.0. Until recently, less emphasis has been placed on phasing out their use because of smaller levels of production than for the CFCs. However, results of more recent scientific research have suggested that tougher measures to control halon production and use are urgently required. Since the Helsinki Declaration, possible reductions in carbon tetrachloride (ODP 1.2) and methyl chloroform (ODP 0.15) have been called for.

The Protocol-controlled substances also contribute in some degree to the greenhouse effect (global warming), so that any cut in emissions has the additional benefit of reduction in greenhouse gas emissions. In developing new options, the greenhouse warming potential of any prospective CFC substitute is a factor which needs to be considered. In Australia, the government has announced that CFC and halon use will be reduced by 95 per cent by the end of 1994, with a total ban on the importation and production of both (except for essential medical uses such as asthma sprays) by the year 2000.

Progress In Eliminating CFCs

In 1988, the British aerosol industry indicated that it would achieve a 60 per cent cut in CFC consumption by the end of 1989. Urethane foam manufacturers have declared their intention to phase out CFC-11 use by 1993.

In Australia, the Victorian Environmental Protection Authority (EPA) reported in 1989 that the foam manufacturing industry planned to halve use of CFCs by 1992, and achieve total phase-out by 1998 in the dry-cleaning industry. The use of CFCs in aerosols has already been cut by 70 per cent since 1974, and the industry claimed to have achieved almost total phase-out at the end of 1989. The EPA report also proposed more stringent regulatory measures and tighter labelling requirements, together with recommendations on ways in which emissions of CFCs can be reduced at each of the three stages of CFC use (manufacture, use and disposal).

New Alternatives

The five CFCs and the three halons named in the Montreal Protocol as controlled substances are all carbon compounds whose four hydrogen atoms have been replaced by chlorine, fluorine or bromine. Partially substituted carbon compounds which retain some of the hydrogen-carbon linkages have been developed and are less able than the CFCs. However, these hydrochlorofluorocarbons (HCFCs) still retain ozone depleting properties (see Table 5.17). Also under investigation are hydrofluorocarbons (HFCs), but these appear to have less value for industry.

Other Atmospheric Pollutants

Lead: Lead in the form of tetraethyl lead and tetramethyl lead is added to petrol to increase its octane rating or anti-knock value. The addition of lead compounds is a cheaper alternative to the extra refining necessary to raise the octane rating by a similar amount. After combustion, the lead deposits are removed by reaction with 1,2-diberomomethane and 1,2,-dichloroethane, and the lead is expelled to the atmosphere as lead bromide, lead chloride or as small particles of metallic lead.

Lead is classed as a cumulative poison mostly affecting the central nervous system. Growing children and pregnant women are especially sensitive. Small amounts of lead are present naturally in some soils and waterways and certain foods may contain very small accumulations of lead. High levels of discharge may occur in streams and water outflowing from mining operations. These will result in the death of some riverine fish species and the accumulation of toxic levels of lead in marine plant algae, shellfish and fish tissues.

Concern about lead gained international prominence following hospital reports linking blood levels of lead in young children with residence and/or diet. A number of studies of the children of families living in towns with lead smelters in the United States, and more recently in Whyalla in South Australia, have confirmed that low intelligence, poor muscle coordination and a short concentration span can be blamed on lead in the atmosphere. It appears that children are particularly susceptible to the effects of lead, and that their developing nervous system tissues tend to concentrate lead absorbed in air they breathe or through the skin.

Air-Borne Cyanides: On the morning of 3 December, 1984 a white cloud of methyl isocyanate, a toxic chemical used in pesticides, escaped from a storage tank at the Union Carbide India chemical plant in Bhopal and wafted over the town. Over five years later, more than 2300 people have already died and more than 200,000 injured, many so seriously that they can no longer work and have permanent lung and sight damage.

Despite the fact that Union Carbide India Ltd was 49 per cent Indian-owned and staffed entirely by Indians, the American parent company has been blamed and become the barb for criticism of the role of large and wealthy multinational companies in the economies of smaller Third World countries. It also became the plaintiff in a complex and continuing court case for settlement of damages to the thousands of affected workers.

While accused of exploitation, a lack of adequate safety precautions and of siting dangerous chemical manufacturing in extra-territorial waters rather than at home, the multinationals involved in hazardous chemical production have tended to gravitate largely towards countries with advanced industrial economies and stiff pollution controls.

A 1984 study by the International Labour Organization determined that multinationals generally maintain health and safety standards well above those of the developing countries in which they operate. No multinational wants to pay millions of dollars of damages and compensation such as those involved at Bhopal, nor to become the focus for protests, sabotage, anti-capitalist propaganda, and scapegoats for deteriorating social conditions.

Have Things Improved?
Most developed countries now police the pollution of the atmosphere from industrial and domestic sources. Consumers are active in demanding more stringent controls, reporting of breaches and imposition of penalties. Conservation groups in developing countries attempt to monitor environmental destruction and large clearing of forests through fires which pollute the atmosphere. But have the actions produced results? Is our air cleaner or are the responsible parties simply becoming more adept at avoiding detection?

TABLE 5.16: Composition of the Earth's Atmosphere

gas	formula	% volume	% weight
nitrogen	N_2	78.09	75.51
oxygen	O_2	20.95	23.15
argon	Ar	0.93	1.28
carbon dioxide	CO_2	0.03	0.046
neon	Ne	0.0018	0.00125
helium	He	0.00052	0.000072
methane	CH_4	0.00015	0.000094
krypton	Kr	0.00010	0.00029
hydrogen	H_2	0.00005	0.000004
xenon	Xe	0.000008	0.000036

TABLE 5.17: CFC Alternatives

Possible replacement	Applications	Ozone depleting potential	Comments
HCFC-22	As a refrigerant, for blowing polyurethane foams (alone or in blends and for leak-testing fire	0.05	Readily available, but more expensive than CFC-11 and CFC-12
HCFC-123	As a refrigerant, for blowing foams and for solvent cleaning	less than 0.05	Toxicity testing is under way and needs to be completed before HCFC-123 becomes commercially available — this could take about 5 years
HFC-134a	As a refrigerant in domestic and commercial refrigerators and in automobile air conditioners	0	Toxicity testing and trial manufacture are under way, with the product expected to be on the market in about five years
HCFC-141b	As an aerosol propellant, for blowing urethane foams and as a refrigerant	less than 0.1	Toxicity tests are under way, but will not be on the market for at least 5 years
HCFC-142b	As a blowing agent in foams and an alternative refrigerant in air conditioners and refrigerators	less than 0.06	Is already commercially available

Taken from the Victorian Environmental Protection Authority paper, Ozone Protection Towards A Safer Future. Note that these compounds are being considered as short-term alternatives only, pending development of more satisfactory solutions.

Notes
1. Report to NSW State Government on tests performed on NSW school heaters under the auspices of a State-Government appointed committee which included representatives from Departments of Education and Health, the Australian GasLight Company (AGL) and the State Pollution Control Commission (SPCC).
2. As reported by Paul Bailey, *Sydney Morning Herald*, 29 December 1989, p.6.

References and Additional Reading

Graedel, T.E. & Crutzen, P.J. 'The Changing Atmosphere' from *Scientific American*, September 1989, p.28.
Stolarski, R.S. 'The Antarctic Ozone Hole' from *Scientific American*, January 1988.

5.4 Water Quality

Water is the most important and most abundant substance on earth. It may well be the most precious resource the earth provides for humankind. Water is essential to sustain life for all natural species and for the functioning of human societies. It is also becoming increasingly hard to find — ocean pure water is a scarce resource in today's world. The fresh water in the world's lakes and streams, creeks and rivers represents less than 0.01 per cent of the earth's total store of water. The oceans represent 97.41 per cent and land or ground water 2.59 per cent. However, of this land-based water, the vast majority is sequestered in the form of ice and snow and is largely unavailable. Only a tiny amount, 0.014 per cent, is readily available for consumption by humans and other living beings.

Water is used by people for washing, cooking and cleaning and is essential to the diet. Water is used in many industrial processes as a raw material, as a cleansing or washing agent and as a coolant. Unfortunately the fresh water which is so highly prized is afterwards returned to the environment dirty, smelly and discoloured. To date, humans have relied on the environment to recycle and cleanse, trusting it will return to them yet again fresh, clean and pure. Global water consumption is increasing largely in response to a growing population and increasing per capita use by agriculture and industry. Although sufficient fresh water is currently available, sound water management is necessary to ensure an adequate supply for the future.

The World Resources Institute estimates that 41,000 cubic kilometres of water per year return to the sea from the land, counterbalancing the atmospheric vapour transport from sea to land. Some 27,000 cubic kilometres, however, return to the sea as flood runoff, which cannot be trapped, and another 5000 cubic kilometres flow into the sea in uninhabited areas. Of the 41,000 cubic kilometres returning to the sea, some amount is retained on land, where it is reabsorbed by vegetation or land forms.

This cycle leaves above 9000 cubic kilometres readily available for human exploitation. In principle, this amount is sufficient to sustain 20 billion people. However, both the world's population and usable water are unevenly distributed and local availability varies widely. When the evaporation and precipitation balances are worked out for each country, water-poor and water-rich countries can be identified. Iceland, for example. has sufficient excess precipitation to provide 68,500 cubic meters of water per person per year. The inhabitants of Bahrain, in contrast, have virtually no access to natural fresh water and are dependent on the desalinization of seawater. Further-more, per capita consumption or withdrawal rates also vary drastically: the average American, for example, consumes more than 70 times as much water as the average resident of Ghana.

Modern times have seen an expansion in the use of water so that it has an even more pervasive influence on daily life. It has become an essential solvent and coolant in industrial and chemical processes, flowing and dripping through mining equipment and manufacturing plant to mould, blend and wash most of the things handled in daily life. It takes 100,000 gallons of water to produce a motor car and 1000 to put a pound of beef on the table. Western households flush around 100 gallons of water down the drain daily, a profli-gacy not shared in developing countries, where for over half of all households every drop of water used must be carried from its source.

The Politics of Water
But water, its amount, distribution and use is also a political issue, both between different groups of people in a country and internationally, between rich and poor nations. The fundamental contrast between the way water is seen in the developed and developing world is the contrast between quantity and quality. In most parts of Asia and Africa, it threatens either absence or over-abundance and sometimes one after the other with bewildering rapidity.

Water is also a communal asset. It makes involuntary neighbours of all who use it. For thousands of years, legal systems have accepted that there can be no ownership of running water.

Nonetheless because access to water is precious, it has to have custodians. Some individuals may claim privileges because their land abuts the lake or stream. Government must say what those privileges may be and regulate rights to use water or to interfere with it. As pressure both on water's quality and its quantity builds, the bargaining over water's role in economic and political life will likewise become more vociferous.

This point was clearly made by Sandra Postel, in her book *Water: Rethinking Management In An Age of Scarcity* (Worldwatch Institute, 1984). In referring to areas of the world already lacking adequate water, Postel pointed out that Egypt's 55 million people are dependent almost entirely on agricul-ture and the Nile River which irrigates farming lands. However, none of the waters of the Nile originates within Egypt's borders. Food and drinking water needs are rising there along with population growth; but Ethiopia, the Sudan and Uganda, controlling the headwaters, use, divert or plan to exploit more water for their own people, crop and development goals.

Conflicts Over Water

Conflicts over water are multiplying. Bitter wrangles have developed between India and Bangladesh over the flow of the two Himalayan rivers, the Ganges and the Brahmaputra, in the alluvial plains of their confluence above the Bay of Bengal. In the Middle East, Israel, Jordan and the West Bank all face a critical shortage of water by the mid 1990s. Observers have suggested that future issues of war or peace in the Middle East will be governed more by access to water than religious differences.

Water holds other hidden threats: contaminants. One of water's most extraordinary qualities is its capacity for change. It dissolves, absorbs, dilutes, concentrates and distributes the range of substances and experiences of washing, mixing, cleaning and other uses made of it. Its makeup alters during the course of its journeys in rivers, its accumulation in reservoirs above and below ground and in the course of its use by living species, human and animal.

In the developed world, engineers tame and treat water into uniformity. About 150 years ago, when industrialization first produced immense urban squalor, medical scientists identified water as the carrier of deadly disease. Since that time, the role of water in industry and in the home has become formalized and engineered. The architects of public health set about tunnelling and flushing, creating an underworld full of drains, sewers, pipes and faucets to contain, distribute and dispose of the water, soon recognised as an essential service.

While the achievements of the sanitary engineers warrant praise, the waterways which diverted and sanitized are again under pressure because of the increasingly complex uses to which they are now placed. Discharges of effluent push their self-cleansing properties to the limit, threatening not only human health but the life systems of plants, birds and aquatic creatures dependent on their natural regime.

Detritus, solid and liquid wastes and the quantities of other pollutant loads in water are increasing yearly in every country, and the results are often visible to the naked eye as rubbish, dead fish, stagnant waters, algal overgrowths, foam, detergents, floating oils and fats. As risks to human health increase, policing and penalties become harsher and infringements more grave in their implications. Maximum admissible concentrations of toxic substances such as nitrates, aluminium, lead, mercury and other elements, tests for their detection and ways of monitoring and controlling levels are constantly changing. A recognition of the need for greater care is dawning in the developed nations and the more profligate of users.

Water Uses

Although the uses to which water is directed differ from country to country, the principal drain on water supply is agriculture. Averaged globally, 73 per cent of water withdrawn from the earth is directed towards agriculture. Almost three million square kilometres of land are irrigated and more is being added at the rate of 8 per cent per year.

Increases in human population and population movements, locally and globally, influence water use, quality and distribution. Coastal developments, building of marinas, coastal and harbour protection works also affect water use and runoff patterns and upset local ecosystems. Human activity in a river basin can often aggravate flood hazards. Deforestation and excessive logging lead not only to increased soil erosion but also to increased runoff: in addition, navigation canals are sometimes dug, which may exacerbate flooding by increasing the amount of water that reaches the flood plain.

Another problem created by over-irrigation is salinization. As water evaporates or is taken up by plants, salt is left behind in the soil. The rate of deposition exceeds the rate at which the salt can be removed by flowing water and so a residue accumulates. Currently more than a million hectares every year are subject to salinization: in the US alone more than 20 per cent of the irrigated land is thus affected. In Australia, major salinization problems occur in 40 per cent of the lands bordering the lower reaches of the Murray River, the major source for water for irrigation of highly productive agricultural land for fruit and vegetable growing, wheat, barley, and oats and for a thriving rice growing industry further upstream.

A projected sea level rise of between 0.5 and 1.5 meters as a result of greenhouse effects would pose a coastal flooding problem but also would lead to salinization of water resources, create new wetlands and increase the ratio of salt water to fresh water on the planet. Precipitation increases of between 7 and 15 per cent in the aggregate have been referred to by climatologists.

Water Pollution and Quality

Fresh water runs off the land and on its way to the oceans becomes laden with particulate and dissolved materials — both natural detritus and the wastes of human society. When the population density in the catchment area is low, waste matter in the water can be degraded by microbes through a process known as natural self-purification. When the self-purifying capacity of the catchment area is exceeded, however, large quantities of these waste substances accumulate in the oceans where they can harm aquatic life.

A water pollutant can be defined as a substance or agent that is added to the water, either directly or by seepage, and that in some way alters the quality of the water thereby rendering it less suitable for use by humans and the environment. Pollutant agents range from heat, radioactivity and discharges or 'leaching' from chemical and mineral waste dumps to polluting substances such as sewage, industrial wastes, agricultural chemicals and urban drainage. This definition thus embraces many manufactured and refined substances and the products of industrial, animal and human use.

Since water is part of the total environment, water pollution is a form of environmental pollution. The concentration of most pollutants is small and has previously been measured in parts per million (ppm). This unit is the fraction of a pollutant present in a body of water, based on its weight relative to the weight of water. However, some countries refer to milligrams per litre as a measure of pollution; mg/L for water pollution, or mg/cubic metre for air pollution.

Water pollution can disrupt the food chain on which living beings depend. If raw sewage is released directly into a river it encourages the growth of large numbers of bacteria. These so deplete the available supply of oxygen that the animals living within the water die off and the natural vegetation is killed. It is often replaced by vegetation which covers the surface of the water.

Non bio-degradable detergents when released in vast quantities into the waterways cause foaming problems, depletion of oxygen in the water and the death of marine life such as fishes, algae and crustaceans. Blue-green algae can replace and re-colonise the water in place of other life forms, rendering the water unfit for drinking. In summer drought conditions, cyanobacteria species bloom in low volume, semi-stagnant waterways, posing a major risk to humans, grazing stock animals and native wildlife, especially in inland Australia.

The Principal Water Pollutants

The main groups of water pollutants are summarised in Table 5.18. Undis-solved solid materials, being solids, make the water look turbid or murky. The larger particles may be removed by gravity settlement and the smaller ones by filtration. The very fine particles, called colloids, will pass through the most elaborate filtration processes and are normally removed by electrophoresis — the use of micro-electric currents between two battery terminals.

Dissolved inorganic materials, mainly salts, are of metal or non-metal origin. Their presence imparts a certain character to the water and gives it properties peculiar to that chemical. If calcium and magnesium salts are present, the water is said to be hard. Sea water contains 3.5 per cent salts (3.5 grams per 100 grams of water) — even lower amounts of salts will render the water saline. Water should contain less than 0.5 per cent salt to be drinkable. The heavy metals, which include mercury and lead, are included in this group since it is in their soluble form that these elements are able to be incorporated and metabolised into the bodies of living organisms.

Dissolved organic compounds contain carbon and are frequently derived from living substances. Their presence in waterways is through the addition of sewage and from industrial processes and industrial wastes such as petrol and oil. Their concentration is estimated by the amount of oxygen that is used by the water. This is called the chemical oxygen demand. Modern chemical equipment can identify the chemicals present in an oil spill such as that occasioned by the oil-tanker *Exxon Valdez*, which hit an iceberg in December 1989. As a result, large areas of the Alaskan coastline became heavily polluted by a huge oil slick which killed millions of marine animals and native birds dependent on the pristine natural environment.

Living micro-organisms, simple animals and plants such as the algae enter and flourish in water from sewage, decaying vegetable and garden refuse, animal and human faecal wastes. The indicator species is *Escherichia coli*, a bacterium which normally lives harmlessly in the human gastric tract and is present in sewage. Its presence in large quantities, however, tends to indicate a heavy concentration of sewage and therefore of more dangerous micro-organisms such as pathogenic bacteria capable of causing typhus, hepatitis and gastroenteritis.

In certain parts of the world, mining for uranium, metallic minerals and rare metal earths results in the entry of dissolved and colloidal metal ions. Nuclear wastes which produce radioactivity may also be present in the water. Radioactive elements can be absorbed by living organisms and are incorporated into their bodies where tumours may result from the harmful effects.

Water is an ideal cooling agent. It takes 4.2 joules of heat to raise the temperature of one gram of water by one degree Celsius. This is called the heat capacity and is higher for water than for most other liquids. This high heat capacity, coupled with its abundance, means that water has become the most widely used cooling agent. Unfortunately the addition of heat to natural waterways can result in a reduction of animal life and a change in character of the stream or waterway. The addition of heat is called thermal pollution.

Suspended solids are usually mud, clay and soil particles and occasionally algae and bacteria. Some minerals including silica (sand, SiO_2), calcium carbonate ($CaCO_3$) and iron oxides may be present. The yellow colour of floodwater and runoff is often due to suspended mud, colloidal clay and iron minerals. Collectively, the iron minerals are called ochre.

The increased turbidity of water results from erosion, industrial discharge of dirty water and sewage effluent. A large population of algae can also cause turbidity: recreational activities such as the use of powerboats in shallow water can stir up bottom sediments and have the same effect. The rate of settlement of a special disc, the Secci disc, is widely used to determine the turbidity of water. Another popular method measures the light that passes through a given amount of water. Turbid or murky water has a low transmittance of light. The actual amount of suspended solids present is measured gravimetrically following removal by filtration and weighing of the dry material present on the filter paper.

Special Problems from Water Pollutants
Traditional organic waste from human and animal excreta, and agricultural fibre from harvested plant debris, present a significant problem. While fully biodegradable, excessive amounts of organic waste can cause oxygen depletion of lakes and rivers. Human excreta contain some of the most pathogenic and persistent organisms as the water-borne agents for cholera, schistosomiasis and dysentery.

In areas of intensive animal farming, ammonia released from manure is partly introduced into the atmosphere and partly converted by soil microbes into soluble nitrates which enrich the soil. Nitrates are soluble in water and do not bind to soil particles. Hence they have high mobility and can become one of the main pollutants of groundwater. Levels of nitrates in soil often breach guidelines established by the WHO, especially in areas where there is heavy use of nitrate and urea-based fertilisers.

Industrial wastes can include heavy metals and considerable quantities of synthetic chemicals such as pesticides. These materials are characterised by toxicity and persistence: they are not readily degraded under natural conditions or in conventional sewage treatment plants. On the other hand, such industrial materials as concrete, paper, glass, iron and certain plastics are relatively innocuous, because they are inert, biodegradable in time and at least nontoxic.

Wastes can enter lakes and streams in discharges from sources such as sewers or drainage pipes, or from diffuse sources such as pesticides and fertilizers from runoff water. Wastes can also be carried to lakes and streams along indirect pathways; for example when water leaches through contaminated soils or minerals tailings dams, transporting contaminants to a lake or river. Dumps of toxic chemical wastes on land have become a serious source of ground-and surface-water pollution. The incidents at Lekkerkerk in the Netherlands and at Love Canal in the US, are indicators of such pollution going on worldwide in thousands of chemical-waste dumps.

Some pollutants enter the water cycle by way of the atmosphere. Probably best known is the acid that arises from the emission of nitrogen oxides and sulphur dioxide by industry and motor vehicles. Acid deposition, which can be 'dry' (as when gases make direct contact with soil or vegetation) or 'wet' (when dissolved in rain water), is causing acidification of low alkalinity lakes throughout the industrialised world. Acid precipitation also leaches certain positively charged ions from the soil and in some rivers and lakes ions can

reach concentrations sufficient to kill fish. Wind can also carry pollutants —
fly ash from coal-burning plants or sprayed pesticides. These can be borne
great distances to be deposited eventually on the surfaces of lakes and rivers.

Another recently recognised aspect of water pollution is the accumulation
of heavy metals, nutrients and toxic chemicals in the bottom mud of deltas
and estuaries of highly polluted rivers. Here, the high pollutant content of
sediments means the marine life is at risk, particularly those sand and sediment feeders, when dredging of sediment for landfill exposes them to people
and animals.

People in the Third World

People in the Third World face different problems in their need for and use
of water than those in the developed world. Most cities in Africa and many in
Asia, such as Dakar, Kinshasa and Chittagong, lack sewerage of any kind.
Human excrement and household waste tend to end up in streams, gullies and
ditches from which people may be drawing their drinking and washing water.
At the end of the 1970s, UNICEF was reporting that a total of 1.2 billion
people in developing countries alone were without a safe supply of drinking
water and 1.6 billion without any proper means of waste disposal.

Piped water, when available, is frequently from centrally located stand-
pipes or pumps which operate for only a few hours daily. In Nairobi, Djakarta
and Bangkok, families are forced to purchase water from a vendor, paying ten
times the rate charged to houses with mains connections. In 1987, a team from
the London School of Tropical Hygiene and Medicine reported that, in the
Sudan, half the average household income was spent on water.

As city populations expand, water and sanitation services are being put
under further pressure. But lack of funds, together with graft and corruption,
delay improvements. Ever present fears of epidemics such as cholera and
typhoid alone maintain pressure for action among bureaucrats.

Pollution in the North Sea

Discharges of fertilisers, detergents and sewage entering the North Sea are
causing serious concern among European nations. The oceanography of the
North Sea is such that pollutants are trapped along the continental coast
where they encourage massive proliferations of algae. Findings from the
North Sea Project, a major study commissioned by the Natural Environment
Research Council for Europe, suggest that significant blame lies with the
nutrients such as nitrates and phosphorus.

In the southern part of the North Sea, a large proportion of the nutrients
in the water comes from sources on land. Close to the coast, more than half
the nutrients are discharged from the great rivers of Europe: the Rhine, Elbe,

Weser, Scheldt, Thames and Humber. In the inner German Bight between 70 and 80 per cent of nutrients result from human activities.

Where the great European rivers discharge into the sea, the water is relatively fresh and clear and does not mix with the saltier water that is further offshore. Algae thrive in this water, and in the presence of high concentrations of nutrients they reproduce at a phenomenal rate. Some of these species are poisonous to fish. At the same time, the algae growth is unpleasant for summer bathers. When the algae die, bacteria break down the corpses and this process consumes the available oxygen at the expense of fish and other species dependent on that oxygen supply.

Britain does not suffer from the same problems as the European countries because currents carry pollutants away from their source. The circulation of water in the North Sea carries some of Britain's pollution across to the Continental coast, particularly into the Dutch and Belgian waters. Persistent movements of currents from the coast of East Anglia towards the Netherlands carry sediments which are often contaminated with heavy metals from industrial waste and sewage, and organic pesticides as complexes bound to particles of mud and silt. Recent studies from the universities of East Anglia, Essex and Liverpool have found that in summer the algae in the North Sea are responsible for almost 25 per cent of the acid rain falling over Europe.

Algae give off dimethyl sulphide (DMS) which oxidizes in the air to form sulphur dioxide, the principal cause of acid rain. Measurements taken along the tracks of vessels chartered to monitor river pollutants, show that the highest concentrations of DMS correspond to the largest populations of algae. The addition of phosphorus encourages the species of plankton that produce the most DMS.

Guarding Water Quality
Few are fortunate enough to live where there is access to fresh, pure water from a bubbling mountain stream free of leached minerals and trace metals, natural plant debris, animal wastes and droppings. In less developed societies and in many non-urban areas of all countries, a regular supply of water may depend on use of rainfall tanks, ponds, small earth dams and weirs. Underground stores may offer water brought to the surface by artesian bores, in wells or by natural phenomena such as mineral springs and geysers. Often of an indifferent quality, this water may require extensive treatment prior to drinking. Dissolved calcium, sodium and magnesium salts contribute also to taste, 'feel' and appearance.

Problems with Britain's Water Supplies
In late 1989, supplies of drinking water in Britain began to cause considerable

public alarm. This coincided with government reports warning that *Crytosporidium*, a parasite in animal waste which causes a form of diarrhoea, is much more common worldwide than previously suspected in industrialised as well as developing countries.

These findings followed upon Britain's largest recorded outbreak of crytosporidium-caused disease early in 1989, and which affected hundreds of people living in and around Oxford and Swindon in southern England. A panel of scientists and environmental health specialists has since been set up to consider the wider implications of the outbreak. The panel will also recommend measures which responsible authorities such as Thames Water and Yorkshire Water can institute to ensure such bacterial infestation does not recur.

Crytosporidium is a parasitic organism that lives in the guts of animals, commonly calves and sheep, and is passed on in bodily wastes. Its spores, called oocysts, remain active for up to eighteen months. People can catch the disease through contact with animals' excrement and by eating raw meat.

There is no known cure for the disease. The oocyst attaches itself to the wall of the gut where it reproduces, causing stomach pain and diarrhoea. The disease can make children and babies very ill with an acute and often pro-tracted form of gastroenteritis.

People with weak or damaged immune systems, including people with AIDS or leukaemia, are also at serious risk. At least one person with AIDS in Britain has died of crytosporidiosis. Information from the Rhyl Public Health Laboratory in Wales suggests that the disease is probably the most common gut infection to affect AIDS patients worldwide.

The micro-organism is ubiquitous. The disease that it causes is now emerging as one of the most ignored causes of severe gastroenteritis. Scien-tists have only recently discovered that water supplies are an important medium for transmitting the disease and that slurry from farm animals is often to blame.

However, the problems of *Crytosporidium* in industrialised countries are not confined to Britain. Scientists in the US are presently monitoring water courses and supplies following an outbreak which affected 13,000 people in Carroltown, Georgia. Microbiologists from the University of Arizona report finding oocysts of *Crytosporidium* in quantities ranging between 14 and 57 per cent of all the samples of surface water taken from western US. Scientists from the US Environmental Protection Agency and from Thames Water are using genetically engineered monoclonal antibodies which, cloned and purified, are marked. These bind with proteins on the oocysts and the com-plex so that they can be detected and its levels monitored.

Blue-Green Algae — Another Major Concern

A potential threat exists to human health through the contamination of water supplies from the overgrowth of cyanobacteria species commonly classed as blue-green algae. These simple organisms are rudimentary plants capable of fixing nitrogen and dependent on energy produced by photosynthesis. They thrive in still water and favour warm humid conditions such as occur in temperate climates, including those of central and southern Europe, parts of Britain in mid-summer, of Africa and large parts of North America.

The risk to water quality arises from high levels of phosphorus and nitrogen present in the nutrient medium. This occurs as a result of runoff from the application of phosphate and nitrogen fertilisers in agriculture and from products in daily use by domestic and industry consumers. Examples of the last include detergents and cleaning agents which contain phosphates. These wastes re-enter the water circulation either through catchment rain-water or after water treatment.

The problem with cyanobacteria or blue-green algae lies in their association with acute problems such as diarrhoea and vomiting, and their link with the development of birth defects and the acceleration of cancers of the gastrointestinal tract.

Dr Ian Falconer of the University of New South Wales, Sydney has been conducting research on the cancer risk. Using water collected from New South Wales inland farm dams, of which at least two-thirds are believed to be affected, Dr Falconer has found that laboratory mice will die within twenty-four hours of drinking affected water and that the survivors are more prone to develop cancer. When tested on mice with small tumours, the tumour growth has been increased by a statistically significant amount over that shown by mice given water from sources free of algal contaminants.

The threat to human health results from deliberate or accidental drinking of affected water, possibly as a result of swimming or bathing in a dam or river which is heavily polluted and carries large quantities of blue-green algae. These simple plants are not always immediately apparent or detectable though present in potentially dangerous amounts. A characteristic smell and colour is apparent when a sample of the water is left to stand.

Concerns to date have been focused on the acute gastric effects induced by drinking affected water, usually as a result of swimming. These findings bring new urgency to efforts in Europe to reduce blue-green algal contamination of large riverine systems. In Britain almost all of the inland canals are also now said to be affected by the overgrowth of algae, which are flourishing in the warmer conditions of the last few years and the rich nutrient mix of the less than pure canal water.

Preventive action has been initiated in many countries. In Europe, attempts are being made to publicise the problem and to alert consumers. Pressure is being exerted on cleaning agent manufacturers to produce new products such as household washing-up detergents, free of phosphates. The use of minimal quantities of super-phosphate fertilisers rather than medium and maximum doses and of crop rotation techniques is being promoted to farmers. Another initiative includes discussions between horticulturalists and biotechnologists. They are talking about the feasibility and likely costs associated with cloning specially designed plants to remove the phosphates and nitrogens from water supplies.

Despite the problems of blue-green algae and *Crytosporidium* in countries such as Australia, not only is there a lower level of awareness, but the problem of water quality generally ranks for low priority with the National Health and Medical Research Council, the country's premier regulating body for public health matters. Presently no controls restrict the quantities of agricultural nitrates and phosphates which enter the water supply. Their presence stimulates algal growth leading to expanses of thick, lush green algae called an 'algal bloom'.

Threats to Water Supplies in Cities in Developed Countries
An example of the threats to water supplies posed by blue-green algae and coliform faecal bacteria can readily be found in Australia. In that country, consumers and recreational users are only now mobilising to exert pressure on companies and authorities to enact tougher guidelines concerning water quality. This has been directed at two main areas: first, reducing pollution of bathing beaches and popular swimming spots along the eastern coast, particularly those centred on Sydney's important surfing beaches which are popular with tourists. The second has arisen in Adelaide in central South Australia.

In the first instance, the disposal of primary treated sewage from ocean outflows within five kilometres from Sydney's shoreline has meant that, when onshore breezes prevail during summer months, there is a serious risk of faecal material and other sewerage debris being washed onto the beaches. In recent years, workers' strikes at local sewerage treatment plants have resulted in raw untreated material being discharged for short periods, exacerbating the pollution problem.

Swimmers and surfboard riders not only experience visual pollution as faeces and rubbish float on the surface, but are likely to suffer stomach upsets, vomiting and/or diarrhoea from accidental swallowing of the water. In Australia also, the problem is added to by the discharge of stormwater drains onto beaches and streams which enter the ocean. Domestic and street refuse, garden refuse and stormwater from suburban areas which are closely settled pollutes water supplies further.

In Adelaide, the problem is different. Unique among Australian cities, Adelaide lacks quality underground water and is not located on a source of fresh river water. The nearest river is the Murray which enters the sea some 130 kilometres east of Adelaide. At this stage it is carrying a heavy burden of mineral and agricultural pollutants gathered in its 1500 kilometre journey to the sea. The water is also very salty and irrigation along the Murray has meant that land beside the river is encountering salinity problems. As a result Adelaide depends for its drinking water on supplies from Lake Victoria further to the east. Yet a new threat emerged to that supply in summer 1990.

Eutrophication often follows an algal bloom. When there are too many algae growing in a confined space, the lower plants receive insufficient sunlight, die, sink to the bottom and decompose. The decay consumes dissolved oxygen which kills the oxygen-breathing animal life and the aerobic bacteria. The decay is continued by anaerobic bacteria which produce methane, ammonia and hydrogen sulphide (rotten-egg gas) so characteristic of stagnant streams and polluted waters. Eventually the remaining algae die, but the water retains its anaerobic character and is barren of its normal non-pathogenic simple organisms, of plant and animal life.

Algal blooms in Lake Victoria, which supplies water to Adelaide, continue to threaten the quality of water and algal overgrowth continues to pose a threat. Instances have arisen, following warm humid periods, of increasing levels of agricultural phosphate and nitrogen recorded in the lake's water. Blame for the pollution has been sourced to the Albury-Wodonga township, and attempts have been made to encourage domestic and industrial users to dispose of excess and waste water by re-cycling, including watering systems for parks and domestic gardens. Health authorities are also promoting the value of land disposal of excess water to farmers using phosphate and nitrogen based fertilisers.

The health risk centres particularly on young children and the elderly who are at risk of acute episodes of gastroenteritis. The ideal growth medium and conditions created in the Lake also encourage the growth of other disease-inducing bacteria such as salmonella. Accidental contamination may result in multiplication of the bacteria and a major public health problem.

Redressing Shortages of Water

Local water shortages can be solved in two ways. The supply can be increased either by damming rivers or by consuming capital — by mining groundwater. Known supplies can be conserved by increasing the efficiency of irrigation or by relying more on food imports. Nonetheless and despite efforts at conservation, water is becoming increasingly scarce as population, industry and agriculture expand. Severe shortages occur as demand exceeds supply. Depletion of groundwater is common, for example, in India, China and the US. In the CIS (formerly USSR), the water levels of both the Aral and Lake Baikal have dropped dramatically as a result of agricultural and industrial growth in these areas. Contentious competition for the water of such international rivers as the Nile, the Jordan, the Ganges and the Brahmaputra is a symptom of the increasing scarcity of water. Decrease in water quality further exacerbates the problems of scarcity.

Action on Water Quality Control and Management
To a large extent, the task of making water drinkable depends on the cause. Skimming removes floating debris: standing over time followed by decanting and filtration removes suspended particles and soil material. Boiling will vaporise any toxic dissolved gases and kill disease-carrying micro-organisms. Resins remove additional organisms and, depending on their nature, a wide range of minerals and trace elements which give the water a characteristic flavour and taste. Further treatment can involve, for example, bubbling chlorine gas into the water or adding the chemicals calcium hypochlorite or sodium hypochlorite.

Thus, in modern industrialised countries, the taste and appearance of water piped to the home depend on its source, storage and treatment. The extent of treatment depends not only on the quality of the initial water but on local community demands.

Quality of inland waters poses a special problem which depends not only on the amount of waste generated but also on the decontamination measures that have been put into effect. The degree of success in the battle for water quality differs from country to country, but it can be generalised into a conceptual formula proposed by Walter Stumm and his co-workers of the Swiss Federal Institute for Water Resources and Water Pollution Control in Zurich. The formula holds that the contamination load of a river basin is based on the population in the basin, the per capital gross national product, the effectiveness of decontamination and the amount of river discharge.

Consequently, actions have been taken to improve the quality of water, both locally and on a country-to-country basis. Some action has crossed national boundaries. A particular example is evident in Europe, where the Rhine River presents a unique example of industrial, agricultural and human waste disposal through a water system. The Rhine runs 1320 kilometres from the Alps to the North Sea and drains a vast basin in four countries — Switzerland, West Germany, France and the Netherlands. The basin is heavily industrialised and the river accumulates and transports into the Netherlands a heavy load of pollutants. Since 1980 the amounts of some pollutants have been reduced. Now the four countries are cooperating in a Rhine Action Plan intended to improve the quality of the river's water. The primary effort will be to institute recycling within industry as a substitute for after-the-effect 'end of pipe' treatment.

Suggested measures range from the sale by governments of pollution licences, to increasingly stringent regulations and policing backed by heavy fines for incursions. Revenue collected would be directed into research into recycling and more efficient and non-polluting methods of production. The system is intended to encourage industry and commercial users of water, including farmers and local authorities, to move to the adoption of more efficient water screening and cleansing measures and to adopt alternatives to non-degradable and toxic water pollutants.

Water Chlorination

The most noticeable chemical added to water is chlorine, which has a characteristic smell and flavour. It is added to minimise the risk of disease associated with water-borne bacteria such as those responsible for cholera and typhoid. Recent evidence, however, suggests that unless there is a significant risk of epidemic and poor hygiene and sanitation, human health is adversely affected by chlorinated water.

The process of water chlorination involves a relatively simple chemical process. When added as a gas or as hypochlorite salts, free, available chlorine is formed which possesses bactericidal and bacteriostatic activity. This is believed to be associated with the ability of the free chlorine to bind the structural proteins of bacterial cell walls, thereby resulting in their breakdown. Alternatively, oxidation of essential bacterial enzymes may occur, resulting in death.

Either an excess or a deficiency of adequate levels of chlorine is responsible for the odour and taste commonly associated with chlorinated water. In the latter instance, the reaction of chlorine with organic materials in the water, such as algae and leaves, will result in the characteristic odour and taste of chlorine. At higher levels of concentration, there is complete breakdown of the organic material with the result that the water is almost free of odour. The impression is given that, in contrast, the chlorine level is below that required and additional chlorine may be added. It is for this reason that close monitoring of chlorine levels is required prior to routine addition of the chemical. Filtration also accompanies chlorination in order to reduce the presence of breakdown products and substances which will create an added demand for chlorine.

Health Risks Associated with Chlorination: Health risks associated with chlorine compounds range from reports of allergies to atherosclerosis, infertility and cancer. They include claims that mutagenic and carcinogenic damage may follow dependence on chlorinated water.

As early as 1951, the *Journal of the American Medical Association* cited cases of asthma relieved after patients had returned to drinking distilled or unchlorinated water. In the same year, Sinclair suggested a possible involvement of chlorine in etiology of heart disease. Epidemiological evidence provided by Pataki from New Jersey, USA, and Passwater referring to the contrast in the levels of heart disease in north and south India further confirmed that view.

In 1974, Price, working in Michigan, USA also linked an increased incidence of heart attack and stroke with the introduction of chlorinated water in 1904. Finally, Benditt provided clinical evidence to support a link between plaque formation in the arteries with chlorinated water. His research suggested that chlorine-induced mutations and the presence of potentially mutagenic and carcinogenic substances such as chloride ions in the blood, could be responsible or associated with the proliferation of cells which resulted in the formation of arterial plaques.

By the mid-seventies, public concern had been raised so that US health authorities began to monitor more carefully the levels of chlorine in water to determine if there might be levels below which its presence and routine use might be regarded as free from risk. At the time, concentrations in different American cities were said to range from a low of 0.1 ppm to as high as 311 ppm. Yet opinion concerning the possible risk of infection and disease in the absence of chlorination continued to hold sway.

Since that time, additional research and evidence has documented the wide range of toxic substances in drinking water which can be directly traced to chlorination. Further weight was added to these findings with reports that laboratory animals on chlorinated water supplies had showed decreased fertility, an increased failure of early implantation of conceived ova, low birth weight and decreased perinatal survival of the young. Short-term tests for the carcinogenic effects of several chlorinated compounds were also shown to yield positive results, confirming earlier findings linking chlorination with cancer.

Chlorine and The Skin: Prolonged exposure to chlorinated water for long periods of time is experienced daily by elite swimmers who may spend up to four hours a day training in a chlorinated pool. Others living in areas where normal water supplies are chlorinated use it for washing and showering. Essentially they are exposing the skin to large volumes of dilute chlorine solution. Some of this chlorine reacts with the oils in the skin to form chlorinated compounds and these may then be absorbed by the body. Recent research suggests that chlorine absorbed through the skin may even cause depletion of vitamins and essential fatty acids.

The strong oxidizing power of the chlorine is responsible for the odour which lingers on the skin. Regular exposure to chlorinated water also serves to promote ageing of the skin not unlike that caused by extended exposure to sunlight. Moreover, chlorine also enhances the ageing affects of ultraviolet radiation by reinforcing the process of cell deterioration.

A further final factor for consideration is the role of chlorine as an antibacterial agent on the skin surface. When hot water is used in bathing, greater amounts of free chlorine are released meaning that a concentration as high as 600 ppm chlorine can be experienced during a ten-minute hot shower. This destroys the natural bacteria necessary to maintain the normal pH and ecological balance important to skin integrity.

Water Fluoridation
The assumption has been that fluoridation is effective and safe. 'Irrational' opposition to the procedure is explained as resulting from lack of education, prejudice or feelings of alienation and powerlessness. However, it is now believed by many that the harmful effects of water fluoridation are more real than is generally admitted, while the claimed dental benefit is negligible and most recent study supports the observation that fluoride ingestion is more strongly related to fluorosis (areas of softened and porous enamel) than to dental caries.

Fluoridation of mass public supplies of water is routine in developed countries and endorsed by most authorities. The procedure is based on research which established that in those towns and population centres where natural water supplies contained high levels of fluoride, a low level of tooth decay (dental caries) was the norm. This research was confirmed by the North American Fluoridation trials conducted during the 1930s on hundreds of communities in the United States.

The New Evidence against Fluoride
One undeniable side effect of fluoridation is dental fluorosis, a kind of teeth mottling caused by ingesting excess fluoride during the ages when teeth are developing. Recent studies report that in fluoridated communities dental fluorosis affects 25-50 per cent of children, some so severely that their teeth are discoloured and pitted. Only 10-12 per cent with barely detectable mottling had been predicted.

In some places, parents or authorities prefer to subsidise or encourage regular supplements of low-dose fluoride tablets, particularly by infants and children under 5 years of age. Reports of excessive tooth mottling in children who had received fluoride supplements have led to drastic reductions in the recommended doses of such supplements. However, the fluoride concentration of fluoridated drinking water has not been similarly reduced.

Critics of fluoridation have difficulty believing that only tooth-forming cells are damaged by fluoride and suspect that other adverse effects have been similarly underestimated. There is evidence of more general systemic damage concerning development of long bones and the integrity of the skeletal bones.

Furthermore, re-examination of the original research and material on which fluoridation procedures were based suggests that the positive results were obtained in small selected samples and that there was also exclusion of less supportive data from other centres.

Toxic Chemicals and Water

Sydney University Ocean Sciences Institute researchers who conducted tests on prawns, the sea bed and sediments in the Torres Strait and Papua New Guinea's Fly River, have detected levels of cadmium and copper which far exceed those recommended by the National Health and Medical Research Council. The amount of copper in the prawns was more than twice the NHMRC standard of safety. The cadmium concentration was considered to be more moderate but still of a level sufficient to affect the reproduction and growth of prawns. While other heavy metals such as mercury and lead remain to be studied, the research team predict similarly high levels.

Cause of the pollution is traced to the commercial mining operations of the giant Ok Tedi gold and copper mine. Unwanted by-products and washings are swept down the Fly River and enter the Strait. In October 1989, the PNG Government set environmental concerns aside to allow Ok Tedi Mining Ltd to continue to dump waste from the mine into the Fly River. The company had claimed that it could not afford the cost of building an interim or permanent tailings dam for the mine wastes. A temporary dam had been washed away in 1984.

Dr Peter Harris, a research fellow attached to the Institute, emphasised that the pilot research had indicated that further investigation was needed to prevent damage to the food chain, the marine environment, and the prawn industry in the Torres Strait which is valued at $14 to $18 million a year. Since the diet of Torres Strait islanders and people living in the Fly River catchment area consists largely of seafood, their survival could be jeopardised if there was any threat to the food source. In addition, large quantities of seafood come from the Torres Strait and are sold in Australian fish markets, air-freighted to Japan and sold in other parts of South-East Asia.

Presently there is no official or regular monitoring for pollutants. The presence of toxic levels of heavy metals represents a potentially serious problem to health through nervous system damage, birth deformities and effects on the muscular coordination mechanisms.

TABLE 5.18: Common Pollutants of Water	
Class	*Members*
animal and vegetable wastes	faeces, food debris, leaves, garden refuge, soil, spent flowers, branches, twigs, etc.
synthetic chemicals	pesticides, polychlorinated biphenyls, detergents, industrial waste and effluent, accelerated eutrophication
petrochemicals	
minerals, metals and salts	mercury, heavy metals and mining wastes, silt, radioactive wastes, salinity problems
solid wastes	
thermal pollution	
living organisms	blue-green algae, especially cyanobacteria, salmonella spp.
human recreational wastes	garbage bags, plastic foam containers, glass bottles, splinters

References and Additional Reading

Ashton, J.F. & Laura, R.S. 'A Hundred Years of Water Chlorination' from *Nature & Health*, 1989, vol.10, no.4, pp.44-9.

Benditt, E.P. 'Atherosclerosis May Start with Cell Proliferation' from *J.Amer.Med.Assoc.*, 1974, vol.227, no.7, p.734.

Bruchet, A. *et al.* 'Characterization of Total Halogenated Compounds During Various Water Treatment Processes in Water Chlorination' from *Chemistry Environmental Impact and Health Effects*, vol.5, Levis Publications, Michigan, 1984, pp.1160-74.

Cutress, T.W., Sucking, G.W., Pearce, E.I.F. *et al.* 'Defects in Tooth Enamel in Children in Fluoridated and Non-Fluoridated Water Areas in the Auckland Region' from *NZ Dental Health*, 1989, vol.85, pp.2-8.

Dean, H. T. 'Mottled Tooth Enamel in the United States' from *Public Health Reports*, 1933, vol.48, pp.703-34.

Dean, H.T., Arnold, F.A. & Elvove, E. 'Additional Studies of the Relation of Fluoride Domestic Waters to Dental Caries Experience in 4,425 White Children aged 12 to 14 Years, of 13 Cities in 4 States' from *Public Health Reports 1942*, vol.57, pp.1551-679.

Dean, H.T. & Elvove, E. 'Some Epidemiological Aspects of Chronic Endemic Dental Fluorosis' from *Am.J.Pub.Health*, 1936, vol.26, pp.567-75.

Dean, H.T., Jay, P., Arnold, F.A. & Elvove, E. 'Domestic Waters and Dental Caries II: A Study of 2,832 White Children aged 12-14 Years, of Eight Suburban Chicago Communities' from *Public Health Reports 1941*, vol.56, pp.761-92.

Diesendorf, M. 'The Health Hazards of Fluoridation: A Re-Examination' from *Int.Clinical Nutrition Rev.*, vol.10, pp.304-21.

Maurits La Riviere, J.W. 'Threats to the World's Water' from *Scientific American*, September 1989, p.48.

New Scientist, 29 July 1989, vol.1675, p.6.

Pain, S. 'Pollutants Collect on Continental Side of North Sea, Study Finds' from *New Scientist*, 11 November 1989, p.6.

Passwater, R.A. *SuperNutrition for Healthy Hearts*, Jove Publications, New York, 1978, pp.155-6.

Price, J.M. *Coronaries, Cholesterol, Chlorine*, Pyramid Publications Ltd, Llanridloes, Wales, 1984, pp.32-3.

Sinclair, H.M. as cited by Clark, L. in *Get Well Naturally*, ARC Books, New York, 1971, p.327.

Smith, M.K. *et al.* 'Developmental Toxicity of Halogenated Acetonmitriles: Drinking Water By-Products of Chlorine Disinfection' from *Toxicology*, 1987, vol.46, pp.83-93.

Szupunar, S.M. & Bart, B.A. 'Dental Caries, Fluorosis and Fluoride Exposure in Michigan School Children' from *J.Dent.Res.*, 1988, vol.67, pp.802-6; Leverett, D.H. *J.Public Health Dent.*, 1986, vol.46, pp.184-7.

Trehy, M.L. & Bieber, T.I. 'Detection, Identification and Quantitative Analysis of Dihaloacetomitriles in Chlorinated Natural Waters' in Keith, L.H. (ed). *Advances in Identification and Analysis of Organic Pollutants in Water*, Ann Arbor Science Publications, Michigan, 1981, pp.932-44.

Wicke, L. *Die Okologischen Milliarden*, Kosel Verlag, Germany, 1986.

6.0 INTERVENTIONS

6.1 Interventions for Environmental and Human Health

The World Health Organization defines health as a state of total physical, social and emotional well-being. Perceived in the same comprehensive and holistic framework, the health of individuals and of human communities is an indication of its relative success, or failure, in coping with and adapting to the environment. The extent to which the human species is healthy will depend on conditions within the environmental niche it has carved out to maximise its survival. Thus, human health now and in the future will reflect our potential, as a species, to continue to accommodate climate change, depletion of the ozone layer, and a decline in air quality. These issues are equally as important and critical to survival as the existence and renewal of basic productive resources such as timber, oil and soil.

Intervening for Health
Human beings have a propensity for 'tinkering'. Others call it investigating, enquiring or seeking solutions. Scientists recognise these activities as 'interventions' which are justified in the course of a disease through a desire to improve chances of human survival. Sometimes they succeed; other times they fail. At still other times, the results are inconclusive.

This Section called 'Interventions' describes areas where humans have intervened in the past in an effort to improve their health. In the first instance, this intervention has occurred through public health practices focused at the community level. To a significant extent, these measures have improved life chances and the quality of life lived. In the second instance, intervention, ostensibly for benefit, has been more ambivalent and controversial. Two examples are cited: pharmaceuticals and pesticides.

Finally, a form of intervention which may, ultimately, be necessary to ensure not only the survival of the human species, but the health of those living and the future of their descendants. The issue in this instance is population control or management: one on which concerted and collaborative action is vital on a global and local scale, and one with political, moral and religious overtones that may be beyond our capacity to resolve.

Suggested Interventions for Addressing Threats to the Environment
In looking to the environment, the context in which individuals experience good or bad health, some workers have proposed intervention of a technical nature. Measures such as deliberately spreading dust in the upper atmosphere would reflect sunlight, but no one can be sure of the implications of such action. Firing rockets into the stratosphere to release vast amounts of ozone has been suggested as a way to reduce ultraviolet radiation reaching earth's surface.

Adaptive strategies recognise the risks of climate predictions and measures proposed to meet them, and point to the potential of spending large sums to solve situations which may never arise. They argue that adaptation, in contrast, is cost effective and forces responsibility for action into the hands of those people who caused the problem. They suggest that infrastructure can be modified, water supply systems, coastal structures, buildings and air quality standards can be altered, to meet changes in the environment as they occur.

Passive adaptation, however, relies mostly on reacting to events as they unfold. Anticipatory or preparatory action, such as detailed by Pearce (his chapter on Policy), could be undertaken to minimise the need for reactive steps in the future when a need is apparent and action required.

Preventive policies are also a form of intervention between cause and effect. In an environmental sense, they tend to focus on measures to limit the effects of greenhouse gas buildup. In human health, they take the form of public health measures. Environmental interventions can be: energy conservation, increased encouragement and use of alternative sources, halting production of CFCs, tax policies to encourage use of cheaper non-polluting sources of fuel, and measures to police and halt deforestation, especially in the Third World. However, these measures need to be undertaken globally and involve commitments and actions by all nations. Public health initiatives have the advantage in that they are best taken at the global level, for example through WHO setting the agenda, for subsequent application at the local level by individual countries and governments.

Intervention as a Public Health Concern
Medical care and public health measures are concerned with intervening between disease and its inevitable outcomes, such as ill-health, disability and death, and to improve health practice by individuals and communities to minimise their future susceptibility.

Medical care can focus on disease control, lifestyle, the increasing distribution and use of health resources, the environment, or all of these. It involves a wide range of measures, from physical and mental fitness, nutrition, contraception, drink-driving laws, teenage pregnancy, abortion and *in vitro* fertilisation, to seat-belt legislation, toxic waste controls, housing policy for the elderly and air pollution controls. These reflect community responses to historical trends and statistical data concerning different communities at varying levels of social, economic and political development.

The Health of Human Communities

The sanitation of the environment, control of communicable infections, organization of medical and nursing services for the early diagnosis and prevention of disease, the education of the individual in personal health and the development of the social machinery to assure everyone a standard of living adequate for the maintenance of health, are matters essential for the survival of modern human communities.

Public health is the science and art of preventing disease, prolonging life and promoting health and well-being through organized community effort. Accordingly, public health practice is centred on environmental health and disease control. It also overlaps aspects of medicine, nursing, school health, personal health and health promotion. Indeed, public health tends to be associated with the work of official government health agencies and is centred on population-based strategies for improving environmental conditions to maximise human health.

Interventions to improve the health of individuals focus on four distinct areas. High-risk individuals, families or groups, decision makers and whole communities are the target for education programmes, which are conducted by the mass media, schools, industry, governments and other organisations. Social interventions take the form of economic, political, legal and organizational changes including the organisation of health care services designed to support actions conducive to health. Environmental supports include the structure and distribution of physical, chemical and biological resources, facilities and substances required for people to protect their health.

The health behaviour of a community includes the actions of the people whose health is in question and the actions of community decision makers, professionals, peers, teachers, employers, parents and others who may influence health behaviours, resources or services in the community. A number of other authors have also described the series of public health initiatives taken over time, to improve human survival against disease.

Public Health Measures Over Time

Excavations have revealed that the Egyptians had community systems for collecting rainwater and for disposing of sewage. Herodotus, in the fifth century BC, described the hygienic customs of the Egyptians. Personal cleanliness, frequent baths and simple dress were emphasised. Earth closets were in general use.

Personal health and well-being was also of concern to the Greeks and Romans, and alcoholic intoxication was considered particularly loathsome, socially disruptive and the cause of personal injury and abuse. One of the oldest temperance tracts advocating moderation was written in Egypt *c.*1000BC. Similar sentiments later appeared in Greek, Roman, Indian, Japanese and Chinese writings. The Old and New Testaments reveal a tolerance of alcohol but denounce excessive drinking and drunkenness.

Early Hebrew society extended Egyptian concepts of disease but their approach centred on personal health-care practice, viewed as a community duty and to ensure its well-being. To Hebrews, observing a weekly day of rest was a health measure equally as much as a religious duty. Family relations and sexual conduct were directed to the best interests of the person, the family and the community. While the Hebrews had rather crude concepts about disease, they did make concerted efforts to prevent its spread. They were among the first to segregate lepers, a practice which is recorded in Leviticus. Recognition that eating pork sometimes resulted in illness led the Hebrews to regard pork as unclean and to forbid it in the diet.

The Mosaic law provided for personal and community responsibility for health, maternal health, the control of communicable diseases, and the decontamination of buildings, for example by fumigation. Water supplies were protected and wastes disposed of through community-organised measures. Lepers were segregated and food protection measures instituted in times of disease or war. Campsites were always scrupulously cleaned to remove wastes from the immediate environment.

The Greek era in history extended over many centuries, but the Classic Period, *c.*460-136BC, saw the time when the Greeks benefited from their approach to personal health, particularly its physical aspects. Games, gymnastics and other exercises were directed toward their ideal of physical strength, endurance, dexterity and grace. Harmonious development of all facilities was the guiding philosophy, and exercise was supplemented by measures in personal cleanliness and dietetics. The Classic Period focused on the individual, thus very little attention was given to environmental sanitation. Yet it could be claimed that it was Hippocrates who wrote the definitive treatise on environment and health, the trilogy *Airs, Water and Places*.

With the destruction of Corinth in 146BC the health knowledge and practices of the Greeks became centred on Rome but changes occurred in approach. To the Romans, the individual existed merely to serve the state and the casualty was Greek hygiene practice. The Romans were expert tradesmen and engineers and their administrative skills were reflected in their many community health projects. The registration of citizens and slaves and the taking of a periodic census served to help in planning community health measures. Regulation of building construction, plumbing and drainage requirements were enforced, including laws to prevent sewage in the streets. Decaying goods were destroyed and poorly maintained buildings demolished.

Building regulations provided for ventilation and even central heating. Town planning served sanitation as well as other needs. Public sanitation was promoted through the construction of paved streets with gutters. Street cleaning and repair were standard procedures in the interests of sanitation, although modern health officials regard such measures as being of aesthetic rather than of direct health importance. Drainage networks carried off rain and other water, which was of some health significance. Removal of garbage and rubbish, although desirable in any society, was not of great health significance as the Romans contended. Public baths were promoted as community health measures.

Although street cleaning, garbage removal and public baths were of minor health value, several other measures promoted by the Romans contributed significantly to health as well as to the growth of civilization. Roman officials had sufficient understanding of health to provide a protected water supply for their cities. Water was brought to Rome from great distances via aqueducts, some of which are still incorporated into the city's water system. City sewer systems were built, portions of which are still operational.

The ability of its citizens to design and construct public water and sewer systems enabled Rome to grow to a city of 800,000 during the reign of Julius Caesar. The Greeks, who depended on family wells and private refuse disposal, were more limited in the size to which their cities might grow. Corinth at the pinnacle of its greatness had a population of only 35,000.

The sixth and seventh centuries saw the rise of Muhammadanism. Following the death of the prophet Muhammad, a series of pilgrimages to Mecca began, each of which was followed by a cholera epidemic.

Leprosy spread from Egypt to Asia Minor and then to Europe. Most nations decreed lepers unacceptable and stripped them of their rights as citizens. They were required to wear identifying clothing and to warn of their presence by a bell or a horn. This isolation, however, together with the early death of lepers, succeeded in virtually eliminating leprosy in Europe.

The Middle Ages: The early years of the medieval period of history (AD476-1000) are usually referred to as the Dark Ages. Western civilization was in a chaotic, almost formless state. The clergy were the only educated class and virtually the entire emphasis of the time was on the spiritual aspects of life. Rejection of the body and glorification of the spirit became the accepted pattern of behaviour.

Sight of the naked body was immoral and sinful. People seldom bathed, and they wore dirty garments. The use of perfumes appears to have stemmed from the attempt to conceal body and other unpleasant personal odours. The more neglected and abused the body, the more likely the individual to be wealthy, in high position or possessing of power. A legendary example of body neglect was St Stylites, who sat on top of a pole for sixteen years to expose his body to the abuse of the elements.

Food crops were often diseased. When harvested they were frequently stored improperly and riddled with vermin. By the time it reached the larger towns, the food had begun to rot. The poor quality food consumed, its foul smell and taste were ineffectively masked by the heavy use of spices. Gastro-intestinal diseases, poor health and chronic diarrhoea were rampant.

The later medieval period, from AD1000 to c.1453, is notable because of the severe pandemics of the time and the attempts made to deal with the spread of disease. The six great crusades to the Holy Land were staged between AD1096 and 1248, and they had a major effect on the health of Europe's population. In the first instance, to provide crusaders who were fit for the long journey, attention was directed at improving the individual's health.

While this approach resembled that of the Greeks by emphasising physical prowess, the general result was to improve the well-being of only one segment of the population. Adult males in the upper social echelons were generally very fit, healthy and robust. Unfortunately, many of them returned from the crusades weakened not only in spirit and strength, but suffering from the ravages of poor diet, bad living conditions and diseases. In addition, they carried with them diseases such as cholera and dysentery. Returning home, the poor sanitation measures at the domestic level tended to spread the disease and the subsequent death rate was correspondingly high among crusaders and families alike.

In 1348 bubonic plague, the Black Death, hit Europe. It traversed a devastating path from Asia to Africa, the Crimea, Turkey, Greece, Italy and into Europe and proved so virulent that it claimed the lives of one-quarter of the population. The death statistics in several of the large cities are sobering: Paris 50,000; Seine 70,000; Marseilles, 16,000 in 1 month; Vienna, 1200 daily; Florence, 60,000; and Venice, 100,000. The Italian writer Boccaccio (1313-75) reported that in the terrible outbreak of plague in Florence in that year, pity and humanity were forgotten. Families deserted their sick. In England, 2 million died, representing approximately half the total population of the country. London recorded 100,000 deaths, and for many years London's deaths exceeded its births. Had it not been for the influx of people from the rural areas, London's population would have declined still further.

While some believed these epidemics and pandemics were due to comets, famines, droughts, crop failure, insects and poisoning of wells by witches or Jews, discerning officials were somewhat better informed. Many recognized that overcrowding, poor sanitation and migrations played a role in the outbreak and spread of disease.

As a result, control measures were established in some cities. In 1377, at Rogusa, it was ruled that travellers from plague areas should stop at designated places and remain there free of disease for two months before being allowed to enter the city. Technically, this is the first official quarantine method on record. In 1383, Marseilles passed the first quarantine law and erected the first official quarantine station. In Venice the government appointed three guardians of public health, and in 1374 all infected or suspected to be infected travellers, ships or freight were denied entry to the city. In 1403, a quarantine of forty days was imposed on anyone suspected of having the disease.

Unfortunately, the measures to control spread of disease were not highly effective. The need for a scientific understanding of the cause and nature of diseases and their spread was evident. Scholars of the time who favoured the scientific approach to pestilence suffered surveillance and public persecution. As a consequence of such resistance, little progress was made in understanding plague and other diseases.

Renaissance: To the historian, the Renaissance in western and northern Europe applies to the period AD1453 to 1600, beginning with the fall of Constantinople to the Turks. From the standpoint of community health, the Renaissance was particularly important because of its movement away from scholasticism and towards realism, with the encouragement and revival of individual scientific endeavour. It ushered in a spirit of inquiry that would lead to the understanding of the cause and nature of infectious diseases. The fifteenth and sixteenth centuries produced such distinguished figures as Copernicus, Vesalius, da Vinci, Galileo and Gilbert.

By the middle of the sixteenth century, scholars had differentiated influenza, smallpox, tuberculosis, bubonic plague, leprosy, impetigo, scabies, erysipelas, anthrax and trachoma. Diphtheria and scarlet fever were not recognized as separate diseases but were understood to be different from all other diseases.

Fracasstorius (1478-1553), a physician from Verona, theorized in 1546 that micro-organisms caused disease. He recognized that syphilis was transmitted from person to person during sexual relations. Knowledge was advancing, but the increase in social concentration, expanding trade and movement of populations tended still to spread disease. Knowledge of communicable disease control lagged behind disease spread, and great plagues still harassed Europe.

Between 1600 and 1665 Europe suffered three severe pandemics of bubonic plague. The plight of London indicates the severity of the outbreaks. In 1603 a sixth of London's population died of plague. In 1625 another sixth was destroyed by the plague, and in 1665 one-in-five of London's residents died from the same disease.

This same era produced Descartes (1596-1650), Voltaire (1694-1778) and Boyler (1627-91). William Harvey (1578-1657) described the circulation of the blood through the heart; Thomas Sydenham made a differential diagnosis of scarlet fever, malaria, dysentery and cholera. Athanasius Kircher (1602-80), using the microscope with a magnification of 33 diameters, examined the blood of plague victims, and not long after, Anton van Leeuwenhoek, using a microscope with a 200 diameter lens, identified bacteria in scrapings from his teeth and found that these could be killed with vinegar.

Edmund Halley, the astronomer, compiled the first vital statistics of births and funerals, the Breslau Tables, in 1693; Plenciz, a Viennese physician, concluded that an organism of some kind was responsible for infectious disease, thereby predating the discoveries of Pasteur and Koch by a century. Edward Jenner, the British physician, showed scientifically that inoculation with cowpox virus can produce immunity to smallpox. Reports indicate that some form of smallpox vaccination had been practised in Turkey, but it took the observations of Jenner to further refine and develop an acceptable technique.

Colonial Period: It is significant that throughout history migrations have served as a vehicle for the spread of disease (Green and Anderson). As Europeans colonized the rest of the world from 1600 to 1800, the health of peoples living in the countries to which they migrated reflected the health problems of Europe. In many instances, native populations were ravaged because they had little resistance to these foreign diseases.

Large numbers of indigenous people in North America, Australia, Africa, Asia and South Africa were struck down by bronchitis, pneumonia, smallpox, influenza, and measles. However, interest in the health of people in the British, French, Spanish and Dutch colonies only attracted attention once the epidemics began to affect all social strata, including the privileged classes. At the time, standards of health were so poor that sanitation consisted merely of community tidiness or general housecleaning. Isolation, quarantine and harsher measures were instituted to contain the diseases, where regulations and standards had previously been ignored or were non existent.

Ironically, however, smallpox aided the European settlers of America.

Introduced to the east coast by Cabot and Gosnold, the disease eliminated so many Indians that the new settlers were able to colonize with little or no opposition. The white settlers seemed relatively immune. Nonetheless, some notable pandemics occurred in the Massachusetts Bay colonies in 1633, the New Netherlands settlement (New York) in 1663, and in Boston in 1752. Of Boston's population in 1752, only 174 citizens completely escaped the pox and 15,684 died.

During George Washington's lifetime, 90 per cent of the people who reached the age of twenty-one had had smallpox, and 25 per cent of those affected by the disease died. Smallpox also decimated the aboriginal populations of Australia, which had no resistance to this and other diseases such as influenza, measles and chickenpox introduced by their colonial white masters. Yet in 1980, just over two centuries later, the World Health Organization was able to announce that smallpox had been totally eradicated from the earth .

In the eighteenth and nineteenth centuries yellow fever became a bigger scourge than plague or smallpox. In 1793 Philadelphia had the greatest single yellow-fever epidemic in America. Of a population of about 37,000, more than 23,000 had the disease and over 4000 of these died. A citizen's committee appointed to deal with the problem drew up a set of regulations which included directions to avoid contact with persons with yellow fever, to placard all infected houses, clean and air sickrooms, provide hospital accommodation for the poor in order to prevent its greater spread, keep streets and wharves clean, and encourage general hygienic measures such as rapid burials, intemperance, adaptation of clothing to the weather, and avoidance of fatigue (Green and Anderson). Vinegar and camphor were used on handkerchiefs to prevent infection. Gunpowder was burned in the streets to combat the disease.

Simple and perhaps ineffective as these measures may have been, they served to increase awareness of the severity of the threat and represented a sincere attempt to combat the disease despite the inadequacy of available knowledge. However, history records that two frosty nights on October 17 and 18 spelled the end of the epidemic. (It is now understood that the low temperatures on those nights killed the *Aedes aegypti* mosquito, temporary host or vector for yellow fever.) As a result of the yellow fever outbreaks, Boards of Health were established in New York and Massachusetts in 1797.

Most advances in environmental health in colonial times occurred during the eighteenth century. Occupational hygiene and safety and well-being of the worker were addressed. Infant hygiene was not founded on any scientific basis but was represented in a humane attempt to give better care to the child. Mental hygiene was limited to a sympathetic understanding and care of the mentally disordered.

In 1639, the Massachusetts colony passed an act stating that each birth and death must be recorded, and the Plymouth colony soon followed. In 1647 the Massachusetts Bay colonies regulated to prevent the pollution of Boston Harbour. Between the years 1692 and 1708 Boston, Salem and Jamestown passed laws dealing with nuisance and offensive trades. In 1701 Massachusetts enacted legislation providing for isolation of smallpox cases and for quarantining of ships.

The superstitions expressed in witchcraft and other practices of the time indicate that the colonial period in history was hardly one in which to expect any great advances in health science. Measured by today's standards, the men who wrote the American Declaration of Independence and drew up the US Constitution were considerably younger than today's legislators and politicians. During George Washington's time, the average duration of life was about twenty-nine years; decidedly few people lived beyond the age of forty.

Early Nineteenth Century: By the first half of the nineteenth century, health measures in Britain were in advance of those occurring elsewhere. However, conditions among the lower socio-economic groups remained poor for some time. Public health was officially recognized in 1837 when legislation relating to community sanitation was enacted. This confirmed the awakening interest in and understanding of public health responsibility within a more educated, prosperous and humanitarian community.

A Factory Commission was set up to study the health conditions of the labouring population, with particular emphasis on the conditions of child employment. Edwin Chadwick, a civilian with a special interest in social problems, was made secretary of the Commission. In 1842, his 'Report on the Inquiry into the Sanitary Condition of the Labouring Population of Great Britain' contained colourful descriptions of the deplorable conditions of the time. Chadwick's report pointed out that half the children of the working classes died before their fifth birthday. It aroused great concern among the new middle class, who began to work to improve the conditions of factories and mines throughout Britain (Table 6.1).

While the death rate and infant mortality data cited by Chadwick did not indicate the complete profile of the nation's health or that of any single community, the length of life and the infant death rates he reported indicated the standards of health applying where survival was the principal goal.

Chadwick's report led to the establishment of a Board of Health in 1848, but England was not yet ready for the sweeping reports advocated. While John Simon was appointed first medical health officer of London, the resulting measures proved slow to enact and difficult to promote. Nonetheless, and despite its brief life of four years, the Board of Health and Chadwick's report remain landmarks in the history of public health.

The Role of War in Public Health
During the nineteenth century, wars were rampant on both European and American continents. Direct loss of life was limited compared with the losses from epidemics and modern warfare, but the death rate among those in battle, especially among the wounded, was high. Ironically, however, these wars contributed to medical and community health advances. Many hundreds of casualties developed gangrene in the conditions applying in hospitals on the battlefield. In 1815, the Napoleonic Wars saw Delpech describing the cause of infection in wounds. Fifty years before Lister developed antiseptic methods, Delpech theorized that 'animal-like matter jumps from one object to another'.

Late in the nineteenth century, the Crimean War led to such high casualties that nursing had to invent the method of 'triage' to sort the injuries of battlefield victims into categories of severity and urgency, so that valuable time and resources would not be lost on those who would die anyway. The method and term have persisted to the present day.

Finally, the medical examination of men being inducted into the US Armed Services during World War I served as the first broad-scale barometer of the health status of American people. With standards lower than had previously been required, the Armed Forces still found it necessary to reject 34 per cent of the men examined because of physical and mental disabilities. The nation was appalled to learn that one-third of its youth were unfit for military service. Professional health personnel analysing the data concluded that public health resources needed to be re-directed. While communicable diseases seemed to be coming under control, other health hazards and problems had been neglected.

Furthermore, many of these defects could have been prevented and most could have been corrected. It was clear that public health programmes had neglected the citizen as an individual and that it was necessary to build up and maintain the highest possible level of health resources for each individual citizen. The measures against communicable diseases were insufficient in themselves.

Private organizations were encouraged to provide financial support to community health research and practices. Voluntary health agencies were established and organized to focus attention on public health education and medical research. It was the latter, however, which benefited most from the resulting influx of funds and the husbanding of human resources to improve the nation's health status.

The Modern Era

The period from 1850 to the present can be divided into five distinct phases representing the different theories dominant at the time. In this, the last period, emphasis has turned to lifestyles and behaviours of individuals and communities as principal determinants of illness, disability and death, and measures to reduce risks from these causes.

During the first phase, the miasma phase (1850-1880), the approach was based on the misconception that disease was caused by noxious odours or miasma. Hippocrates, for example, observed that people who ventured out at dusk invariably contracted malaria. The term 'malaria' literally means 'bad air'. Similarly, diphtheria was thought to be caused by gases associated with putrefaction. Disease control efforts were directed entirely toward general cleanliness. Garbage and refuse collection became important and general cleanliness measures were initiated. Street cleaning was pursued relentlessly. Unfortunately these cosmetic measures were not aimed at specific causes of disease, and consequently proved of little value in disease control.

Public health teaching began in this period. An English manual of hygiene by E.A. Parkes, a professor of military hygiene in the army medical school of England, was published in 1859. In 1879, A.H. Buck edited a pioneering text called *Hygiene and Public Health*, dealing with environmental sanitation, housing, personal hygiene, child and school hygiene, industrial hygiene, food sanitation, communicable disease control, disinfection, quarantine and infant mortality. The volume is notable for appreciation of the growing influence of medicine in public health. For example, it recognised that county lines are not natural boundaries for diseases and have no influence on when and where a disease appears in a community. In other words, that disease and its occurrence are not determined by place and climate.

The bacteriology phase (1880-1920) began with the work of Louis Pasteur, Robert Koch and other bacteriologists who demonstrated that a specific organism causes a specific disease. This knowledge made it possible to change from general disease-control measures to specific measures for protecting health by blocking the routes over which causative agents would travel. As evidence implicating disease-carrying organisms accumulated, attention was directed to those areas where measures could be undertaken, as in protecting water supplies by filtration and chlorination, the pasteurisation of milk and proper disposal of sewage. The definition and subsequent refinement of laboratory procedures further assisted the advance of scientific research and the productivity of researchers of the era.

Apart from demonstrating that a specific organism caused a specific disease, Louis Pasteur was notable in his discoveries of the cholera bacillus and of a method for inoculation against rabies. Robert Koch (1843-1919) discovered the tubercle bacillus and the streptococcus; he also discovered the cholera vibrio which he demonstrated was transmitted by water, food and clothing. During the same period, Joseph Lister (1827-1912) developed the practical use of phenol (carbolic acid) as an effective antiseptic.

This period also saw the staffing of official public health departments with bacteriologists, laboratory technicians, sanitation inspectors and engineers, quarantine officers and others who specialized in disease control measures. With the initial emphasis placed on isolation and quarantine, the public soon came to treat with considerable derision the hideous placards and rigorous policing undertaken by the large numbers of departmental officials.

Nonetheless, and despite their limitations, the health and environmental protection measures undertaken did succeed in reducing the death rate. A comparison of the figures for 1930 with the average for the years 1881 to 1885 indicates a pronounced and continuing trend throughout western Europe and the United States.

In Britain, World War II brought the issue of health and the care of all British people — in and out of the armed services — to the attention of the government. Making the advances of health science available to all posed a pyramid of problems because of the various individuals and groups unable to acquire health knowledge or to obtain health services. The poor and the socially and educationally disadvantaged had clearly missed the benefits of community health programmes.

In Britain, the social levelling process begun with World War II gathered new momentum. Because of its importance to the nation, health and the provision of medical and hospital services became the responsibility of the government. The country's doctors, hospitals, pharmaceutical and health services were nationalised. The British National Health Service came into being in the years immediately following the war and continues to this day, a socialised system of health care delivery to which all British residents are entitled by virtue of their residence and work status in that country.

The Era of Social Medicine and Preventive Care
In the United States, the social engineering phase during the period 1960 to 1975 was marked by the realisation that technical advances in health care were often beyond the capacity of an individual's means, particularly among the poor and in nations where resources were considerably smaller than those of the industrialised countries. The husbanding of human resources required that the products of technological developments be made available to every world citizen. The social aspects of health were given a new priority.

Since economic barriers were considered the most urgent, Medicare and Medicaid legislation was passed to put purchasing power in the hands of the poor and medically indigent to enable them to receive needed health services. The educational and social isolation of the poor necessitated carrying health resources and services to those who apparently were not receiving the benefits of community-directed programmes. The 'outreach' services of public health nurses and indigenous community workers thus became a mainstay of local health departments.

By the early 1970s the resources of health facilities, manpower and research developed in earlier decades had been more equitably distributed to the poor and the medically indigent. In Europe, Canada, Australia, Asia and Africa even more sweeping social and medical reforms were made. Citizens in most communities began to take a more active role in allocating health resources at a local level.

However, the accomplishments were not always reflected in the mortality and morbidity statistics. The differences between death and disease rates of rich and poor, white and black, urban and rural populations persist. It is becoming apparent that increasing expenditures on health are mainly directed into the provision of medical care. These, in turn, are not yielding proportionate improvements in the health of nations as a whole. The rapidly escalating costs of medical care have been attributed to the expensive medical technology and facilities created by earlier investments in medical resources, now made more universally accessible by new health insurance coverage and related programmes of distribution. The search is now on for ways to contain the costs of medical care.

The search for cost-containment strategies in the United States led first to the 'health planning' acts of the late 1960s and early 1970s. New forms of health-care delivery, such as 'health maintenance organizations', were instituted, designed to encourage physicians to keep their patients healthy and away from hospitals rather than admitting them unnecessarily for operations, procedures and tests likely to achieve little real benefit to that individual.

Today, the emphasis in disease control centres on policies aimed at prevention of disease and the promotion of health. The focus has shifted to the individual and is directed at the local, community level. Individuals are now encouraged to actively maintain and build their health and to be more aware and conscientious in their responsibilities to other members of the community, by refraining from smoking in public areas, disposing of wastes in a responsible manner, caring for their local environment and assisting in policing of atmospheric pollution.

TABLE 6.1: Mean Age of Death and Infant Death Rates
(England, 1842)

Class	Mean Age of Death London	England	Infant Deaths per 1000 Births
Gentry, professional persons and their families	44	35	100
Tradesmen, shopkeepers, their families	23	22	167
Wage classes, artisans, labourers and their families	22	15	250

Based on Chadwick. See Richardson, B.W. The Health of Nations: A Review of the Works of Edwin Chadwick, vol.2, p.2, Longmans Green & Co., London, 1887.

References and Additional Reading

China Reconstructs, vol.XXXVIII, no.9, September 1989, p.66.

Green & Anderson. *Community Health*, 5th edition, The C.V. Mosby Company, St Louis, 1986, pp.3-25.

The Environment Digest, 1989, Issue no.27.

Krause, F. & Bach, W. *Energy Policy in the Greenhouse*, Vol.One, International Project for Sustainable Energy Paths, El Cerrito, CA94530, US, 1989.

New Scientist, vol.1691, 18 November 1989, p.55.

New Scientist, vol.1893, 2 December 1989, p.8.

Pearce, D., Markandya, A. & Barbier, E.B. *Blue Print for a Green Economy*, Earthscan Publication, London 1989.

6.2 Chemical Hazards to Health

Intervention can sometimes go wrong. Pesticides were developed to minimise crop loss from insects and reduce diseases in agricultural plants. More healthy crops would produce more reliable yield. With more food and better quality produce, the spectre of famine would retreat and infant mortality rates decline, health standards would improve and socio-economic conditions begin to rise throughout the world. At least, that was the way the argument went and, to a certain extent, some of the predicted gains have been achieved. But not without a price. Pesticides and associated agricultural chemicals, fertilisers, herbicides and weedicides are not without their drawbacks. They bring new problems while solving old ones: and sometimes the new problems are more intractable than their precedents.

This is certainly the case with the second example of human intervention to be discussed: the use of pharmaceuticals. Once hailed as the salvation to infectious disease and premature death, antibacterials have side-effects and can be misused and abused. Their distribution can be selective, so that in one instance they might be held out as 'scarce goods' attracting a premium price and available only to the rich. At other times, these same drugs might be discarded and 'dumped' on Third World nations, ostensibly for humanitarian reasons. In reality, the donor pharmaceutical manufacturers are attempting to attract goodwill overseas, without conceding that consumer legislation in their parent country has changed to prohibit distribution of the dumped pharmaceutical because of new scientific evidence of risk and toxicity.

Chemicals manufactured in and developed by industry pose a further problem. While industries might attempt to improve the quality of life by developing and introducing new synthetic products or finding new techniques for maximising benefits from raw materials, dangers may sometimes arise in the course of this work. Those exposed and at greatest risk are frequently the workers themselves. Occupational diseases are one particular aspect of environmental health and human intervention which is likely to become a focus for future concerns.

Finally, food toxins and chemicals associated with the use of pharmaceuticals, pesticides and other agents can be responsible for alterations in body hormones and the emergence of new diseases.

Pharmaceuticals: Ally or Foe?

Pharmaceuticals are a class of chemical agents often ignored by writers on environment and health but which are valuable tools in the process of intervention in the natural disease process. Many exert adverse effects on a wide range of body systems, particularly when over-prescribed, old or misprescribed as often occurs in developing countries. The dumping of out-dated antibiotics and products which fail to meet strict western government standards of safety has resulted in large-scale toxic poisoning caused by out-of-date tetracyclines. Incorrect labelling and poorly translated directions of common, freely available products such as aspirin have caused minor outbreaks of illness.

Thalidomide was found to be responsible for the birth of children without properly formed arms, hands and legs; women developed massive thromboses (clots) following use of the oral contraceptive pill; and patients given anti-cancer drugs in the absence of firm evidence of tumour subsequently developed leukaemia from their use.

The development of drug resistance in humans and animals is a further complication. Antibiotics are perhaps the most reckless example of the misuse of modern medicine, and of promotion of the 'magic bullet' or 'panacea' for all illness.

Antibiotic Resistance

Fifty years after their introduction to modern medicine, the sulphonamides, penicillins and newer antibiotics are losing their potency. For example, some antibiotics succeed by undermining and interfering with the synthesis of the bacterial cell wall. Others inhibit essential enzyme metabolism or block the sites of protein replication inside the bacterial cell. Many bacteria have become resistant, evolving alternative pathways or new strains to overcome the source of attack by these antibiotics.

When antibiotics fail to do their job the consequences can be disastrous. Between 1968 and 1972 an outbreak of bacillary dysentery caused by an antibiotic-resistant strain led to thousands of deaths in Central America. Ironically, those at greatest risk of infection are often those in medical institutions where antibiotics are widely used.

Early in 1986, British doctors were alarmed to discover a germ resistant to virtually all common antibiotics. Known as 'Super Staph', it was found in at least thirty-two London hospitals (*British Medical Journal*, 1986 editorial). A methicillin-resistant strain of *Staphylococcus aureus* was finally blamed for the incidents. However, the hospitals of eastern Australia had earlier recorded hundreds of deaths of in-patients affected by resistant strains of a similar *S. aureus* strain. At the time, the organism was found susceptible only to vancomycin.

The Developing World

The WHO has warned that the problem of antibiotic resistance is global, but often worst in developing countries. It blames the misprescribing and over-use of antibiotics for coughs, colds, fevers, diarrhoea, and a host of other minor conditions with which natural immunity could cope without recourse to drugs. Due to non-existent regulations, gross and irresponsible promotion and people's unreal expectations, antibiotics are often given incorrectly.

A particular example applies to the use of antibiotics for diarrhoea, an acute condition where the drug industry continues to promote antibiotic use. This is contrary to medical practice where routine treatment does not involve antibiotics. Nonetheless, almost 65 per cent of anti-diarrhoeals on the market in eight regions of the world contain antimicrobials, usually in irrational combinations with other ingredients.

In Peru during 1983, the Italian firm Carlo Erba marketed a drug contain-ing chloramphenicol and tetracycline in a formulation with chocolate flavour-ing as a treatment for children for diarrhoea. The drug, Quemiciclina, was so popular that it became known as 'Erba' or candy for diarrhoea. Yet tetracyc-line is contraindicated for children under 12 years of age and is reserved for use only in life-threatening infections.

Unfortunately there is no sign that the practice of antibiotic promotion in the Third World will change in the near future. With sales in excess of $20,000 million yearly and rising fast, incentives are weak. As fast as commonsense pharmaceutical controls are imposed in some markets, the corporation sales teams transfer their efforts elsewhere. For instance, while the tetracycline market was expected to decrease between 1980 and 1990 from $70 million to $47 million in the US and from $65 million to $51 million in Europe, in the rest of the world an increase was projected from $47 million to $75 million.

Widespread emergence of resistance has also prompted further research into stronger and more potent antibiotics. These often have unpleasant and undesirable side-effects. In addition, they further lead to the emergence of more resistant strains. It is a matter for concern that some scientists secretly confess that these agents may be undermining the body's own immune defence mechanisms and hampering its capacity to fight a disease threat. The rash of new diseases and conditions suffered post-infection, and the pro-longed recovery now required by many following infection and treatment with bacterial and viral diseases, raises questions about drug use. Clearly, these ideas are challenging and need further research and development of hypoth-eses for testing before any conclusive statements can be made.

The Emergence of Resistant Micro-Organisms

Antibiotics are used to treat microbial infections in animals as well as humans, and are given prophylactically to healthy humans to prevent infections. Antibiotics are also given in low doses to food animals to improve their growth rate and feed-conversion. However, use of an antimicrobial in humans or animals is often followed by the appearance of resistant micro-organisms. This effect leads to treatment failures and the need for newer, often costlier antimicrobials.

Many investigators have concluded that the phenomenon of resistance is related to the amount and pattern of use of antimicrobials and have recommended more prudent use in animal as well as human health management. Animal growth promotion also accounts for a substantial portion of all antimicrobials used in the United States and parts of Europe, and this use has become controversial.

Defining the significance of antimicrobial resistance to human health and identifying the source of such organisms has proved difficult. The application of molecular biology techniques for sub-typing of bacteria or identifying specific genes has assisted in providing some epidemiological clues on antimicrobial-resistant organisms. However, these studies suggest that the antimicrobial drugs to which food animals are exposed provide selective pressure that leads to the appearance and persistence of resistant strains (Cohen and Tauxe).

Salmonellosis
In the United States, Salmonellosis poses a major problem to public health. Annually more than 40,000 cases are reported, with 500 deaths, and financial costs estimated in the region of $50 million. *Salmonellae* are gram-negative bacteria that can infect both humans and animals. The organisms most frequently cause a self-limited gastro-enteritis, ranging from asymptomatic colonization to major illness involving meningitis or osteomyelitis. Most uncomplicated Salmonella infections do not require treatment. Patients with bacteremia, meningitis or other extra-intestinal infections require effective antimicrobial treatment.

Although more than 1500 different forms exist, only about ten have been linked to over 70 per cent of the human cases identified each year in the US. The most frequent isolate is *Salmonella typhimurium*, which accounts for approximately 35 per cent human disease. Other common isolates are *S. enteridis* (10.0%), *S. newport* (4.5%), *S. infantis* (3.0%). Some forms are highly host-specific and rarely cause disease in other species. For example, *S. typhi*, which causes typhoid fever, only infects humans and there are no known animal reservoirs. *S. gallinarum* and *S. pullorum* are exclusive pathogens of poultry. *S. typhimurium*, however, has a broad host range and causes disease in many species.

Cases of typhoid fever and of non-typhoidal salmonellosis can be traced many years back. Since the early part of the century, the incidence of salmonellosis has changed dramatically. In 1963, several large outbreaks were traced to commercial egg products and this provided the impetus for more intensive monitoring by public health authorities.

Today much more data is available. *S. typhi* has become uncommon, whereas non-typhoid salmonellosis has increased. This has been related to changes in animal husbandry practice said to encourage the spread of Salmonella among animal populations. Industrialization of food processing further facilitates swift and broad distribution of contaminated food items.

The emergence of resistant organisms has been linked to treatment failures with penicillin derivatives used in treating post-operative bacterial infections and persistent chest conditions in the community. As a result of these experiences, a reported outbreak of Salmonellosis in the United States can be associated with the use of pharmaceuticals like penicillin. An outbreak of Salmonellosis following use of penicillin derivatives for persons with streptococcal pharyngitis, bronchitis and otitis was first believed to be caused by contamination of the pharmaceutical with *S. newport*. It was subsequently found that the patients involved had recently consumed ground beef contaminated with the *S. newport* strain. This resulted in asymptomatic colonization of the patients with the resistant strain. The penicillin derivative provided the selective advantage for the organism to proliferate and cause typical symptoms.

Antimicrobial therapy can promote emergence of antimicrobial resistant strains of Salmonella in other ways. Furthermore, an outbreak caused by a resistant strain affects a larger number of people than if it had been caused by a sensitive strain. For example, in 1985 in Illinois, many people who drank contaminated water and developed salmonellosis would not have become ill if they had not been taking antibiotics at the same time, or if the strain had not been one resistant to antimicrobials.

Salmonella strains isolated from healthy food animals show considerable antimicrobial resistance. In 475 outbreaks of salmonellosis reported to the Centre for Disease Control in Atlanta in the 1970s, major sources of the bacteria were foods of animal origin — meats, poultry and dairy products. In outbreaks investigated by CDC, 69 per cent of Salmonella outbreaks were of resistant strains and 46 per cent of sensitive strains. Only a small proportion of either sensitive or resistant Salmonella outbreaks were attributed to transmission from other humans, and the remainder were linked to foods of animal origin.

Antimicrobials are fed to food animals in low doses, referred to as sub-therapeutic because the amount used is less than that to treat specific illness.

The use of these sub-therapeutic doses in animals produces effects in them which are similar to those seen in humans who receive antimicrobials for salmonellosis: they develop resistant organisms. Those animals fed sub-therapeutic doses of tetracyclines excrete resistant organisms in their faeces for long periods, and also transmit the organisms to other animals. Because large numbers of animals have longer and more constant exposure to low doses of antimicrobials than therapeutically indicated, their use may have a greater impact on the rate at which resistant strains of bacteria develop and persist in the environment.

Pests and Pesticides
Pesticides are another form of human intervention which has had mixed benefits and poses serious future problems for human health. Pesticides are natural and synthetic chemicals used to kill pests. Here, a pest is defined as any living organism that, for one reason or another, is unwanted by human beings. The best known pesticides are the insecticides: by definition, substances that kill insects. There are three principal subgroups: chlorinated hydrocarbons, organophosphates and carbamates.

Chlorinated hydrocarbons contain specific arrangements of atoms of hydrogen, carbon and chlorine, and include substances classified as organochlorines, such as aldrin, dieldrin, heptachlor, lindane and mirex. Organophosphates contain carbon, hydrogen and phosphorus. This group includes malathion, methyl-parathion, parathion and diazinon. The carbamate insecticides contain arrangements of carbon, hydrogen and nitrogen, and the most common one is carbaryl.

The chemistry of the major insecticidal groups causes them to behave very differently in the ecosphere. Organophosphates are water soluble and usually break down rapidly in the atmosphere. They are often less persistent but may be more toxic than chlorinated hydrocarbons, particularly the organochlorines. However, they are likely to be more acutely toxic to humans and other mammals.

Chlorinated hydrocarbons and organochlorines are the more persistent and take time to break down in the environment. They tend to accumulate there and in body reservoir sites such as the fatty tissue and the liver. For example, DDT tends to become concentrated in living animals as it passes through the food chain. In the soil it may persist for ten years or more.

Persistence in the Environment
Persistence of a pesticide in the environment, and the extent to which it tends to be concentrated in species which form part of the human food chain, varies with the situation. It depends on the method of pesticide application and

quantity used, the place where applied, be it soil, water or atmosphere, and climatic conditions at the time, such as temperature, humidity and wind.

Pesticides may become broken down by the action of micro-organisms, sunlight or by the physical-chemical process of weathering. In the degradation processes, both inside large organisms and in the environment, conversion to substances that are more toxic sometimes occurs. In living beings, products containing mercury or arsenic may be transformed or metabolised to systemic poisons.

Pesticide Effects on Human Health
Pesticides are possibly more alarming in their adverse effects than pharmaceuticals, which were always recognised to have potential risks as well as benefits. Evidence is now building against the safety of many of the chemicals used in agriculture.

Of special concern are the synthetic analogues of natural pesticides, herbicides and weedicides developed for greater sensitivity and wider application to crops and environmental circumstances. Developed to improve agricultural yields and crop resistance to disease, they have proved to be a mixed blessing for the agricultural sector.

Organophosphates
The effects of organophosphates on living species are short, sharp and nasty. These pesticides affect the nervous system by inhibiting the enzyme cholinesterase, which regulates transmission of nerve impulses at nerve endings (see Chapter 4.6, Sensory and Nervous System, for more detailed description of the role of cholinesterases in nervous transmission). Tremors, paralysis and death follow. In some formulations, organophosphates can act by inhibiting enzymes involved in metabolism and thereby produce a slower, more painful death.

How severe the effects of an organophosphate are depends on many factors, including the synergistic effects. Two or more pesticides acting at the same time are sometimes used to improve pesticide effectiveness while employing lower amounts of each individual chemical. This strategy is intended to improve the safety to humans at the same time as maintaining pesticide potency and effect.

Organochlorines
Among the organochlorines of widest use and greatest potential danger to humans are the combination of n-butyl esters of phenoxyacetic acids 2,4-D and 2,4,5-T; that is, the family containing dieldrin, DDT, chlordane, lindane, heptachlor and picloram.

The main danger of the organochlorines principally used in farming and insect control lies in the length of time they take to break down into their component elements. Since these substances enter the food chain and are ingested by human beings, they can be further concentrated in the body. In the liver, reproductive organs and body fat deposits, they can cause immense damage because of their effect on cell reproduction through chromosomal DNA. In addition, the techniques of use of the organochlorines as well as other potent agricultural chemicals, can induce respiratory, skin and gastric reactions in users or those who come in contact with them.

A History of Organochlorine Use: The organochlorine pesticides were first developed and selected for special attention in the late 1940s. It was apparent that a large number of them were useful in controlling pests affecting crop and farm produce yields, and were easy to use. But it took researchers in the United States less than ten years to realise that they offered opportunities, not just for agriculture, but also for chemical warfare.

In particular combinations, the organochlorines penetrate the waxy cuticle of leaves. By coating the leaf surfaces, they affect water loss through transpiration and at the same time permeate the leaf tissues to enter the plant's system. The result is loss of leaves and eventual death of the tree or shrub. The US Defence Department recognised that by spraying these agents from low-flying aircraft they could clear thick jungle, removing the dense foliage of the leafy camouflage thereby forcing into the open any enemy guerilla troops (such as those which were proving very effective against Americans in North Vietnam).

Unfortunately these chemicals also strip the crops of the local farmers and enter the streams, wells and drinking supplies of the villagers. Furthermore, because their actions tend to persist, these organochlorines induce damage which is permanent and long term. In Vietnam, the threat of famine became very real for the villagers living in areas near jungle sprayed by American aircraft. The farmers' crops failed to grow and when their children and wives drank the water or ate the shrivelled grains, they would suffer nausea, vomiting or start behaving strangely. Farmers working on defoliated crops would suffer skin burns, smarting eyes and continual vomiting.

However, these effects were not restricted to just one or two of the organochlorine family of pesticides. It did not take long for the chemists to realise that the most potent of the group was 2,4,5-T. This chemical produces a contaminant called dioxin (also known as TCDD), so potent that one kilogram would kill one billion guinea pigs. It is considered to be the most potent nerve chemical ever produced.

Dioxin

Dioxin acts as a powerful carcinogen and teratogen — it causes cancer and deformations. It is also a mutagen and alters the chromosomal DNA, the cell's hereditary material. Dioxin causes cows to give birth to stillborn calves, chickens and ducks to become sterile, and women to cease ovulating. It attacks the lymph glands, damages the sweat and thymus glands, and generates skin diseases, including extreme acne pustules. It also induces liver cancer and scarring of the conjunctival surface of the eyes. As early as 1957, German scientific journals began to report the toxicity of dioxin to humans. In 1962, the *Journal of Investigative Dermatology* discussed methods of testing its harmful effects by using the skin from the ears of laboratory rabbits.

In 1962, an accidental explosion at the Duphar factory in the Netherlands released 200 grams of dioxin into the atmosphere. Twenty workers plus the inspectors who investigated the damage developed severe acne, as did the nine cleaners who came to clear up after the accident. Within two years four men were dead. The plant was sealed off for ten years then finally dismantled, brick by brick. The rubble was embedded in concrete and dumped into the Atlantic — hopefully where no further damage might ever result.

According to the American Environmental Protection Agency, samples of 2,4,5-T produced by manufacturers between 1956 and 1968 contained as high as 40 parts per million. Until the late 1960s, manufacturers did not bother about the TCDD levels in their batches of 2,4,5-T, despite the fact that it had been identified following industrial accidents similar to those which occurred in Holland.

Dioxin In the Bleaching and Cleaning Industry: Dioxin is not confined to the agricultural section. It is also a residual chemical in the manufacture of bleaches used for giving a snowy white appearance to many items of regular daily use, ranging from paper goods and cleansers to toilet paper, disposable baby nappies and women's sanitary pads. It is because of the presence of dioxin in these last three items that consumers have become concerned.

Extensive United States research in animals has confirmed that, when ingested even in small amounts, dioxin is responsible for birth abnormalities and changes in the chromosomal material of the genes. It enters the human body through the skin and on sensitive areas can cause local irritation, such as redness and a rash. In very sensitive skin such as that of a new-born baby, there are isolated reports of blistered and eroded skin surfaces around delicate nappy areas. Women with sensitive genital areas have complained of local irritation after using particular brands of white paper sanitary pads.

Against this background, the sale of unbleached toilet paper and other household items has proven a surprisingly successful initiative by some

Australian firms. Whether the initial interest and positive response generated by the commencement of sales of unbleached goods in 1989 will continue should prove telling — especially if the price of these items is more than that of their bleached rivals.

Around The Home
Lately it has been realised that the use of organochlorines is not restricted to farming. The more potent of the insecticides such as DDT, dieldrin, lindane, chlordane and heptachlor have regularly been used in normal household pest prevention and eradication. Conventional practice is to use an organochlorine pesticide prior to building, by treatment of the soil surrounds when concrete foundations are laid. Pest exterminators making routine inspections treat small infestations of ants, termites and other insects using these substances.

This means that in modern suburbia there is residual organochlorine present in the soil near most homes, in store-rooms, cellars, playrooms, etc., and under the house. Entry of residual chemicals such as organochlorines can occur through the hands, skin and the mouth: e.g. babies sucking fingers, children not washing their hands after playing outside or under the house.

So great is the alarm about the adverse effects of the organochlorines that national government committees are constantly monitoring reports about their use and toxicity. At the same time, it appears that much of the initial American work which failed to confirm the risks of organochlorines used in household insect eradication has proved faulty.

Pesticide Use and The Health of Farm Communities
Dr Colin Keenan works with the Health Department in Northern New South Wales and is particularly concerned at the level of use of pesticides by the farming and fruit-growing communities in the area. The reasons for his concern stem from common complaints about nausea and feeling unwell, to skin and respiratory allergies, hay fever, asthma, and the numbers of babies born with developmental disabilities to wives of farmers and agricultural workers. And it is not merely a case of careless or improper use of agricultural chemicals.

Workers with pest compounds do generally follow the manufacturers' guidelines on safe use by wearing breathing masks, thick protective gloves and overalls when applying the chemicals, but the discarding of gloves when mixing or in cleaning equipment, not wearing the covering overalls when the weather is hot and failing to take note of prevailing wind direction undermine best efforts at safety.

Tests reveal that most penetration of pesticides occurs through the skin and the mucous membranes of lungs and mouth. These immediate sites of contact are the first to display the acute effects of a chemical. Itchiness, rashes, allergies and large wheals can occur on the hands and exposed skin areas. Comparable effects on the fragile mucous membranes result in hay fever and breathing problems.

Pesticides, particularly those of the organochlorine family, do not degrade readily in the soil or break up in the air after use. Nor do humans or animals readily metabolise them — it treats them as foreign substances. In grazing animals, crop sprays with pesticides can result in high residual concentrations. At the same time, airborne and surface residues of pesticides themselves are further concentrated in liver and fat deposits. Studies in British birds and the fish of waterways have recorded concentrations at least several hundred times that of the surrounding air or water. Some species of petrel showed organochlorine residues 600 times above atmospheric readings.

The Mode of Action

The long-term effects of pesticide use are not easy to prove. Monitoring of pesticide levels in the tissues is dependent on the level of cholinesterases in the blood. These are a class of enzyme whose job lies in breaking down the chemical acetylcholine (the neurotransmitter, or messenger chemical) into its components after it has carried a nerve message from one nerve cell to the next. Its importance lies in the fact that once a message is sent from one nerve to the next, a new message can be sent along the same fibre only if the nerve has recovered and completed transmission of a previous message. This is where cholinesterase comes in.

Once cholinesterase has broken the acetylcholine into its component parts, these can be reabsorbed and used again. If there is insufficient cholinesterase, the acetylcholine remains and effectively blocks the nerve from sending or receiving any more messages. Eventually this brings spasticity or paralysis of the nerve. The result is the plethora of nervous system effects produced by organochlorines.

Tests reveal that between 5 and 8 per cent of farmers and agricultural workers have depressed levels of cholinesterase enzymes. Researchers are now following these men and checking their health, and that of their families, at regular intervals. Thereby it is hoped that some definite and conclusive links may be found between use of organochlorines and other farm chemicals and any subsequent medical problems.

Of the effects of organochlorines on the hereditary DNA of the nucleus much less is known. It is believed that the molecule causes breakages in a small 7-segment area of the DNA helical strand, but research is still in its initial stages.

The local skin damage and irritation caused by organochlorines is explained as a normal reaction of a sensitive surface (the skin) to a harsh foreign substance (the pesticide). The degree of irritation and subsequent penetration is related to the medium in which the organochlorine is present — that is, whether it is present in an aqueous liquid, as a gas or in solid form.

Aerial Crop Dusting with Chemicals

The Wangaratta area of Australia's northern Victoria, a predominantly farming area, relies heavily on production of citrus and stone fruits, tobacco and some rice growing. Regional health surveys of birth abnormalities, cancers and respiratory problems suggest that the local community is far from healthy.

It was in 1985 that the area's high level of birth abnormalities was first recognised. For example, the town of Myrtleford with a population of approximately 3-4000 listed 49 children suffering various degrees of birth abnormalities. These ranged from Down's syndrome to anencephaly, which is a defect in development of the neural tube in which the uppermost part of the brain and skull of the foetus are missing or incorrectly formed. Hole-in-the-heart babies, a number with improperly formed internal organs and defective nerve-muscle coordination, were also reported. In adults, malignant tumours of brain and internal organs are well above national levels. Skin allergies are wide-spread and common. Non-specific symptoms such as nausea and dizziness, breathing difficulties including bronchitis are common. The sufferers invariably report no previous history of illness or similar condition.

By comparison, in Mansfield and Bright which have similarly sized communities, though less oriented to farming, the local health authorities recorded only eight or nine children who were seriously affected. The level of allergies, asthma and cancerous tumours in family members was also lower.

When the concerns of the local Myrtleford community failed to produce government action and further enquiries by the Health Department, an independent consultant was called in by the local council. His results indicated levels of organochlorines and minerals in the water supply well above recommended levels at times of peak flow. Despite initially dismissing these results, the Environmental Protection Agency, the Agriculture and Health Departments in Victoria have recognised the urgency of the health problems in the Wangaratta area, and have marked them to a high priority for study. Further action is being anxiously awaited by the local communities.

Fertilisers

In southern parts of the Commonwealth of Independent States (CIS, formerly USSR), cotton production is virtually a monoculture with most food and agricultural items being imported from nearby provinces. Heavily dependent on a high water supply to support the lush growth in a dry, hot climate, the cotton crops are also dependent on the application of large quantities of fertiliser. Despite the benefits of mechanisation, much of the crop is still harvested by hand. This poses a major problem because the chemicals used to deter insect pests and weed growth are potent defoliants and members of the organochlorine family.

Yet despite the imposition of bans on their use, organochlorines are only being phased out gradually. Moreover, because of its persistence in the soil, in locally drawn water and human tissue, the effects will continue to be evident for years after its use has ceased. The following describes a distressing situation in the CIS due principally to overuse of pesticides and fertilisers in agriculture.

Health in the Farming Communities of the Aral Sea
Health is a major problem in the Central Asian Soviet Republics bordering the Aral Sea. In 1961 this Sea, the fourth largest inland stretch of water in the world, contained 1064 cubic kilometres of water, with depths to 53 metres and a surface area of 66,000 square kilometres. Fifty-six cubic kilometres of river water flowed into it each year, maintaining average salinity at 10 to 11 per cent. Today the surface of the Aral is reduced to 36,500 square kilometres and it has lost 69 per cent of its water. Its contributory rivers have been plundered for irrigation and the waters which drain into it are heavily contaminated with pesticides and fertilisers used in agriculture.

The drying of the sea is being blamed for a rash of epidemics in the farming communities. Nine out of ten babies born on the coastline suffer anaemia due to frequent child-bearing, poor diet and weakness in the mothers. The salt content of breast milk in many areas is four times the norm.

In Karakalpak, the incidence of paratyphoid is twice as high as in the rest of Uzbekistan and twenty-three times higher than the Soviet Union (CIS) as a whole. In the past ten years, cardiovascular disease has increased by a factor of 1.6. Tuberculosis has doubled, gallstones are five times as common and cancer of the oesophagus seven to ten times.

In the Kzyl-Orda region of Kazakhstan, the rate of gastric typhoid has increased twenty times in just five years. Dysentery is rife throughout the Aral region, hepatitis and infectious diseases are prominent. Infant mortality is higher than in Paraguay and twenty times higher than in Japan.

These effects are the long-term results of excessive use of pesticides and fertilisers. In the Aral area the use of these chemicals has become so excessive that it has contributed to the degradation of the soil due to over-reliance on irrigation, over-cropping and erosion.

Alternatives
And there are alternatives to these chemicals. Arguably they are less potent and their effects are not long-lasting, but they avoid the long-term negative implications of use associated with organochlorines. Many natural non-residual chemicals once used in farming can be re-examined: new biodegradable products are available. Effective pest eradication can be achieved in the home by use of degradable chloro-compounds.

Chemical Industries and Cancer
Many modern industries are involved with manufacture or use of potentially toxic chemicals. Workers in pharmaceuticals and pesticides manufacture, paint production, petroleum, plastics, rubber and many new synthetic substances industries are continuously exposed to these chemicals during time spent on the job.

Exposure to chemicals in the workplace affects people in different ways. Some develop sensitivity after prolonged use, like a cumulative effect. Others become sensitised only as a result of accident which has exposed them to a sudden large dose. Still more exhibit no signs of disease or developing sensitivity until after retirement from the industry.

One of the earlier studies of disease and occupational exposure to chemicals occurred in Sweden and sought to establish whether professional chemists were at excess risk of developing cancer as a result of their work. Olin (1978) found that of the 857 men who graduated from Sweden's two schools of chemical engineering between 1930 and 1950, by 1974 a significant number had died from lymphatic and haemopoietic cancers. This has prompted further studies and the results obtained have been valuable in tightening legislative controls on protection for workers, including regular testing and research on all chemicals in use by a process.

A long-term study of the Australian petroleum industry commenced in 1980, seeks to establish whether there is a link between occupationally determined exposure to environmental hydrocarbons and adverse health effects. The population for study was defined as all employees working in refineries or petroleum marketing terminals and who had been continuously employed for at least five years. No direct measures of exposure were available so an accurate job description was used as an indirect measure. A personal interview was carried out with all subjects to obtain information on such potential confounders as smoking history and previous employment. Of interest were the cause of any death recorded and the evidence suggesting cancer played a role. By the end of 1985, 30,300 person-years at risk had accumulated and 96 deaths had occurred.

Thus far, the results are inconclusive because of the 'healthy worker' effect which occurs where people who are gainfully employed are, almost by definition, healthier than the general population, which contains those who are unemployable for reasons of ill-health (Table 6.2).

Food Toxins and Food Quality
A few examples serve to highlight problems associated with food toxins in developed as well as under-developed and Third World countries.

Mad Cow Disease (or Bovine Spongiform Encephalopathy, BSE) attacks cows' brains leaving them looking like sponges. The disease is caused by the practice of feeding cows with sheep's offal to boost the protein in their diet. Some offal came from sheep infected with a disease called Scrapie, which researchers believe mutated into a more virulent strain and attacked the cows. Mad Cow Disease was first noticed in Britain in 1985 but it was not until 1988 that sheep's offal was banned as a feed. In the intervening period, 14,000 cows had died or been destroyed.

In 1989, concern was heightened when Professor Richard Lacey, a former British government health adviser, claimed that a danger existed because humans could develop Creutzfeldt-Jacob disease, a virulent form of the virus related to kuru and multiple sclerosis. The human form of the disease lies dormant for about twenty years and only becomes apparent with mid and later life deterioration of the nervous tissues.

Before BSE hit the headlines, there had been outbreaks of salmonella, listeriosis and toxoplasmosis. All these diseases are caused or exacerbated by turning the rearing of animals into a factory process from conception onwards. Much of the problem is caused by force-feeding animals with a processed protein diet made from animal products. Calves, for example, while usually fed on formula at first, are then weaned onto a chocolate-flavoured feed based on pigs' blood. Later, blends of proteins made from meat and bone-meal or soya and fish meal are added to the growing calves' diets.

Chicken litter from battery farms — containing faeces, feathers and dead birds, all of which may be infected with salmonella — is still fed to cattle after processing, despite the 1989 scares. These include an outbreak of botulism in British cattle traced to suffocated and rotting chickens which had been swept from their pens into the cattle feed.

Pigs are also considered a source of food poisoning to humans because they are given large doses of antibiotics to prevent respiratory infections caused by the cramped breeding conditions in which they are reared. Pig meat and products from animals treated with certain antibiotics are banned from sale in the United States and many European countries, due to fears associated with hypersensitivity reactions in susceptible people.

Pesticides in Beef

In 1987, Australia's $750 million beef export market to the United States and Japan was threatened when chemical residues, particularly organochlorines, exceeded the maximum US limit.

Organochlorines were banned from use in agriculture, and efforts made to identify sites where previous use had left residues which were still present in pastures and fodder fed to cattle. The Cattle Council of Australia and the Australian Meat and Livestock Corporation began a $47 million campaign to identify and eliminate the problems. A comprehensive national testing programme was organised and, to pay for its cost, farmers charged $3.50 for each animal slaughtered. Since that time, every one of the 148,000 beef properties in Australia has been tested for pesticide levels. Among the anomalies uncovered was evidence that cattle grazing near telegraph poles treated for termites, had high levels of dieldrin, the organochlorine used to protect the poles against termites. Legitimate spraying on potatoes and subsequent use of the land for grazing animals had also contributed to the levels of residues in beef cattle.

Cadmium in Sheep

Cadmium, a heavy metal, was found in unacceptably high levels in Australian sheep kidneys and livers in a 1988 study. Dr Jim Langlands of the Commonwealth Scientific and Industrial Research Organisation (CSIRO) in Armidale, central NSW, identified cadmium in 1663 liver, 1779 kidney and 2526 muscle samples of stock destined for export for human consumption. Eight per cent of the kidneys and 2 per cent of the livers exceeded the maximum levels set by the National Health and Medical Research Council. A high cadmium intake is associated with increased risk of kidney failure and arthritis.

Older sheep and those from South and Western Australia contained the highest levels of cadmium, while those from NSW were within acceptable levels. High cadmium levels can occur naturally in some soils, but in these instances Dr Langlands traced the cadmium to excessive use of fertilisers.

Contaminated Fish

Heavy industrial pollution has made fish unsafe to eat in many parts of Europe. The Rhine River and North Sea are so polluted that much of the catch is deformed and unfit to eat. In Bohemia, Czechoslovakia, fish caught within an area of 2000 square kilometres in 1988 were banned when they were found to contain up to ten times the permitted level of polychlorinated biphenyls (PCBs), recognised carcinogens. The PCBs were believed to have come from an unidentified industrial leak in October 1986.

In the early 1990s, the British Government issued warnings against eating shellfish, crabs and lobster caught off Britain's north-east coast because of lethal levels of shellfish toxin, a poison that causes paralysis and death. Fish caught off Sydney, Australia, in 1989 were found to be contaminated with levels of an organochlorine that were on average 122 times those recommended by the National Health and Medical Research Council. The worst specimen had levels that were 250 times the limit. The fish were also contaminated with dieldrin and heavy metals. Fishing was banned within 500 metres of the three major Sydney sewage outfalls — Manly, Bondi and Malabar.

In 1988, Coffs Creek, Coffs Harbour was closed after thousands of fish died from toxic levels of aldrin dumped into the water. The aldrin oxidized in the fish to form dieldrin. A pesticide operator was later prosecuted for dumping of aldrin.

In May 1989, 752 people contracted food poisoning after eating oysters grown in Sydney's George's River. The Norwalk virus, a common seafood virus which flourishes in raw sewage washes in waterways, was shown to be responsible. Heavy rain over the previous month had caused sewage overflows from stormwater drains and sewerage in the area.

Conclusions

Thus it seems chemicals may have a range of different effects in the environment and upon other species, such that the final consequences for humans may be unexpected. And we have but touched the surface. There remain many issues, from agricultural chemicals which are carried into streams and rural waterways, to food additives, irradiated and other 'treated' foods, and medical and pharmaceutical interventions as a whole. The lid on Pandora's Box remains ajar.

TABLE 6.2: Deaths, Age-Standardised against the Australian Population and against Federal Government employees

Standard-ised	Deaths Observed	Deaths Expected	SMR
Australian population	94	147	64
Government employees	90	84	107

References and Additional Reading

Cawcutt, L. & Watson, C. *Pesticides: The New Plague*, Friends of the Earth, Collingwood, 1984.

Christie, D., Robinson, K., Gordon, I. & Rockett, I. 'Health Watch: The Australian Petroleum Industry Health Surveillance Programme' from *Med.J.Aust.*, 1984, vol.141, pp.331-4.

Cohen, M.L. & Tauxe, R.V. 'Drug Resistant Salmonella in the United States: An Epidemiological Perspective' from *Science*, 1986, vol.234, pp.964-9.

Harrison, P. *The Greening of Africa*, Paladin, 1987.

Lifton, B. *Bug Busters*, McGraw-Hill, New York, 1985.

Morna, C.L. 'A Maize Miracle' from *New Internationalist*, June 1990, pp.10-11.

New Scientist articles, issues of 29 July and 18 November 1989.

Olin, G.R. 'The Hazards of a Chemical Laboratory Environment — A Study of the Mortality of Two Cohorts of Swedish Chemists' from *Am.Indust.Hyg.Assoc.*, 1978, vol.39, pp.557-62.

Seagrave, S. *Yellow Rain: A Journey Through the Terror of Chemical Warfare*, Evans & Co. Inc., New York, 1981.

Stilling, P.D. *An Introduction to Insect Pests and Their Control*, Macmillan International Editions, 1985.

Sydney Morning Herald. 'A Consumers Guide to the Environment', 8-page supplement, 23 November 1990.

The United Nations Economic Commission For Africa. *African Alternative Framework to Structural Adjustment Programmes for Socio-Economic Recovery and Transformation*, United Nations Publications, 1989.

The World Bank. *Sub-Saharan Africa: From Crisis to Sustainable Growth*, The World Bank, 1989.

6.3 Population Control

There is a further final aspect of intervention which many believe is most important of all, and one which is crucial to human survival and well-being. The issue is the human population, already rapidly exceeding the available resources. Thus far, action in this area has been limited, and may continue in the same vein. However, there is overwhelming evidence that such a situation does not auger well for the future well-being of human beings or for planet earth.

Unfortunately, population management and discussions about controls over numbers are controversial issues which carry political and religious overtones. Initially, it is important to establish the extent to which increases in population are to blame for the environmental problems being faced today, and the depressing cycle of hunger, famine and poverty referred to in previous chapters. This brings us to two other contentious issues, the matter of development aid, and the role of family planning for economies of newly developing countries, particularly those in Asia and Africa.

The neo-Malthusians blame most of the developing world's ills on population growth. They include groups such as the Population Crisis Committee in the US and individuals such as Prince Philip in the UK. Their basic logic is simple and apparently convincing: more people consume more of every kind of resource, from energy to land to minerals, and produce more waste. The anti-Malthusians, such as environmentalist Barry Commoner and Frances Moore Lappe, an American writer on agriculture, blame inappropriate technology, overconsumption by the affluent, inequality and exploitation which squeeze farmers into marginal land and force them to overuse it.

This highly politicised debate is not scientific but ideological, fuelled by politics and religion. The Malthusian side provides support to those who favour drastic population control programmes such as those adopted by India in the 1970s, which after a period of enforced sterilisations alienated people from family planning. The anti-Malthusian arguments back up those who wish to deny women the right of free choice in family planning, and other improvements in their position. Neither side can win the debate because both arguments are oversimplified. Both sides must accept that population is only one of the factors that lead to degradation of the environment.

Population Effects on the Environment
Three key factors determine the effects which the number of people have on

the environment. First is their level of consumption, which is determined by lifestyle and income. Second, the technology needed to satisfy that consumption and dispose of the waste generated. These two factors together decide how much environmental damage is done per person. Multiplying by the third factor, population, arrives at the total level of damage.

Worldwide emissions of carbon dioxide, the most important of the greenhouse gases, rose from 2349 million tonnes in 1950 to 6793 million tonnes around 1985, an increase of 3.1 per cent per year. Over the same period, world population grew by 1.9 per cent per year. Emissions per person rose by 1.2 per cent as a result of changes in technology and higher consumption of goods that involve production of carbon dioxide.

Population growth therefore accounts for almost two-thirds of the increase in carbon dioxide entering the atmosphere between 1950 and 1985. Increases in consumption and in technology together account for a little over one-third. If present trends continue, the future impact of population growth on emissions of carbon dioxide looks alarming.

Population Growth in the Developing Countries
Predicted increases in greenhouse gas emission levels from the developing world cause greatest concern for future planners, and this is associated not only with the recent rapid rise in industrialisation but with the traditionally high birth rate in these countries.

If output per person in newly developing countries such as those of southeast Asia continues to grow at the same rate as over the past forty years, the average person in these parts of the developing world will be producing 1.7 tonnes of carbon dioxide each year by 2025, more than double the current level of 0.8 tonnes. Meanwhile the number of people would have risen from 3680 million in 1985 to 7114 million. The population increase in the Third World alone would therefore produce an extra 5.75 billion tonnes of carbon dioxide, which is not far short of the current world total of 6.8 billion tonnes.

In estimating the growth of the world's population, the United Nations has made several projections giving a low, medium and high figure for the year 2025. According to the low projection, world population will reach 6331 million, 783 million less than the medium projection. It is quite possible to achieve this lower figure: it would not require a rapid decline in birth rate of the type seen in China, Thailand or Cuba, but would require the more modest reductions achieved by countries such as Tunisia or Jamaica. Such reductions require a wide choice of freely available family planning methods. Even more crucial are improvements in mother and child health, female education and women's status — all valuable measures in their own right.

Achieving the low population projection in developing countries would reduce annual carbon dioxide emissions in 2025 by 1330 million tonnes without reducing carbon dioxide output per person. This compares with some 1570 million tonnes currently produced each year as a result of tropical deforestation.

Slower population growth could make an even bigger contribution in the case of methane, another important greenhouse gas. About half of 'man-made' methane emissions come from decomposition in irrigated fields and the guts of livestock. These are not examples of wasteful consumption that could be cut back. The area of irrigated land and the number of livestock have expanded to provide livelihoods for growing rural populations, and to meet the world's increasing demand for cereals and meat. The irrigated area has grown by about 1.9 per cent per year since 1970, about the same rate as the world's population. The number of cattle has grown only about half as fast, averaging 0.9 per cent a year.

Livestock and irrigation will continue to expand in line with populations in developing countries. Slowing the growth of the population is the only feasible strategy for reducing the increase in methane emissions from these sources. When the environmental problem involves deforestation, loss of species, loss of soil fertility and erosion, the influence of increased greenhouse gas emissions becomes more complicated.

Land Use and Population
The United Nations Food and Agriculture Organisation calculates that forests and woodlands in developing countries shrank by 125 million hectares in the fifteen years to 1986. During this period, the area of pasture land in the Third World increased by 7.9 million hectares, which equates with only a 6 per cent loss of the forests. Latin America accounted for the majority of this increase: in Asia the total area of pasture did not increase at all and in Africa it shrank. Increases in non-agricultural land such as dwellings, factories, offices, roads and infrastructure accounted for some 58.7 million hectares over the same period. As towns expand mainly in agricultural areas, most of the increase has been at the expense of existing agricultural lands.

Despite this loss, the total area of farmlands grew (58.7 million hectares). This means that an expansion of approximately 100 million hectares in farmlands must have occurred, over half of which would merely compensate for losses to non-agricultural sectors. Most of this expansion will have been at the expense of forests and woodland. It probably accounts for more than 80 per cent of deforestation. It has been estimated that the rest may be due to degradation through logging or overgrazing.

What share of the blame for the loss of forest to cropland can be sourced to population growth? Between 1971 and 1986 cropland expanded by 0.51 per cent a year. Population grew by 2.2 per cent a year. Food consumption per person grew by 0.58 per cent a year. Technology change, in this case, improved yields so that the area of cropland needed per person decreased by 2.3 per cent a year. So, of the two factors pushing for an expansion of cropland, population accounts for four-fifths of the effect, the increase in consumption accounts for only one-fifth. On this basis, the rough calculation is that population growth was responsible for around two-thirds of deforestation in developing countries.

In southeast Asia, the pressure to earn foreign currency and the lure to entrepreneurs of windfall profits from logging carry much of the blame. Recent massive deforestation in Amazonia is more the result of government policies encouraging land clearance for ranching and farming as an alternative to land reforms. Within the colonised areas, however, population growth will create pressures to clear more forest to provide land for settlers' children.

The population of developing countries should reach a plateau at around 9.1 billion towards the end of the next century (up from a present 4.1 billion). The extra 5 billion people will need roughly an extra 280 million hectares of land for non-agricultural needs. Experience shows that this will come mainly from prime agricultural areas.

If improvements to agriculture through techniques such as intensification do not keep pace, the amount of land given over to crops will have to increase to make good the deficit. Slower population growth would make a substantial contribution to decreasing the inevitable pressure for further deforestation and pasture growth.

The Role of Soil Erosion

Soil erosion is a more contentious issue. According to anti-Malthusian Lappe, inequality and cash-cropping on large estates force the rural poor to farm dry, hilly or infertile areas, which are vulnerable to erosion. Piers Blaikie, a sociologist at the University of East Anglia, UK, believes that exploitation by national and international elites, rich landowners, large companies, etc., pushes the poor below subsistence level. They are then forced to mine the soil simply to survive.

Soil erosion is a function of several factors. The more torrential the rainfall, the greater the erosion. Erosion is also faster if, for example, it has less organic matter to facilitate clumping of soil particles. Erosion is highest on steep slopes where it occurs over longer time periods. It is greatest where vegetation cover is least.

Population growth affects several of these factors. As human numbers grow, the area of open fields expands. This increases the length of any particular slope. As dense, fallow vegetation declines, the overall vegetation cover is reduced. Expanding livestock herds help to thin the vegetation by grazing. There is an associated decline in the amount of organic matter contained in the soil.

Assuming that technology does not change these estimates, conservation techniques can reduce the damage. Terracing or contour hedging can reduce the length and degree of slope. Adding compost and mulch can increase organic content of soil. Feeding livestock in stalls can reduce grazing pressure. However, in only the more advanced of the developing countries have conservation techniques managed to keep pace with population growth.

Poverty, exploitation, misguided government policies, all have significant implications in increasing the pressure for technological solutions. Poverty or low farm prices, for example, starve farmers of funds, prevent investment and inhibit technological change. Inequality in land ownership artificially confines the poor to smaller or more marginal lands. Within those areas, higher population densities will lead to greater erosion unless technology keeps pace.

Will Population Growth Be Worth The Effort?

Population growth aggravates many types of damage to the environment. Slowing that growth reduces the damage but involves a time lag, sometimes of twenty years, before noticeable improvements occur. In the shorter term, other measures have greater impact: reducing consumption, shifting to sustainable technologies, halting deforestation, attacking poverty and inequality, and introducing land reform.

In the medium to long term, reducing population growth can have a very significant impact. To achieve this goal, governments, development agencies and aid donors need to focus their attention on enhancing the rights, education and health of women and children. This will improve both the health of the people and the environment.

Population Control in the Developing World

Aside from moral and religious issues involved in population control, the final ticklish issue may revolve around the question of 'How'. Emphasis has tended to be placed on women, the child bearers. However, birth control and family planning programmes in the developing world have frequently floundered because of a lack of awareness of social factors and the influence of culture. Often the women who are made the focus of such campaigns remain under the nominal control of their husbands, or are restrained from acting according to their own wishes because of traditional values and social pressure. At other times, they themselves may be socialised into traditional ways which may advocate large families. In these instances, considerable education is required before attitudes and values become more receptive to new ideas concerning family size.

In certain societies greater attention may be placed on the male. Vasectomy has proved successful in limiting family sizes only where financial incentives or gifts are offered in exchange for agreement. India, Indonesia and other parts of Asia serve as isolated examples where such methods have achieved mixed success. In Thailand recent emphasis has been placed on encouraging the use of condoms. The AIDS scare and the prevalence of sexually transmissible diseases have further strengthened support for this method of birth control.

Natural methods of birth control such as the Rhythm and Billings Methods, are less likely to inflame traditional religious authorities and proponents of the Right To Life movements. At the opposite extreme are the controversial methods of direct intervention — abortion, the use of pharmaceuticals and chemicals including 'morning-after' pills, surgery or a variety of folk techniques using massage and herbal medicines.

In the middle, but still classed as interventionist, are the intra-uterine devices. These range from Lippes Loop to Copper-7 and the Dalkon Shield, still widely used in the Third World despite their implication in legal suits for birth-related damages in the United States and other areas of the world. Here, too, there is a tendency to use the simpler chemical methods which require least active though continual involvement by the women. Long-acting injectable contraceptives, such as Depo-Medrol, are used to provide contraceptive protection for periods from six months to a number of years. Oral contraceptives, available at greater cost and requiring regular compliance, are less favoured in traditional societies, particularly where levels of education are low and incomes small.

References and Additional Reading

Harrison, P. 'Too Much Life On Earth?' in *New Scientist*, 19 May 1990, pp.10-12.
—— *Inside The Third World*, Penguin, 1987.
The United Nations Population Fund. *The State of World Population 1990*, United Nations Publication, 15 May 1990.

7.0 CONCLUDING SECTION

7.1 Economics and Politics

The final, concluding Section to this book is where we look back over what we have learned in the context of our own experiences in the environment, and with health, at this point in time. It is also the time when we try to reconcile a range of scientific facts and theories with the realities and practicalities of daily living.

To a certain extent, the difficulty of converting scientific findings into political action is a function of the uncertainty of science and the pain generated by the action. Sometimes human intervention brings overwhelming improvement in living conditions and human health. Public health measures continue to substantially improve the life expectancies of whole communities.

On other occasions, gains may be more qualified and benefits may be mixed, as we have seen with the use of pharmaceuticals and pesticides. Sometimes the action needed may be considered necessary, but governments and competing nations are unable to make the commitment because of sacrifices required of them in the race for development and growth. The limiting of greenhouse gas emissions and the banning of CFCs are two such issues. Population is another involving ideologies and beliefs as much as governments, industry and people. In practical terms, decisions also become economic and political issues, dependent on the success of competing political lobby groups and reliant on financial commitments from governments and industry.

The Contentious Matter of Commitment

Given the current uncertainties surrounding just one aspect of the global environmental crisis — the predicted rise in levels of greenhouse gases, with their associated effects on the climate — and the enormous technological and social effort that will be required to control that rise, it is fair to say that responding successfully to the multifaceted crisis will be a difficult political enterprise.

It means trying to get a substantial proportion of the world's people to change their behaviour in order, perhaps, to avert threats that *may* affect a world most of them will not be alive to see. In other words, to forego their own immediate needs in case future generations will suffer.

How, then, can we make changes happen? Scientific studies on individual issues have documented the reality of global ecological threat and they have pointed to some ameliorative measures. The final Section of this book examines policies, the design of programmes and their successful management. The problem for the future is to develop policies and programmes which will convince ordinary people throughout the world to make firm and continuing commitments to permanently alter their behaviour.

Global Action
The 1972 United Nations Stockholm Conference on the Environment marked a watershed in global thinking. For the first time, recognition was given to the important link between earth's ecosystems and the rate of economic development. One argument said that poverty led to destruction of the environment, and crash programmes of western-style development would remove that threat. The other, more radical argument claimed that such crash programmes in backward Third World nations to raise living standards for poor people were responsible for environmental destruction; and that this damage was frequently irreparable and permanent.

A series of conferences, scientific meetings and increasingly more alarming publicity on the state of earth's environment have raised the tenor of the debate and heightened awareness beyond the initially low key, minimally publicised discussions of the first Stockholm meeting. Fifteen years later the well-meaning Brundtland Report, sponsored by the United Nations and entitled 'Our Common Future', was published in 1987. Though intellectually muddled and turgidly written, this document took the prospect of green capitalism as an article of faith and propounded the concept and feasibility of 'sustainable development'. In itself, however, it forges new pathways and has relevance as the basis for concerted and consistent action plans in all member nations.

Steps On the Road To Earth Repair
'Our Common Future' was published by the Brundtland World Commission on Environment and Development. It sought to identify causes for environmental ill-health and to offer solutions. These are:

(i) Regional poverty, which creates a vicious cycle of inequitable distribution of wealth and resources such that people, particularly the poor, abuse their resources in order to survive.

(ii) Uncoordinated economic growth which uses resources and generates waste products in a manner insensitive to environmental limits.

(iii) The process of environmental abuse has brought us to the thresholds of the integrity of many ecosystems, including that of the atmosphere itself.

(iv) The world economic system, with its dependence on Third World Debt and penalties against sharing efficient technology, enforces short-term exploitation of environmental systems.

A most poignant example of these causes and a clue to the difficulties facing those who search for solutions is apparent in many developing countries such as the Philippines. Here, villagers in remote areas of Mindanao have abandoned their marginal farmlands and agricultural lifestyle in order to mine gold in the nearby mountains. They labour on slopes prone to land slippage and regular cave-ins, using crude equipment. Many work underground in tunnels lacking timber support structures to limit collapse, and risk their lives to fossick for the valuable mineral. Aware of the dangers and risks, and despite government attempts to close or prohibit mining in dangerous areas, these desperately poor peasants continue to collude with the owners and jointly defy the authorities.

What makes matters worse and increases the dangers to their health is that these same miners also use toxic mercury and its salts in extracting and purifying their minings. Here, too, and for as long as they are able, victims of mercury intoxication who are suffering degrees of confusion, disorientation, and loss of muscular and nervous control continue to work alongside the healthy.

Sustainable Development

> 'Our Common Future' is not a prediction of ever-increasing environmental decay, poverty and hardship in an ever-more polluted world among ever-decreasing resources. We see instead the possibility for a new era of economic growth, one that must be based on policies that sustain and expand the environmental resource base. And we believe such growth to be absolutely essential to relieve the great poverty that is deepening in much of the developing world. But the Commission's hope for the future is conditional on decisive political action now to begin managing environmental resources, to ensure both sustainable human progress and human survival.
>
> (Australian edition, 1990, page 1)

'Sustainable development' is the key concept and theme of the Brundtland Report. It promises to become the environmental catch-phrase of the 1990s, a political slogan wielded by the different sides of the environmental debate. For this reason, many environmentalists are becoming suspicious of the term itself. Basically, the phrase 'ecological sustainability' underscores the belief that human society can only be sustainable in the long term by conserving the environment. In spite of the uncertainty of meaning, the fact that the concept of sustainability has moved to the centre stage of political, social and economic debate is a very significant measure of the success of the environmental movement on a global scale.

Definitions abound for 'sustainable development'. The World Commission for Environment and Development has defined it simply as 'development that meets the needs of the present without compromising the ability of future generations to meet their own needs ...'. This definition contains two key concepts: 'needs' (in particular the essential needs of the world's poor to which overriding priority should be given); and 'limitations' which are imposed by the state of technology and social organization on the environment's ability to meet present and future needs. The WCED report also emphasizes that sustainable development is a process of change in which painful choices will have to be made. Ultimate success depends on political will.

A cursory reading of newspaper coverage in recent times indicates that 'sustainable development' in most western countries has been interpreted to mean almost whatever suits a particular interest in the least 'painful' way. For example, the Australian Mining Industry Council believes that sustainable development involves the relaxing of mining and exploration restrictions on national parks. At a more sophisticated level, politicians from each end of the political spectrum maintain that sustainability is only possible in the context of continued economic growth. They point to the Brundtland Report to support their views.

Ecologists, on the other hand, believe that economic growth which entails further expansion in resources consumption is unsustainable, and that there is a need to redefine societal objectives in terms of qualitative growth in human welfare whilst achieving a decline in the level of material resource consumption, especially in the developed world. In this context, environmental groups around the world are now asking how do people deal constructively with the sustainable development debate? How to ensure that the focus is on ecological sustainability and that political momentum is not lost before real reforms are negotiated rather than compromised?

The level of debate is increasing in all developed and developing countries throughout the world, as peoples everywhere begin to realise that the threats and grave scenarios of scientists are not imaginary and exaggerated. A huge international process is building momentum for the biggest Conference so far planned to address the imminent problems associated with earth's future survival. Scheduled for 1992 is a full UN General Debate in Brazil, which promises to be of immense if not unparalleled political significance.

Although the importance of the UNCED 1992 debate is not denied by many in the environmental movement, there is a concern that by becoming too deeply involved in the process the movement could lose momentum on issues such as climate change, rainforest destruction and other important issues. Sceptics point to exercises such as the World Conservation Strategy and the National Conservation Strategy for Australia, as examples of processes which produced little fundamental change in spite of the input of much time and energy. On the other hand, however, the opening up of a structured debate on sustainability provides an unprecedented opportunity to develop new social and political institutions and policy frameworks that could assist in fundamentally resolving the world's environmental crises. From this perspective, the failure of the environmental movement to play a central role, either through a conscious decision or simple oversight, would be disastrous.

Beyond Brundtland

A myriad of books and reports on the state of the environment have preceded and followed the Brundtland Report. They have raised the level of public knowledge and heightened awareness of issues concerning exploitation of resources, energy use, industrial development, the quality of air, water and the human environment experienced by individuals.

In the industrialised developed nations, reviews of past errors and indiscretions, evidence of environmental exploitation and damage have resulted in widespread calls of concern. However, faced with commercial boycotts, threats of inevitable business failure, unemployment and curtailed growth, government promises for action and concerted efforts to prevent further denigration have been toned down. Steps to ameliorate damage have been delayed or graduated over years. Frequently, commitments to action have failed to result in concrete and definitive plans.

Similarly, in Third World nations, agreements and commitments once made at international levels are softened or paid lip-service nationally when political will and action is demanded. The complaint is made that the poor countries are being forced to sacrifice their chance of catching up with the developed world, and of sharing in the benefits of economic development, because of the damage done to the environment by these very countries. They argue that they are equally entitled to exploit any development opportunities available to them and to use those of least cost, whether or not they result in damage, emit pollutants or exploit scarce resources. They, too, should be free to harvest their timber for energy and export in the same way the developed countries stripped their own lands.

To date, a few suggestions have been made for individual action and personal behavioural change. Yet, on the whole, these have been confined to western developed nations. Consequently they have failed to inform and galvanise action in countries of the Third World where the new thrust for economic development, accompanied by rampant environmental destruction, is creating widespread concern. In most cases, in First and Third World countries, the central issue of economic growth has been avoided.

Yet this is surely the pressing issue. There are many ways in which governments and people can move beyond the mentality which sees growth and development as necessary ideals. Sustainable development can be viewed as a method for 'empowering people'. In practical terms, in the Third World, this means putting peasants rather than proprietors and absentee landowners in charge of farm management; forest dwellers rather than foresters in charge of plantations and forests. Thijs de la Court, for example, suggests that shanty town dwellers can take control of their lives by growing vegetables in their slum areas (de la Court, 1990). This suggestion provides a powerful image of people, discarded by the development process, re-establishing their contact with the land in a hostile urban environment.

Neglected also have been considerations of public policies — the actions that governments might take to address environmental problems. One of the few such publications, *The Return of Scarcity* by H.G. Coombes, suggests that governments need to be forced to implement effective legislation to encourage individuals and companies to make more efficient use of energy. He suggests that governments can tax petrol and polluting chemicals to provide disincentives for their use. They can provide incentives and subsidies for industries which exploit the natural sunlight and the energy of winds for power generation. These are not new ideas. Coombes concedes that, in the final analysis, commitment and will to make sacrifices are essential.

Economists and Sustainable Development

Economists have also addressed the problems, probably more from compulsion than desire, because the changes required in finance, costing and global markets are enormous. Significant among these have been the ideas of David Pearce, who released a major contribution to the debate in 1989. *Blue Print For A Green Economy*, Pearce's document, sets out to answer two main questions:

(i) Is sustainable development a practical concept?

(ii) Can economics give it structure and substance?

His ideas are worth consideration here because they present a genuine and feasible effort to seek solutions.

Pearce argues that the same economic principles as have guided those development-oriented actions which created the environmental problems being faced today, can be used to solve them. He calls for a synthesis of policy and economics, at a national level, which for Pearce is the United Kingdom, followed closely by action at an international level. Yet in ensuring these goals, economists such as Pearce are seeking to re-define and re-direct the very principles that they have advocated — growth, 'development', increases in capital assets, etc. They are abandoning the form of development which uses and produces a commodity that has no further value and of which no further use may be made. They are seeking a form of development which creates a product equally as usable and valuable as that of the original item from which it was made.

To tempt and attract converts, economists maintain the incentives of wealth, status and power associated with economic development, but add a further higher ideal appropriate to Maslow's final fifth category of basic human needs. These higher wants and needs include the satisfaction of contributing to the philosophical capital of the race and, by action, benefiting all mankind. The environment, other social, philosophical and natural circumstances in which humans live neatly fulfil these goals.

Thus, productive efforts become directed towards restoring damage caused by the earlier actions, now recognised as having been misguided and ill-considered. New capital is created from the use of resources, so that those which are not easily renewed or replaced are being turned into further resources of increased worth and value for the future. Notions of equity and social justice appear. Goods created become new capital which benefits the whole, not simply the élite, privileged few who might have gained from the use of the original resource. At the same time, the time horizon lengthens beyond the life-span of the individual to that of a society, a way of life and a people.

Throughout the work, Pearce's theme is clear:

> 'future generations should be compensated for reduction in the endowments
> of resources brought about by the actions of present generations'
> (Pearce, 1989, p.13)

The policies promoted and the ideas canvassed aim to create more total 'wealth' than they use, and leave more valuable assets or capital for future wealth creation. This represents a fundamental reappraisal of the way development goals have been pursued to the present day.

Pearce's Notion of Capital: Pearce begins by defining two forms of capital. 'Wealth' capital represents the end product of human activities and ranges from buildings, computers and roads to money, human intelligence and institutions of government. 'Natural' wealth or capital he defines as living species, the planet and its basic resources, soil, timber, water, etc. In the past, creation of the former involved denigration and loss of natural wealth. Now, he argues, economists need to appreciate that wealth capital will only continue to grow if natural capital is not lost but also permitted to grow. This dual notion of capital thus forges a link between production, economics and the environment.

Compensation or Revenue?: Pearce talks about the need to compensate future generations for present-day use and abuse of resources capital, and of damage done to the environment by human activities. Compensation is applied to activities by the present generation in anticipation of the needs and rights of their descendants.

Inherently, the notion of compensation implies guilt. In the use of this term, however, Pearce is speaking about wealth capital generated and produced by a generation for its use now and in the future. Why should the producers of that wealth apologise for it, especially if its production does not compromise the total available capital for future descendants? Pearce seems to be hinting that capital in the form of wealth capital is 'not good' — almost an overt anti-capitalist's criticism. In contrast, natural capital is assumed to be 'good' and to carry high value. In this scenario the term 'compensation', with its implication of guilt, fits comfortably and appears consistent .

There is some justification for the use of the term by present generations to cover their uses of finite, non-renewable resources, because their use will lower its stores and down-grade the total remaining resources. Unfortunately, however, the notion of compensation sets the underlying negativity to a work which otherwise might be regarded as positive, optimistic or — an alternative is suggested — the word/term 'revenue'.

Anticipatory or Reactive Planning? David Pearce favours anticipatory planning — action now in expectation that failure to act will mean a more negative future, a worsening of problems all based on existing and admittedly inadequate information. It is a worse-case scenario. It adopts costly, unmeasured strategies without prior resource into cheaper alternatives, targeting better policy and further research.

A reactive policy (a wait for more information before action) approach may be more cost effective, better targeted and planned to an economist, but it is incompatible with sustainable development goals. Should the predicted but unmeasured negative effects result, it can prove more costly and difficult to repair than would have been the case if anticipatory action had been taken.

Throughout the book, Pearce goes on to use the principles of conventional economics (growth, supply, demand, quality and quantity curves and notions) to argue that sustainable development can be and is compatible with capital accumulation; i.e. to overturn these assumptions (anticipatory versus reactive planning). He does this by showing that:

(i) economic quality frequently improves quality of life and fuels economic growth: e.g. improved health of the workforce, job creation in pollution abatement sector and waste treatment industry;

(ii) discussion of sustainable development turns focus away from economic growth to consideration of economic policy due to implicit association of growth with capital with the term 'development';

(iii) a recognition that trade-offs are inevitable between economic growth and environmental quality once a term like 'sustainable development' is adopted.

In doing so, Pearce draws attention to notions of values and prices traditionally associated with trade-offs and rationality involved in such decision-making; and to searches for initiatives on how to achieve goals (planning by objectives).

But, more importantly, Pearce forces traditional economists to realise that, to date, they have been making their assessments of growth and basing their arguments on use of misleading indicators. Use of the term Gross National Product (GNP) has assumed that increased wealth capital is accompanied by improved quality of life. On the contrary, processes involved in creating that wealth may degrade quality of life. In other words, the notion of growth or GNP has ignored 'natural capital' and the loss of finite capital resources that may accompany capital wealth creation endeavours, geared to achieve a maximal response. The result is an emphasis on the environment and the future rather than simply wealth creation and the present.

But what insight can Pearce provide, using economic arguments? Where does his Report disappoint or appear limited? First, Pearce offers economic arguments to bolster the political will of governments and industry. He might be accused of raising no new ideas and offers no new suggestions to solve the pressing environmental concerns faced by the planet. Nor does he attempt to suggest any new ways to make them possible technically and socially. Instead he offers advice using the tools with which he is familiar and skilled to provide new insights. Thus he redefines traditional economic indicators pertinent to national status, such as GNP, to make slower growth appear more attractive to commerce, policy planners and governments. Of greatest disappointment is the inability of Pearce, like those before him, to offer advice on handling of wastes.

The second barrier faced by Pearce hovers throughout his Report: the problem faced by behavioural psychologists and advertisers, campaigners for public health and the ideologues of political parties — changing human attitudes and altering their behaviour. Surveys have established that members of the public claim they are prepared to pay more for goods which are 'environmentally friendly' and for products which are well thought-out, and to buy substitutes for polluting detergents, aerosol sprays and potentially hazardous but tasty food additives. However, when faced with a price difference which may be quite small or a product which is clearly less satisfactory, once used, these consumers return to their old ways. They opt to save money or to carry on regardless of the dangers.

In a similar vein, indications of intent and willingness are equally flawed. Constructive, satisfactory plans meeting the concerns of all interested parties are difficult to achieve. They are equally hard to put into practice and achieve results. A positive response and change in practice from public or industry is rarely maintained, particularly after initial enthusiasm has waned or policing of infringements (for example of laws against polluting emissions) is seen to be ineffective or non existent.

Politics injects a further element, particularly nationally between parties competing for funds or support from wealthy industries. For their part, industries are reluctant to pay for polluting and to admit infringements voluntarily. They are disinclined to spend profits on technologies for waste containment and management, or to clean up the environment they have despoiled.

Pearce makes little headway in giving government powers to enforce unpopular measures on reluctant industries unwilling to cut profits or restrict growth. Their success rests on achieving voluntary support and enlisting public support, making the change from short-term to long-term goals a policy for all.

Earth's Environmental Sinks
Pearce cannot solve the problem of environmental sinks — neither can Brundtland nor other scientists. The finite nature of earth's environmental sinks — the oceans, the atmosphere, the firmament or lithosphere — is really the same for all. Waste is the problem. It is true that many materials can be recycled. However, there are limits to recycling, and energy is the one entity which, once used, is lost totally. Lead in batteries can be recycled but lead in petrol exhaust is dissipated/lost into the atmospheric sink. Human initiatives are required to meet the decline in capacity of earth's sinks and to find alternatives to them. Pearce, importantly, concedes this barrier and attempts to scale it — alas without success. Resources, even finite natural ones, can be substituted but garbage tips are limited in their capacity.

Signs of breakdown in environmental sinks signal major disasters. An immediate example might be ozone but the jury is still out, evidence is still accumulating. It took millions of years for earth's mantle to acquire the ozone barrier which enabled the first living creatures to venture from the seas onto the land. Eventually these creatures gave way to newer, higher forms of life including the animals and the primates which, like us, depend on the integrity of the ozone mantle for continued life on the planet in its present form.

An Answer to Pearce

Proposals for immediate action are controversial because they often entail large and immediate investments as insurance against future events which are far from certain. However, it does make sense to take actions that will lessen the environmental stresses; actions which would unite and 'condition' people and governments to change their behaviours, embarking on more cautious strategies towards new development initiatives, and looking for alternatives to those technologies which result in the environmental degradation.

Pursuing energy efficiency is one component to this strategy. More efficient fossil-fuel use will slow the carbon dioxide buildup in the atmosphere. This makes good economic sense even if the predicted effects of carbon dioxide buildup prove to have been overstated.

A reduction in fossil-fuel use would curb acid rain and urban air pollution, and lessen the dependence of many countries on foreign producers. Independent of any climate change predictions, developing alternative energy sources, revising water laws, searching for drought-resistant crop strains, negotiating international agreements on trade in food and other climate-sensitive goods are measures which would prove beneficial. Nonetheless, these measures are costly and politically controversial. Regulations or incentives to foster energy-efficient technologies will create unemployment for coal miners and timber loggers, leading to the collapse of coal-based industries and producers. Poorer people will be least able to respond to the changes: poor countries will be unable to switch to more expensive and sophisticated forms of energy, de-commission coal-burning power houses and adopt solar, wind or nuclear alternatives.

Actions to prevent a greenhouse warming will have to be coupled with domestic and foreign policy measures that attempt to balance fairness and effectiveness. On the positive side, direct investment in traditional and people-based forms for energy production, such as the more sophisticated use of methane digesters in rural China, has been argued to be more valuable and effective in the long-term fight against poverty and backwardness. These policies would act by stimulating initiative and sponsoring self-sufficient industries within each individual country. They would also develop environmental understanding and sensitivity in the people. These arguments, however, fail to recognise the perceived advantage already gained by developed, First World nations now enjoying the fruits of fossil-fuelled development.

Others argue with David Pearce that the 'true' costs of polluting industries and products on the community, and the loss of environmental capital through mining and timber use, are not incorporated into the end price of the commodity. They place a price and a value on the environment and its natural renewable and non-renewable resources, and include any additional costs for environmental repair, recycling. The philosophy behind such policies — sustainable development — is based on the notion that the present generation should not deny its descendants any fewer wealth goods or environmental capital than it received from its forebears. They use the free market, not government regulation or tax incentives, to dictate increases in energy efficiency and the elimination of CFCs.

The Watershed — Government Action
Implementing sustainable development and economic principles requires comprehensive and uniform national policies. Governments have begun looking at the ideas proposed by Brundtland and Pearce. Areas of concern to them are financial and legal; in other words, what it will cost them, how to get the money, from whom and how.

In the financial sector, success lies with resolving key problem areas. First, that goods and services and natural resources be at their true cost to the community, and at market values. To date, environmental resources have been treated as 'free goods', and their true value is not reflected in their price either at source or in the marketplace for the finished product. Second, government policy needs to reflect more truthfully the 'user pays' principle. Making the polluter pay, particularly at the industry level, is designed to encourage conservation in the use of scarce resources, and increased use of recyclable, renewable and environmentally friendly alternatives.

These policies require the polluting industry to compensate communities and governments and to repair damage (where possible); e.g. purchase of a 'licence to pollute' and granting of permission to discharge wastes, at a cost per unit of discharge. Such rights need to be realistically costed and priced so that use of alternative methods and techniques is considered to be worthwhile investment by these industries. Prohibitive pricing and penalties must ensure that individuals and industries are conscientious in meeting guidelines and complying with legislation. Policing and monitoring must be adequate to discourage non-compliance and dishonest practices.

Also suggested as a government strategy is the concept of 'conservation bonds'. Under this scheme, industries and firms wishing to exploit natural resources are required to purchase conservation bonds, shares or commercial paper. These require the company in question to reinvest a certain minimum proportion of the value of resources used by them in the restoration of the resource which they are using; e.g. costs of re-forestation to be met by timber export and logging companies.

A final suggestion relates to the use of more imaginative accounting principles to monitor environmental costs and use of resources. For example, credit and debt principles can be applied to the balancing of capital resources use budgets, profit and loss statements, etc.

Legislating For The Environment
Additional legislation aside from that which provides for the above financial initiatives, can also be suggested. An example lies in the operation of Forest Accords.

The concept of Forest Accords legislation was initially conceived as a series of principles or policies to control use of forests by government groups and individuals, industry, researchers and recreationists. These Accords were to be followed by full negotiated trade-offs and bargaining between conservationists and the timber industry. The idea was that the initial round of Accords would commit the timber industry to an agreement which involved their admission that forests needed to be conserved for posterity, not just future logging. Once 'locked in', the industry would be forced to accept steadily increased apportionment of forested areas for conservation purposes. Eventually, the industry would be forced to abandon any claims to native timbered areas and to rely entirely on plantation and marginal forest for its needs.

The Forest Accord concept was first developed by the New Zealand Labour Government after the Second World War. The idea was first conceived as a way of resolving continuing conflict between the woodchipping industry and conservation groups. The further extension of the idea to embrace the concept as now viewed, occurred at a later date.

The Forest Accords worked in the following way. They were first implemented in government-owned forests. Conservationists surveyed the forests and moved to ensure that public forest was divided into zones which could and could not be logged. This was followed by gradual extension of the zones for non-logging. Eventually the situation became such that it is uneconomic to log the forests at all. Stage 2 extended the policy from public to private forests. The Accords thus proved to be a form of negotiated agreement to gradually phase out all logging.

Flora and Fauna Guarantees are a similar idea proposed, in this instance, to conserve endangered species. They involve granting special status to a species which has been shown to be in danger of extinction or whose environment is threatened by imminent development. The Guarantee thus provides an opportunity for special areas to be designated to ensure protection and regeneration of the species. At the same time, conservationists could organise programmes to encourage increase in numbers of species and knowledge about the essential environmental requirements for their continued survival.

It is envisaged that a guarantee and associated status of endangered species be granted only by a special expert committee of scientists. This committee would regularly report to the Environment Minister on listed species, and act to ensure that endangered species received adequate protection. They would also lobby local authorities and inform the community about threat to the species.

Other suggestions have been raised concerning the value of a form of national service or conscription for all young people. This would involve:

(i) mandatory/obligatory service following completion of secondary education — no abstention;

(ii) two years in length involved in physical fitness training, basic survival skills, training and work in administrative, technical (machinery and servicing) or national reconstruction core — housing, soil regeneration projects, tree planting, emergency services for natural disasters work;

(iii) special voluntary units for national defence and policing of nation's boundaries against illegal entry, drug smuggling and infringements of entry and customs regulations.

Popular support for this idea is not common and would require concerted government promotion and major policy and attitude shifts at community level.

The Present and The Future
Most rivers in the industrialised countries where the population and per capita GNP are stable and decontamination procedures tend to be fairly effective, are nonetheless polluted by both traditional and industrial wastes. Yet some stabilisation and improvements in pollution levels have been reported since the early 1980s. Less promising have been measures to improve and degrade inorganic pollution, mineral and toxic chemical loads.

Where increasing industrial activity in a river basin has been matched by increasing waste treatment, a decent level of water quality can be maintained. Yet the balance between contamination and decontamination is precarious. A serious accidental discharge, such as occurred in 1986 following a fire at the Sandoz pharmaceuticals factory on the Rhine in Switzerland, was sufficient to wipe out large numbers of aquatic organisms and force drinking-water purification plants to close their intakes downstream from the accident.

In most newly industrialising countries, both organic and industrial river pollution are on the increase, since the per capita GNP is rising quickly together with population, and decontamination efforts tend to be neglected. Major water quality problems are apparent in Eastern Europe, Czechoslovakia, East Germany, Bulgaria, Rumania and Poland. In these countries, rapid industrialisation has been a high priority. In Eastern Asia and South America, the same policies are being followed and pollution reduction and minimisation measures are being avoided to keep costs down and increase the speed at which development can occur.

In less developed countries where the population is growing and where waste treatment is practically non-existent, water pollution by organic wastes is widespread. As a result, millions of people, especially children, die from water-borne diseases that could be prevented by proper sanitation measures. These countries still suffer from diseases eradicated in the West long ago. Although the United Nations declared the 1980s to be the International Drinking Water Supply and Sanitation Decade, and instituted a programme to provide safe drinking water and appropriate sanitation for all by 1990, the programme's ambitious goals have not been met. The most notable progress has been made in Mexico, Indonesia and Ghana.

The Future for Water Resources
Improved water management practice, however, should be but one aspect of a more comprehensive approach to the hydrosphere. For example, ecological and toxicological studies of marine life are required to improve husbandry of the oceans and gain a better understanding of their role in the water cycle. Many aspects of the hydrological cycle, including the fluxes between its compartments and the extent of groundwater reserves, are not accurately known. These problems and others are currently being addressed by the International Hydrological Program of the United Nations Education, Scientific and Cultural Organisation (UNESCO). Major international research programmes studying interactions between climate and the hydrological cycle have recently been launched by UNEP, as well as WHO and the non-governmental International Council of Scientific Unions.

Land Management
Land management principles are required to back better use of water resources. Restoration measures should embrace:
   * re-planting depleted forest areas, e.g. in and around west German cities.
   * re-introduction of wildlife species and recolonising of forests with wildlife which had become depleted through disease and predators; accompanied by protection measures and support through vaccination programmes and supplementary feeding, e.g. red wolf in Canada, small rodents and reptiles in north-eastern Australia.
   * re-stocking of inland rivers and estuaries with fish lost through environmental accident or over-fishing.
   * re-cycling of formerly non-degradable wastes such as plastics through use of new technologies.

Global Politics and The Legislative Opportunities
While much must be done at the national level, by individual countries, initiatives and co-ordination on a global scale are essential. This is particularly the case in matters such as the Greenhouse Issue and Climate Change. The remaining pages in this Section detail progress in achieving success in this arena.

Without pre-empting the issues, it is hard to avoid cynicism when considering the extent to which money and domestic politics undermine the level of commitment countries such as the United States are prepared to make, when the time comes for them to be signatories to an effective Agreement.

A Bit of Background
One of the more important recent proposals for an international framework to resolve the problem of greenhouse gas levels in the atmosphere has been focused on emissions of CFCs. These chemicals are among the most negative in their impact upon earth and human health, through their role in temperature management and ozone level depletion.

The suggestion to work towards reducing levels of CFC was first raised in 1976 by the anthropologist Margaret Mead, and William W. Kellogg of NCAR. Together they proposed a 'law of the air' which would keep emissions of carbon dioxide below a global standard by assigning polluting rights to each nation.

Although it is only in the past few years that climate change has risen to the top of the international political agenda, the possible impact of changing atmospheric composition on global climate has attracted much international scientific study and cooperation for more than fifty years.

The present major impetus was initiated in the United States General Assembly in December 1961, and followed regular meetings during the 1950s of the World Meteorological Organisation's (WMO) Commission for Climatology. The General Assembly was responding to a call by US President Kennedy for closer collaboration between WMO and the non-governmental International Council for Scientific Unions, over development of satellite and computer technology to address the challenges of weather prediction and climate control. A Resolution 1721 (XVI) set in train a period of greatly accelerated development of atmospheric science and technology in support of human needs and wise use of global environment.

The twin birth in 1967 of the WMO World Weather Watch and the WMO-ICSU Global Atmospheric Research Program (GARP) laid the foundation for two decades of productive international cooperation and rapid development in the monitoring and understanding of global weather and climate. Although the major effort under GARP through the 1970s was directed to its so-called first (weather) objective culminating in the Global Weather Experiment of 1978-9, much of the scientific and technological progress in both observing systems and model development was equally supportive of the second climate objective.

Concern at the prospect of human interference in climate developed rapidly during the 1970s, and despite a brief period of preoccupation with the prospect of an imminent ice age, the threat of greenhouse-induced global warming received substantial attention at the First World Climate Conference in 1979. The Conference Declaration embraced the following commitments: exploiting existing knowledge on climate, taking steps to improve that knowledge and also anticipating and minimising potential climate change caused by human activities.

The next major development and initiative on climate change occurred in 1985 at a special conference held at Villach in Austria to provide an ongoing assessment of data on increases in atmospheric greenhouse gases. The Statement arising from this meeting recognised that many important economic and social decisions are being made today on long-term projects — major water resource management activities such as irrigation and hydro-power; drought relief; agricultural land use; structural design and coastal engineering projects; and energy planning — all based on the assumption that basic climatic data, without modification, are a reliable guide to the future. However, this is no longer a good assumption since increasing concentrations of greenhouse gases are expected to cause a significant warming of the global climate in the next century. Hence there needs to be a renewed emphasis on establishing estimates to improve such decision-making processes.

The Villach Conference participants also acknowledged that climate changes and sea-level rises were realistic and were being monitored, for example in the Pacific Island states; that these changes were closely linked with other environmental issues such as acid deposition and threats to the ozone layer, mostly arising as a result of man-induced atmospheric change. The possible moderating effect of reduced coal and oil use, energy conservation measures and prohibitions on manufacture and use of CFCs were unknown.

A Global Climate Programme

In 1979, a major international, interagency and interdisciplinary effort was mounted to establish a World Climate Programme. This sought to coordinate scientific research and data on climate to provide the means of foreseeing possible future change. Its work has been further aided by that of the United Nations World Commission on Environment and Development (the Brundtland Commission). Today, the prospect of major climate change within the lifetimes of the present generations, its possible impacts and the strategies through which nations might minimise and adapt to it, are widely seen as among the most important issues facing humanity in the closing years of the twentieth century.

The 1988 Toronto Conference on The Changing Atmosphere recommended a 20 per cent cut in emissions by the year 2005 and a 50 per cent cut by 2020. Following the severe winter of 1989-90 initial results suggest that the Americans have failed to cut back on continued increases in carbon dioxide emission levels. It appears that a government committed to a policy of environmental protection and efficient energy use is necessary, as are legislation backed by sanctions and severe penalties against offenders, improved community education and individual consumer initiatives.

Yet China is attempting to minimise its contribution to greenhouse effects and to preserve natural resources. In 1988, 210,000 methane gas pits were built for rural families and 12 million energy-saving stoves installed, saving 25 million tonnes of standard coal each year. In addition, the straw saved was available for use as fertiliser, forage or fuel for other purposes, and trees were not stripped of their branches in the customary manner. The installation of 20,000 wind-driven power generators on grasslands in the Inner Mongolian autonomous region of northern China has brought electric power to 20 per cent of the area's herdsmen and provided the opportunity for improvements in lifestyle.

According to official data provided by the Ministry of Agriculture, rural China is responsible for over one-fourth of the country's energy use. Most of this energy is obtained from the burning of coal and other carbon dioxide producing fuels. Township enterprises alone use energy equivalent to 100 million tonnes of standard coal. Poor equipment means that the efficiency of coal production and use, and low controls on emission, are lacking and beyond the available funds.

Global Action To Limit Greenhouse Gases
Early in 1989, the world's leading industrial nations, meeting in a specially convened conference on global warming and the greenhouse effect, committed themselves to developing strategies to reduce their emissions of greenhouse and other related atmospheric gases. Despite increased funds for development of socially acceptable national policies, few plans have followed on the Toronto Summit.

The first detailed formula for reducing carbon dioxide releases by 20 per cent by the year 2005 has emerged from research funded by the Dutch Government. It argues for determination of a global 'carbon budget' for distributing fossil carbon 'cuts' equally between industrialised and developing countries: industrialised countries would have to reduce their emissions of carbon by 20% by 2005, 50% by 2015 and 75% by 2030. Developing nations could increase their emissions over the next twenty-five years to provide them with the opportunity to industrialise their economies at least cost. This would permit these countries to meet international loan obligations, improve per capita income and begin to attack the social problems of poverty, health and education. However, by 2030 these nations too would need to begin reductions to achieve the present level of emissions.

The Dutch plan suggests that national budgets be based on rates of release of carbon dioxide per head of population. This would allow Britain to continue emitting carbon dioxide at current rates for forty-two years before 'using up' its emission quota under the formula. Yet, the United States has only twenty-five years left and eastern Germany only twenty-two.

In order to arrive at their carbon budget, the researchers set strict ecological limits on the acceptable level of global warming. Using data from the California-based International Project for Sustainable Energy Paths, they limited the average rate of global warming to 0.1° C per decade, with an absolute ceiling of 2.5° C. The authors of the study argue that earth's environment, its trees, plants and animal species would be unable to adapt to greater increases. They substantiate their claims by drawing on research on increased and rapid die-back by trees growing in industrialising areas, and compare the rate of loss with the time taken for evolution of any new, more resistant species. Beyond a level of 1° C they claim that the risk of major 'forest breakdown' is unacceptably large. Furthermore they point out that the current levels of human activity, the burning of fossil fuels, clearing of forests and farming breach this limit. Thus, on present trends, the risk threshold will be exceeded by many times.

Using this information, the Dutch Formula proposes a budget of 300 billion tonnes of emissions of fossil carbon to cover the period from 1985 to 2100. At current rates the budget would be used up by 2030. At the same time, it assumes sweeping progress in other areas influencing greenhouse effects. These include the phasing out of CFCs by 2000, returning the amount of carbon stored in forests and soils to mid-1980s levels through several decades of reafforestation, reducing overall emissions of methane from agriculture (partly by replacing beef farming with forestry), and a slowdown in the growth of concentrations of nitrous oxide in the atmosphere. The study also suggests that any international agreement on global warming should include separate figures for emissions of biotic carbon, including that from burning trees or straw and firing of sugar-cane crops prior to harvest.

Action By Member Countries: Expressions and Realities
It seems that global concern about greenhouse warming and the destruction of the environment has become a major force in international and national politics. Individuals, lobby groups and whole communities have pressed governments worldwide to address the imminent problems and concerns. They are demanding action, not merely words. But as time passes and new interests attract attention particularly of the international community, the pressure for change has less effect. Will it eventuate in long-term substantial change, in new measures to ensure a better health for all?

The industrialised nations have committed themselves to embarking on policies which will ensure that their own societies and economies make the necessary adjustments to limit damage and reverse that already done. But will these commitments be met by the individual countries and against powerful vested interests able to dislodge them from power? Thus far, it is only the smaller nations directly affected which have developed strategies and are implementing definite plans.

The first tentative steps were made, for example, at a November 1989 Conference in Male, the capital of the Maldives. Attending the meeting were representatives and politicians of fourteen small island states from the Caribbean, the South Pacific, the Mediterranean and the Indian Ocean. By virtue of their status as islands, these small countries are among those which will be more dramatically affected by rising sea levels which would accompany increased temperatures and consequent melting of polar ice-caps. Available for discussion and consultations were scientists and technical advisers who could contribute realistic ideas and suggest practical measures to be undertaken.

During the discussions the Maldives revealed that it had already instituted its own National Action Plan for Environmental Management and Planning, specifying limits for atmospheric contaminant emissions, measures for improved waste management and strategies to maximise productivity of land by methods which conserved soil quality by natural means. Plans for building on lands threatened by rising sea levels need prior approval; the gradual movement of people onto higher land and an appreciation of the need to consider abandonment of threatened sites all attract mention.

The Male conference concluded with a declaration signed by all representative countries calling on the larger developed nations to urgently limit their outputs of greenhouse gases and to become more efficient in their use of energy. It concluded by making a further call for an international convention on climate change, along the lines of the 1987 Montreal Protocol which focused on greenhouse effects. This convention would further bind the signatories to limit the production of fluorocarbons and halons that are responsible also for destruction of the ozone layer.

Commitment and The Developed World

In stark contrast was the outcome of a similar international meeting held in late 1989. Representatives of 72 nations were told that limiting the worldwide use of fossil fuels would cost almost 200 billion pounds over the next fifteen years. Horrified by the cost and the tough deadlines, the meeting failed to agree on specific measures to finance the switch to less polluting forms of energy and to commit themselves to definite measures to limit or reduce their emissions of greenhouse gases.

The US, the CIS (former Soviet Union), Japan and Britain vetoed Scandinavian and Dutch proposals for a freeze on carbon dioxide emissions by the year 2000 and a 20 per cent reduction in emissions by 2005. The dissenting nations, which between them emit 50 per cent of the world's carbon dioxide, insisted that the proposed targets were unrealistic and the wording of the final communique be less specific on a date.

Thus, it remains difficult to be optimistic, despite the early enthusiasm. There is still great reluctance to view environmental problems on a global scale and to actually begin concerted efforts locally to decrease levels of carbon dioxide, phase out chlorofluorocarbons and turn productive agricultural land to forest. Economic declines, opposition from commerce and business, fears of consumer resistance to change and increased associated costs to pay for the use of more expensive biodegradable replacements are a major concern.

A number of the larger countries, such as the United States and Great Britain, maintain that they carry significant debt burdens, and are reliant on the business and profit opportunities of wealthy multinational companies, many dependent on oil revenue. Their populations are accustomed to luxuries such as air-conditioning, easy-to-use spray-can paints, deodorants and fly-sprays and fast, petrol-guzzling automobiles. Any unpopular government measures which reduce the standard of living and constrain populist needs would result in electoral defeat.

While less familiar with higher standards of living, the poorer nations maintain that they have not contributed to global warming and that they have far greater debt burdens to service, principally to the developed countries. Hence, they argue that in order to meet their financial commitments, rapid and minimal cost industrialisation is essential.

Realities of the Toronto Target

The world can certainly adapt to some of the adverse impacts of imminent greenhouse warming. Current 1990 estimates suggest that the atmosphere is heating up at around 0.2-0.5° C per decade. According to the US Environment Program and the WHO, this must be reduced to 0.1° C per decade and contain total warming to less than 2.5° C. Beyond this level the delicate balance of the total ecosystem would be placed in serious jeopardy.

For any reductions to work they must be worldwide. No one country will have much effect if it acts alone: hence the Toronto target of a 20 per cent reduction by the year 2005. Since adoption of this target by the major OECD nations, additional goals have been developed by a joint Australian and New Zealand Environmental Council (ANZEC).

ANZEC recommends first, the phasing out of CFCs completely by 1998. Many nations, including Australia, have already signed the Montreal protocol which undertakes to cease production of all CFCs and like ozone-depleting substances by that date. An associated reduction in greenhouse warming is expected to accompany elimination of CFCs. The second goal of ANZEC is to stabilise carbon dioxide emissions at the 1988 level well before 2005: in other words, to prevent an increase in emissions in the interim period as well as aiming for a 20 per cent reduction by 2005. The third goal is to cut total emissions of greenhouse gases by 40 per cent, measured as carbon dioxide equivalent; that is, to reduce carbon dioxide emissions by 20 per cent and to reduce emissions of methane nitrous oxide and CFCs by an amount equivalent to a further 20 per cent reduction in carbon dioxide emissions.

In defence of these goals it can be argued that energy-intensive industries account for 29 per cent of total energy consumption. About 23 per cent of that 29 per cent (or 6.8%) could be saved. Corresponding figures for low energy industry and residential use respectively are 5 per cent and 4 per cent. The total for these 'end-user' sectors is 15.8 per cent.

In addition, 39 per cent of this energy is supplied in the form of electricity from coal-fired power stations, and apparently carbon dioxide emissions from these could be halved by switching to more efficient combustion technology or natural gas. This would give a further saving of half of 39 per cent of the 15.8 per cent (or 3%) which, together with the end-user savings, would be 18.8 per cent savings in all.

The petroleum-burning transport sector accounts for 39 per cent of total energy consumption. A written report for the mining giant CRA in 1989 suggested that 60 per cent savings are possible for private cars through greater fuel efficiency, though this would increase the initial purchase price of cars by 25 per cent. This would amount to a conservative estimate of a total 40 per cent savings for the transport sector overall; and 40 per cent of 39 per cent is 15.6 per cent. So the overall potential savings in energy consumption and consequent carbon dioxide emissions, purely by technological improvements in efficiency, is 18.8% + 15.6%: a total 34.4 per cent.

This is well above the Toronto target of a 20 per cent reduction even allowing for errors in the estimates. Furthermore it does not involve any switch to nuclear energy, hydro-power or large-scale solar or wind energy. It also does not include possible savings from revegetation and soil conservation. Nor does it involve a general economic slowdown or loss in material standards of living — just great efficiency in generating and using energy from existing fossil fuel sources by innovation.

ANZEC's broader goal is a 40 per cent reduction in all overall greenhouse gas emissions by 2005. At present, greenhouse warming induced by human activity derives about 44% from carbon dioxide, 19% from methane and 18% from CFCs. CFCs are due to be phased out in 1998 so that 18 per cent is already marked for saving. As calculated previously, it is possible to cut carbon dioxide emissions by 34 per cent: so that is 34% of 44%, or 15 per cent of the total.

More conservatively, the Toronto target of a 20 per cent reduction in carbon dioxide emissions would be equivalent to 20% of 44%, or 9 per cent of the total (between 27 and 33 per cent just from reduction in carbon dioxide and CFCs).

Methane gas is emitted from ruminant animals, landfills, leaks in natural gas distribution, burning off, rice paddies and mining. Emissions of methane and nitrous oxide contribute to greenhouse warming. A total potential cut of about 4.3 per cent is possible for methane. A possible 3 per cent in nitrous oxide could be saved from reduced use of nitrogenous fertilisers and reduced burning off. Thus, overall, a reduction of between 34 and 41 per cent in total greenhouse emissions is feasible.

Implications of the Toronto Target
If greenhouse gas targets are not met, the implications are serious. Worldwide economic disruption and a severe drop in living standards are likely. Yet national and international resolve is hard to achieve, especially in the depressed economic and financial state in which many nations now find themselves.

Why is this so? Opinion polls suggest that people express their willingness to make sacrifices to preserve the planet and minimise environmental damage. Unfortunately, industry and consumers balk when the true costs involved become evident. An Australian industry report in 1989 suggested that to meet the Toronto target, a 40 per cent rise in electricity tariffs, a 60-120 per cent rise in car fuel prices and a 25 per cent rise in car prices would be necessary. The net effect on GNP would amount to a $30 billion loss, for the period to 2005.

Expressed in annual terms, however, the suggested increases appear less catastrophic. An annual 2 per cent increase in electricity prices could be managed with the introduction of more efficient appliances to reduce consumption. A 2.8 to 4.7 per cent increase in petrol prices is relatively small in view of rapid speculative increases that occurred mid 1990 during the Gulf crisis. A 60 per cent increase in fuel use efficiency in new cars could balance the large increase in prices. On the positive side, if the Toronto target is reached, an increase in private consumption of 2.24 per cent per annum is possible compared with 2.29 per cent if the target is ignored.

The main argument used by countries against adoption of the Toronto target is the fear that they might lose any competitive edge in international trade if other countries fail to adopt the target. The developing nations are also arguing a special case, claiming that they should not be disadvantaged because the present crisis is the fault of the developed nations. Thus, persuading the developing nations to cut greenhouse gas emissions is not easy.

Interventions To Conserve The Environment
Finally, it is important to recognise that suggestions other than those applying to greenhouse gases and CFCs have been made. They can be summarised in the series of policies listed in Table 7.1.

The Environmental Agency of Japan has suggested a system of internationally tradeable carbon dioxide emission leases. The initial allocation of rights raises a number of possibilities. Developing countries might argue on a per capita basis, forcing the developed countries to bear the brunt of the reductions. Alternatively, emission rights might be contested proportionally on an area basis, putting pressure on Japan, Hong Kong, Singapore and the United Kingdom.

To persuade developing countries to act, the developed countries would first have to show their commitment by adopting the Toronto targets unilaterally and universally. Herein lies the first test of global commitment and political resolve: the classic case of the 'prisoner's dilemma'.

A Novel Solution
Extensive forest plantations have been proposed to absorb the approximately 5 billion tonnes of carbon generated annually in the world. Another suggestion canvases the use of microscopic algae. Toha and Jaques of the University of Chile, Santiago, argue that the quantity of algae to digest a comparable amount of carbon as a forest occupies only one-tenth the space, will grow in fresh or saline water and be available for immediate action. With an adequate addition of nutrients and with average climatic conditions, the scientists contend that it is not difficult to achieve concentrations of 0.3 grams per litre, with doubling in three days. This means that in a one-hectare pond with a

depth of one metre, 1800 kilograms of carbon dioxide are absorbed daily. The approximately 22 million hectares needed to absorb the excess carbon dioxide could be distributed along the coastline of some sixty countries.

TABLE 7.1: Preventive Policies for Australia

* increase the efficiency of energy production and end use

* develop alternative energy systems that are not fossil-fuel based

* plant trees to increase the sink for CO_2 and to combat soil degra-dation and salinity

* tighten emissions controls on vehicles and stationary sources

* reduce oxides of nitrogen — necessary both to control ozone levels (smog) and as they are pollutants in their own right

* reduce hydrocarbons, carbon monoxide, sulphur compounds and carcinogenic particulants

* recycle deposit on all recyclable containers — also reduces litter pollution

* carbon tax on fossil fuels with the tax corresponding to the amount of carbon in each field. This would most affect coal as it produces more carbon dioxide per unit than oil or natural gas. Energy prices would rise and efficiency improve. Renewable energy sources, increased insulation would be adopted.

* disposal tax on all non-recyclable non-biodegradable packaging, e.g. plastic fast-food containers, to encourage the use of recyclable or biodegradable packaging

* redeploy ex-coalminers and those employed in fossil-fuel dependent industries to conservation schemes, e.g. in production of environmen-tally friendly products, learning to grow seedlings in nurseries and other forest management skills

Source: In Touch, *Newsletter of the Public Health Association of Australia (Inc.), November 1989, vol.6, no.4, p.4.*

References and Additional Reading

Brundtland Report. *Our Common Future*, United Nations, 1987; Australian edition, Oxford University Press.

China Reconstructs, vol.XXXVIII, no.9, September 1989, p.66.

Coombes, H.G. *The Return of Scarcity*, Angus and Robertson, Sydney, 1990.

de la Court, T. *Beyond Brundtland: Green Development in the 1990s*, Zed Books, UK, 1990.

Krause, F. & Bach, W. *Energy Policy in the Greenhouse*, Vol.I, International Project for Sustainable Energy Paths, El Cerrito, CA94530, US, 1989.

New Scientist, no.1893, 2 December 1989, p.8.

New Scientist, no.1691, 18 November 1989, p.55.

Pearce, D. *Blue Print for a Green Economy*, Earthscan Publications, London, 1989.

Starke, L. *Signs of Hope: Working Towards Our Common Future*, Oxford University Press, 1990.

The Environment Digest, 1989, Issue no.27.

Zillman *et al*, Lecture on Climate Change delivered to the Australian Meteorological Organisation, 1989.

8.0 RESTORING THE BALANCE

What does the future hold? Can human beings survive in the short term as the species we know, or does their level of evolutionary sophistication already mean that, in time, they will pass into extinction in the long process of evolution and natural selection? Will this be the result of normal evolutionary processes, or of the disturbances which humans have caused to earth's ecosystem and the balance of nature? Do they face a new wave of disease and ill-health because they have tampered with the delicate balance of relationships on which all life depends, and from which they themselves evolved? The human species is now omnipresent. It also believes it is in control of nature, with power over all other species. Is this the case, or does nature have in store a salutary lesson for this arrogance?

The impact of human activities on their own health should be cause for concern. Hopefully, this book will provide an insight into the dangers and risks involved in the haste towards maximising opportunities, competition and exploiting available resources. In the developed world, the probable and imminent impact of environmental change on human health can only be compared with two previous large waves of public health: the infectious diseases of the nineteenth and early twentieth centuries severely curtailed life chances; in the later, twentieth century the lifestyle diseases of cancer, cardiovascular disease, and self-initiated abuses associated with drugs, alcohol and behaviour combine with longer life-spans to lessen the quality as well as the quantity of many lives.

The sad reality is that even in comparatively affluent countries those at greatest risk are the very people least well equipped to manage their illnesses, and with a decreased capacity to change the contributing circumstances. In Australia, for example, of a total population estimated at around 19 million, more than 5 million individuals are regarded as disadvantaged and unable to enjoy full access to available health services for socio-economic, racial, occupational or educational reasons.

Yet the new environmental diseases bring graver problems and greater risks for the under-developed world. Here the poorest countries lack the capacity to deal with even small incremental stresses caused by land degradation, increased population, inadequate food production, and depressed prices in world markets, often caused by the more developed nations. Indeed, the policies of the developed countries erode the traditional markets of the poor countries, dependent on agricultural exports and commodities. Export subsidies and incentives force under-developed countries to compete, at a disadvantage, against wealthier, better resourced and capital-rich nations. To achieve sales, they must sell their goods at prices which do not provide an adequate return for the individual producer or boost substantially the moneys returning to the parent country. Per capita incomes decrease, survival pressures increase and health standards fall further. Famine, warfare over territory, disease and death eventuate: Eritrea, Mexico, South America and South-East Asia come to mind.

The greatest concern lies in the lack of appreciation that new health threats are real, that the first signs exist and that relevant health policies must be developed. We have already seen that few governments have implemented preventive policies, public education campaigns or begun planning facilities to handle increases in the number of patients. These can be anticipated to be suffering from skin cancers resulting from increased ultraviolet radiation following depletion of the ozone layer, from asthma and lung diseases with higher levels of atmospheric pollution, and even from occupational cancers such as those associated with coal mining, the asbestos fabrication industry and minerals processing.

There is certain to be a range of newer diseases thus far not viewed as resulting from environmental changes initiated by humans. Human numbers and high fertility rates place great pressures on existing and future resources. While auto-immune diseases and the new scourge, AIDS, may prove to limit human population growth in ways similar to the infectious diseases of the past, there is no reason why modern humans will not be equally successful in adapting themselves or developing effective treatment measures. Times of population stress seem to trigger in mice the release of unidentified hormones which reduce fertility, delay maturation and cause death of developing young *in utero*. Crowding in other mammals causes emotional disturbances in the population, resulting in a breakdown of order. Rats have been observed to kill their young and exhibit abnormal sexual behaviour. Will humans react in the same ways when population pressures become extreme? Are the first signs apparent in those areas where numbers are already beyond the capacity of available local resources?

Glossary

acid rain: the precipitation of dilute solutions of strong mineral acids from the atmosphere. These acid solutions are formed by the mixing of various industrial pollutants with atmospheric oxygen and suspended water droplets. These acid substances then fall as rain, snow or fog on soil, vegetation and buildings. The pH of the resultant 'rain', however, is affected by the pollutant substances suspended in it (e.g. sulphur dioxide, nitrogen oxides, hydrogen chloride) resulting in an acidic or low pH precipitant. This weak acid damages the surfaces of leaves, the bark of trees and leafy crops, and within the soil mobilises aluminium thereby inhibiting healthy absorption of nutrients by the roots. In time, it erodes the limestone and sandstone surfaces of buildings and is responsible for visible signs of 'pitting' on bricks and window sills.

aetiology: cause of disease.

aquifer: an underground water storage area. Water that falls as rain or snow on land may seep into the soil where it becomes groundwater. Some rock layers are more permeable than others and if groundwater reaches a rock layer that is relatively impermeable, the water tends to collect above the layer and may form an aquifer. The upper surface of such underground water is called the water table.

attributable risk: the difference between the rates at which disease occurs in people who are exposed and in people who are not so exposed.

bacteraemia: term used to indicate the presence of bacteria in the bloodstream, but usually taken to mean that they are present in small numbers which are not seriously affecting the health of the host organism.

biogeochemical cycles: dynamic processes in which atoms are recombined and rearranged and the starting materials are regenerated. Progress through portions of all such cycles requires energy, and the principle energy source driving biogeochemical cycles is the sun. Solar energy powers the processes of evaporation and condensation that move water in cycles.

biomagnification: As living things break down the complex molecules of dietary carbohydrate, fat and protein, they assimilate some components, excrete others and still more they store in body fat and bone. Chemicals that are stored in this way, such as pesticides, accumulate over time and are passed up the food chain in higher and higher concentrations. At each link in the food chain, the amount of pesticide in the organism's tissue is increased. The process is referred to as biomagnification.

biomass : the quantity of living plant and animal matter in a given area. Biomass values represent the amount of organic matter which has accumulated within an ecosystem. The more conducive the environmental conditions, the higher the biomass which can be built up within the system. Biomass values are usually quoted as dry weight in tonnes per hectare or kilograms per square metre.

biotic potential: maximum possible growth rate of a species, given the provision of optimal environmental conditions for its growth and reproduction.

Body Mass Index (BMI): the index is calculated by dividing the individual's weight in kilograms by their height in metres squared.

cancer: an uncontrolled growth of a group of cells in the body. The cells lose their functional differentiation and begin to reproduce themselves indiscriminately, overgrowing their natural boundaries, distorting their surroundings and destroying the normally ordered structure of the tissues in which they are growing. Secondary growths occur elsewhere in the body, disseminated from the primary growth by cells which separate off and are swept away to new locations by the bloodstream or in the flow of lymphatic fluid through lymph channels and lymph nodes. Sooner or later the growing masses of cells interfere with the function of a vital organ of the body, or the growth itself outgrows its own blood supply and becomes necrotic. The immune system, unable to effectively target and fight cells which so closely resemble the body's own tissues, becomes compromised and the individual's resistance to disease is lowered. Eventually the life of the victim is destroyed by organ failure, infection or massive haemorrhage. Sometimes the sheer inanition caused by the growth and by treatments to contain it, leads to exhaustion and death.

carrying capacity, of the environment: the largest population of a given species that can be sustained indefinitely in a particular ecosystem. This is a theoretical balance point between biotic potential and environmental resistance.

cytotoxic food test: a test which involves placing a drop of buffy coat cells suspended in a mixture of serum and distilled water, onto a slide on which a food extract has been dried, and examining the cells under a microscope to detect changes in motility and appearance. Despite the fact that it has been shown to lack precision, sensitivity, specificity and predictive value, and that the cellular 'changes' do not correspond to any known pathophysiological process, the test remains in widespread use by some practitioners.

degeneration, of cells: occurs where there is damage to cell structure or function which is, at least potentially, reversible. These changes are most readily recognisable in the densely cellular, parenchymatous organs such as the heart, liver and kidneys. Affected organs are pale, soft and slightly enlarged. This type of lesion can progress to frank fatty change. Mild to moderate physiological fatty change can occur depending on diet.

epidemiology: a science which looks for significant 'associations' between the occurrence of specific factors, usually environmental, and that of a given disease. If factor and disease are found to be associated more frequently than would be expected on a purely random or chance basis, a 'significant' association is said to exist. An association may be deemed to be direct or indirect in nature; if it is direct, it may not be causal. The mere finding of a significant association is not, in rigorous scientific terms, proof of a causal association, although the finding is likely to stimulate researchers into seeking to investigate its nature in considerable depth.

eutrophication: the process of nutrient enrichment of an aquatic system. In water environments, occurrence of eutrophication has increased due to the accumulation of nutrient-rich effluents from agricultural production and discharge of sewage and stormwater from towns lining the banks of inland waterways.

extinction: the disappearance of a species from the surface of the planet. The disappearance of a species changes the environmental setting as much as does the appearance of a new one. Extinction has thus been one of the regular elements of change in an ever-changing and dynamic ecosystem. The extinction of individual species or families of wholesale extinctions have created niches that drive evolution further along.

facultative relationship: a relationship between two or more species such that one or the other is dependent on or contributes to the well-being of the other. Includes symbiotic relationships which are mutually beneficial (mutualism), or beneficial to one with the other unaffected (commensalism).

GIT: abbreviation for gastro-intestingal tract

Gondwanaland: the southern protocontinent believed to have resulted from the fragmentation of Pangaea supercontinent in the late Paleozoic or early Mesozoic era.

gram-negative bacteria: one of the principal classes of bacteria. Gram-negative bacteria, unlike gram-positive members of the species, do not 'take up' and retain the gentian-violet stain used to identify micro-organisms. They tend to take up the red counterstain. In comparison with infections associated with gram-positive bacteria, those associated with gram-negative bacteria tend to be more resistant to conventional penicillins and antibiotic treatment; e.g. pseudomonas, proteus, hemophilus, legionella.

gram-positive bacteria: unlike gram-negative bacteria (see above), gram-positive bacteria retain the colour of the gentian-violet stain.

granuloma: circumscribed lesions characterised by the presence of macrophages and lymphoid cells. This type of chronic inflammation tends to form in response to the presence of inert foreign material or moderately pathogenic but persistent organisms. The formation of a granuloma involves immunological as well as inflammatory processes.

Gross National Product (GNP): total monetary value of all goods produced from all sources (industrial, agricultural, services, etc.) within a country in any one year.

groundwater: water which reaches the land as rain or snow and seeps into the soil.

heavy metal: any metal with a high atomic weight, generally over 100. Examples include mercury, lead, cadmium, chromium, plutonium.

humidity: the amount of water vapour in the atmosphere.

hyperplasia: excessive proliferation of normal cells in the normal tissue arrangement of an organism.

immunoglobulin (Ig): a family of closely related though identical proteins that are capable of acting as antibodies. Five major types are normally present in the human adult, and they differ in function and metabolic weight.

incidence: the number of new cases of disease occurring in a defined population over a specific time period.

infant mortality rate: the number of deaths of children less than twelve months of age per thousand population in any one year.

inflammation: a non-specific defence mechanism which operates when any form of tissue damage occurs and is the first major line of defence against invasion by micro-organisms. The process of inflammation underlies many of the visible changes seen at post-mortem examination in the majority of diseases. The process of inflammation is a complex series of reactions characterised by changes in the calibre and flow of blood vessels in the area, formation of a protein-rich fluid exudate, and infiltration of white blood cells into the affected area. These alterations are responsible for the characteristic signs of acute inflammation: heat, redness, pain, itchiness or tissue irritability, swelling and loss of function.

Law of the Minimum: In 1840, Justus Leibig observed that a plant tended to grow only to the limit of the foodstuff available to it in the most extreme minimum quantity. The law of minimum expresses the idea that organisms and the living systems they comprise are held in check by the scarcest of the things which they need.

Law of Tolerance: Shelford's Law of Tolerance states that both too much and too little of various environmental chemical and physical factors can serve as limiting factors or regulators in ecosystems. Thus, if the lowest annual temperature reached in a particular environment is below the lowest temperature that a particular plant can tolerate, that plant is not likely to be found in that environment, even if moisture or soil type are sufficient. Similarly, some organisms have a wider range of tolerance than others, and these differ according to various environmental factors.

leucocyte: white blood corpuscle. There are two types: granulocytes (those possessing granules in their cytoplasm) and agranulocytes (those lacking granules). They form the first line of defence within the bloodstream and are responsible for 'mopping up' dead cells, dead bacteria and debris.

lyme disease: a recurrent inflammatory disorder accompanied by distinctive skin lesions, polyarthritis and involvement of the heart and nervous system. Evidence suggests that a spirochete called *Ixodes dammini*, acquired from ticks, is the causal organism.

mesosphere: the layer of the atmosphere that exists between stratosphere and thermosphere and extends between 50 and 80 km above the earth's surface: it ends at the mesopause.

morbidity (Standardised Morbidity Rate: SMR): the ratio of total observed cases of sickness to total expected cases in a defined population, multiplied by 100.

mortality (Standardised Mortality Rate: SMR): the ratio of total observed cases of a disease to total expected cases, multiplied by 100.

niche: a species niche refers to the unique, functional role or place of that species in an ecosystem. While the habitat of any given species is simply the kind of environment where one would go to find that species, its niche is more complex. According to Odum, the ecologist, a complete description of a species niche should describe: (i) its habitat niche, or physical location within a particular habitat; (ii) its trophic or food niche, or ecological role within the ecosystem, e.g. the species it eats, its predators and competitors; (iii) its multidimensional niche, or its preferences for temperature, shade, pH, humidity, slope, etc.

perinatal mortality rate: the number of deaths of infants less than 30 days old per thousand population in any one year.

polymerise: chemical reactive process by which molecules become joined in long chains, often resistant to breakdown by conventional acids and alkalis.

ppb: parts per billion

ppm: parts per million

prevalence: the number of cases of disease that exist in a defined population at some point in time.

relative humidity: the ratio between the water vapour content of a given amount of air and the moisture it could contain if the air had been saturated at the same temperature and pressure.

relative risk: the ratio of the rate at which disease occurs in people who are exposed to the presumed cause, to the rate among those not so exposed.

replacement rate: in respect of human beings, the proportion of the original residential population that can be accommodated in a redevelopment site, e.g. a slum.

reservoirs: places in the ecosystem that serve as major abiotic storage sites for chemical elements such as nitrogen, phosphorus, carbon. The chemical elements can move between reservoirs, and their size and accessibility at any one time depend on the patterns of movement and the rates at which nutrients travel between reservoirs.

resistance: refers to the collective environmental factors which keep growth rate of an organism below its biotic potential. Much time is spent by ecologists in identifying and measuring the factors or combinations which influence the population growth of important species.

septicaemia: a profuse and persistent bacteraemia in which there may be actual multiplication of organisms within the circulation. Bacterial endotoxins (cell wall components released when bacteria are destroyed) are poisonous to tissue cells generally, so that septicaemia is associated with a number of generative or toxic changes. The classical lesions of septicaemia are widespread petechial haemorrhages, degenerative changes in the liver, kidney and heart, splenic congestion and enlargement, lymph node congestion, oedema and fever. The infective organisms associated with septicaemias in animals include the streptococci and staphylococci, although a number of specific infections (e.g. anthrax) are primarily septicaemic in nature. The widespread dissemination of organisms throughout the tissues or organisms means that when septicaemia is the cause of death of animals used for food, infection can easily be transmitted to humans. This is a major cause of unremarkable death in poor, rural families in the developing world.

stratosphere: the layer of the atmosphere that lies between the troposphere and the mesosphere. The base altitude of the stratosphere varies with the latitude, beginning at 9 km at the poles and increasing to 16 km in equatorial regions. The stratosphere extends for around 50 km and ends at the strato-pause. The stratosphere contains most of the atmosphere's ozone which concentrates at an altitude of 22 km. Air temperatures rise with altitude in this layer and clouds rarely occur.

thermosphere: the layer of the atmosphere that lies above the mesosphere and the magnetosphere, extending from around 80 km to 500 km above the earth's surface. It is sometimes referred to as the ionosphere. Solar radiation causes ionization of the atmospheric gases which leads to the phenomena of the Aurora Australis and the Aurora Borealis.

toxaemia: in some cases the organisms do not spread from the initial site at which they become established, but toxic products may be absorbed and cause severe illness. The resultant morphological changes include 'toxic' degeneration of the parenchymatous organs and capillary damage leading to haemorrhage, so that lesions may be indistinguishable from those of septicae-mia. Specific toxaemias are due to exotoxins, which are poisonous substances produced by living bacteria. Such toxins frequently exert their effects on the nervous system so that characteristic changes include paralysis or muscular spasm (e.g. botulism, tetanus, ciguatera).

troposphere: the lowest level of the atmosphere, extending between a height of 9 to 16 km above the earth's surface.

ulceration: occurs when there is localised damage to epithelial cells, with loss of cells and exposure of the underlying connective tissue. An erosion is a shallow ulcer with only superficial cell loss.

vesicle: (blister) a small cavity below or within an epithelium which may be filled with plasma or blood. A bulla is a similar but larger lesion. A pustule is a vesicle containing large numbers of inflammatory cells. Vesicles arise due to separation of the epithelium from the underlying connective tissue, or to localised cell destruction under an intact surface.

water table see **aquifer**

Index

3,4-benzopyrene, 72

Aborigines 26, 75, 113
acetylcholine 278
acid deposition 35
acid rain 8, 29, 224, 241, 320
acids 115
acid-alkaline balance 61
acne 129
actinic cheilitis 123
Adelaide 245
adverse food reaction 131
aerial crop dusting 279
aerobes 62
aerosols 41
aflatoxin 96
ageing 4, 6, 19, 119, 123
Agent Orange 95
agricultural chemicals 73
agricultural societies 8
agricultural workers 278
agriculture 78, 103
AIDS 242, 291, 319
Ainu of Japan 26
air pollution 166, 219
alcohol 27, 82, 86, 88, 98
Aldrin 78
algae 241; *see also* blue-green algae
alkalis 115
alkaloids 70
allergens 127ff., 140
allergic alveolitis 139
allergic lung disease 139ff.
allergy 130-31, 279
alternative practitioner 161
alveoli 136
American Mormons 26
ammonoid 37
anaerobes 62
angina 81
angiosarcoma 74
animal danders 138
ankylosing spondylitis 65
Antarctica 45, 227
antibiotic resistance 269

antibiotics 55
antibodies 129
antigen-antibody 58, 71
antimony 66, 142, 146
anxiety 164
apathy 164
apples 121
aquifer 50, 320
arable land 209-10
arsenic 31, 63, 69, 115, 146
arterial plaque 81
artery disease 81ff.
arthritis 5
artificial breakways 50
asbestos 65, 94, 103, 108, 143
asbestosis 142ff.
Aspergillus spp. 60
asphyxia 29
asteroids 36, 38, 46
asthma 15, 72, 127, 140, 164, 221, 279
— treatment for 141
atheromatous plaque 82
atherosclerosis 81, 84, 95
atmosphere 13ff., 46
atmospheric humidity 9
atmospheric pollution 15, 140, 169,
    220, 218ff.
atmospheric pressure 115
atmospheric turbulence 46
atomic veterans 75
atopic asthma 127
atopic dermatitis 126-27
Australia 74, 120, 144
Australia & New Zealand Environmen-
    tal Council (ANZEC) 313
Australian Newsprint Mills Plant 67

bacillary dysentery 77
bacteria 25, 58, 61, 72, 129, 148, 173,
    269, 322, 323
bacteriology 264
bagassosis 152
Bantu 87
beauty 167
beef 283
benzene 103, 115
beryllium 64

betel nut chewing 103
betocarotene 73
B.H.C. 79
Bhopal 230
bilharzia 5
biodegradable products 281
biometeorology 163, 176
birth abnormalities 68, 75, 279
birth control 291
birth rate 198, 211
Black Forest 221
bleaching industry 276
blood changes 29
blood pressure 86, 88
blue-green algae 237, 243, 251
body hormone 129
body mass index (BMI) 84, 321
body repair processes 57
body weight 82
bore water 89
botanic medicine 173
Bovine Spongiform Encephalopathy
    (BSE) 282
brain tumour 19
brassica 73
breast milk 73
bronchi 136
bronchitis 138, 221
bronchospasm 29
Brundtland Commission 43
Brundtland Report 11, 293-5
BSE *see* Bovine Spongiform Encephalo-
    pathy
Burkitt's tumour 100

cadmium 31, 66, 89, 115, 146, 283
caesium-137 63
caffeine 70
calcification 81
calcium 89
cancer 16, 55, 58, 63, 92ff., 321; *see
    also* leukaemia
— of bladder 97
— of bowel 144
— of breast 64, 96, 109
— of cervix 101, 109, 110
— of colon 96, 114

cancer of digestive tract 64
— of intestine 73
— of larynx 16
— of liver 96
— of lung 16, 64, 66, 96, 105, 106,
    143
— of nasal sinus 16
— of oesophagus 95
— of ovary 109
— of pancreas 16
— of prostate 66
— of scrotum 97
— of skin 75, 111, 126
— of stomach 16, 96, 112, 144
— of thyroid 64
— of urinary tract 16
— of vagina 109
cancer and chemical industries 281
cancer and locality 96
cancer and occupation 96, 98, 103
cancer causes 102
cancer of skin, protection against 126
cancer, chemical initiatives 104
cancers, women's 109
Candida albicans 131
capacity of the land 201ff.
Cape Coloured 87
capital 298
carbamates 78, 273
carbaryl 78
carbohydrate 69, 115, 207
— complex 86
carbon dioxide 20, 219, 309, 312
— annual emmissions of 54
carbon disulphide 115
carbon monoxide 16, 34, 89, 115, 167,
    219
carbon-based cleaner 41
carboxyhaemaglobin 151, 167
carcinogenesis 93
carcinogens 72, 79
carcinoma, basal cell 14, 28, 111, 123
carcinoma, squamous cell 28, 111, 123
cardiac failure 19
cardiac muscle cells 89
cardiovascular disease 81ff.
cartilage 178

cataclysmic events 35
cataracts 14, 28
causeways 50
cement 142
cereal grains 114, 211
CFCs 20, 30, 41, 126, 220, 226, 254, 292, 310
chemical contaminants 68
chemical hazards 66ff., 268ff.
chemical intervention 268
chemicals 129, 136
Chernobyl 64, 75
chicken litter 282
chilblains 15
China 44
chlamydia 59
chloracne 29, 78, 130
chlordane 78-9, 277
chlorides 6
chlorinated hydrocarbon 6, 273
chlorination 247ff.
chlorofluorocarbons *see* CFCs
chlorpheroxy acids 78
cholesterol 71, 82-3, 96, 103
cholinesterase 278
chromium 31, 108
Chronic Fatigue Syndrome 171-2
Chronic Mono-nucleosis 171
cigarette smoking 4, 72, 82, 88, 98, 103, 108, 115, 137; *see also* tobacco
CIS 44, 280-81
clean water 7
climate 10, 13, 39, 162, 209
climate change 43, 53
climate change response 52
climatologists 34
clonorchiasis 77
clouds 46
coal 103, 219, 224
coal tar 16
coal workers 142
coastal development 236
cobalt 63, 146
cocaine 4
cockroach 77
coke ovens 108

coliform faecal bacteria 244
colonic fermentation 115
colonic irrigation 173
Colonial Period 260-62
Colorado 17
commitment, political 292
Commonwealth of Independent States *see* CIS
competition between species 35
conjunctiva 28
conservation bond 303
contact dermatitis 122
contact urticaria 121
convectional current 39
Coombes, H.G. 297
copper 15, 113, 146
copra itch 128
coprophagy 25
coral reefs 50
corals 37
coronary heart attack 88
coronary heart disease 55, 81, 84
cosmetics 129
cramp 81
credit facility 187
Cretaceous period 35
crime 26, 169
crinoids 37
crop production 209
crop residue 78
crowding 6, 168
Crytosporidium 242
cyanide 230
cyanobacteria 237, 243

daffodil 121
Darwin, Charles 27
DDT 227, 73, 273, 277
death rate 191, 198, 211, 262
defoliant 95
deforestation 181, 236
degenerative processes 4, 6, 19
dengue 30
Denver 17
deoxyribonucleic acid *see* DNA
depression 162, 164
Derbyshire Peak District 24

dermatitis 72, 121, 130
desalination 51
desert 203
detoxification 173
diabetes 82, 95, 168
Diazinon 78
dieldrin 73, 78-9, 277
diet 82, 88, 98, 103-4, 110, 113, 114,
    129, 131, 173, 192, 207-8
dietary fibre 114
dinosaurs 36
dioxin 29, 78-9, 2276
diphtheria 62
dirt 136
disasters 301
disease 3-4, 192, 319
— a holistic approach 5
— degenerative 6
— infectious 9, 20, 192
— vectors of 19, 30
disease control 254
dithiocarbamates 78
dizziness 166
DNA 93, 100
domestic violence 169
Down's syndrome 279
drainage 7
driftnet species 8
drought 41, 237
drug abuse 169
drug resistance 269
drugs 27, 129
dust 72, 136-7; *see also* house dust
dust mites 127

earthquakes 36
Eastern Congo 24
economics 292
economists 297
ecosystem 13
eczema 72
eddies 39
education 4, 88
electromagnetic disturbance 46
electromagnetic force 175
endotoxin 59
energy production 78

enterocolitis 77
environment as holistic system 13
environmental factors 129
environmental hazard 3
environmental health 9
environmental resources 303
environmental sinks 301
enzyme 71
epidemics 6
epidemiology 191
Epstein Barr virus 100, 101, 172
erythema 165
Escherichia coli 238
eutrophication 245, 322
evolutionary process 27
exercise 173
exorphins 71
exotoxin 59
extinctions 35, 322
Exxon Valdez 238
eyes 165

factory waste 7
Familial hypercholesterolaemia 82
famine 184, 207, 319
farmers' lung 152
farming communities 277, 280
fats 69, 83, 103, 207, 208
— saturated 96, 104, 110
feathers 138
fertiliser 89, 155, 210, 211, 239-40,
    244, 280
fertility 319
filiariasis 30
fish 283-84
fisheries waste 78
fitness 173, 254
fleas 77
flies 77
flooding 49, 236
Flora & Fauna Guarantee 304
fluoridation 249-50
fluorosis 249
foetus 4
fog 16
folic acid 25
food additives 4, 68

food aid 183
food allergens 68
food allergy 131
food and cardiovascular health 82
food intolerance 71
food sensitivity 69
food shortage 6
food subsidy 187
food supply 205
food taboo 209
food toxins 68, 268, 282
food treatment processes 72
Forest Accords 304
forestry waste 78
formaldehyde 29
fossil fuel 34, 302
freon 6, 219
fruit 121
fumes 136
fungal disease 59
fungi 140, 148
fungicide 70, 73,151
future planning 10, 299

galvanized iron 66
gangrene 15
gardiasis 77
garlic 121-22, 210
gases 29, 36, 115; *see also* freon,
    greenhouse, halon, methane, radon
gastroenteritis 245
genetic engineering 25
genital herpes 101
geography 13, 24
geology 24
Germany 44
glandular fever 100
glaucoma 164
Global Climate Models 46
global warming 10; *see also* greenhouse
    effect
glycoprotein 127
goitrogens 73
grain itch 128
grain weevil 139
grazing 209
Green Revolution 211

greenhouse effect 34, 51, 228, 236, 302
greenhouse gas 14, 20-21, 23, 40
greenhouse phenomena 43ff.
Gross National Product (GNP) 300, 323
growth regulator 73
gypsum 142

habitat change 35
haemangiosarcoma 30
haematite 66
halon 229
hay fever 72
H.C.B. 79
headache 164-65
health maintenance organization 266
health status 4, 8, 26
health, definition of 1ff.
hearing loss 156-59
heart disease, genetic factors 82
heat 14, 159
heatwave 9
heat-stroke 14
heavy metals 31, 151, 173, 251, 323
Helsinki Declaration 228
hepatitis 77
hepatitis B virus (HBV) 101
heptachlor 73, 78-9, 277
herbicide 73, 155
herbs 64, 173
heroin 106
herpes simplex virus 111
herpes virus 100
high-fat diet 114
high-rise buildings 169, 214
histamine 70
hoarseness 166
Hodgkin's disease 100
house dust 140; *see also* dust
housing 4, 15, 169, 212
human B Lymphotrophic virus 172
human papilloma virus 101
human T lymphocyte virus 1 (HTLV-1)
    101
humidity 13, 72, 127, 128-29, 323
hunger 206
hunter-gatherer 5, 8
Hunter River Valley 24

hydrocarbons 34
hydrofluorocarbons 229
hydrogen cyanide 115
hydrogen sulphide 115, 167, 221
hypertension 81, 95
hypothermia 15
hypoxia 29

ice age 38, 45
Icelandic disease 172
ill health, primary prevention 3
— web of causation 4
immigration 190
immune system 174
immunisation 55
immunoglobulin 71, 323
Impact Theory 38
Industrial Revolution 6, 15, 225
industrialisation 6
infant mortality 198, 262, 323
infectious mononucleosis 100
inflammation 57, 71, 323
inorganic chemical 115
insecticide 73, 151
insomnia 164
insulin 168
International Monetary Fund (IMF) 186
interventions for addressing threats 254
iodine-123 63
ionosphere 40
iron 31, 65, 142, 146
irrigation 203, 236, 245
irritability 162, 164
irritant hairs and spines 121
ischaemia 81, 88
isotopes 78
itching 166

Japan 20, 26, 66

kerato-conjunctivitis 28
kidney 73
Kiribati 49

lactose 71
land degradation 8
land management 306

land surfaces 46
landfill 22
large families 26
latitude 39
leachates 78
leachings 19, 89
lead 4, 15, 31, 65-6, 115, 146, 173,
    229, 237
lead monoxide 167
Legionellaceae spp. 147ff.
Legionnaire's disease 139, 146, 150
legislation 304-5
leptospirosis 77
leukaemia 16, 29, 64, 100, 242
life expectancy 198
lifestyle 6, 88, 98
life-span 6, 76
Lignin 114
limestone 142
Limits to Growth 191
Lindane 79, 277
lipoproteins 84, 86
lithosphere 301
liver 73; *see also* cancer of liver
living conditions 98, 192
logging 236
longitude 39
Los Angeles 222
low-fibre diet 103-4, 114
lungs 136, 139, 150, 152; *see also* can-
    cer of lung
lyme disease 324
lymphoma 100

Mad Cow Disease (BSE) 282
magnesium 89, 146
magnetic influence 39
malaria 5, 30
Malathion 78
Maldives 49-50, 311
malnutrition 206-7
malt workers' lung 152
Malthus 190, 205
maltose 71
manganese 19, 89
mantle 40
maple bark lung 152

marble 142
marijuana 4
mastectomy 110
ME *see* Myalgic Encephalitis
Medicaid 265
Medicare 265
melanocytes 120
melanoma 14, 28, 111, 123
Melbourne 16, 18
mental deterioration 19
mental fatigue 165
mercury 31, 115, 146, 173, 237, 251
mesophilic organisms 61
mesosphere 40, 324
mesothelioma 143
Metal Fume fever 146
metallathonien 67
metals 65-6, 115; *see also* heavy metals, non-ferrous metals
metastasis 59
methaemoglobinaemia 29
methane 314
methyl isocyanate 230
methyl mercury 66
miasma 264
micro-organisms 4, 62
Middle Ages 257-59
migraine 71
migration 192-94
— Australian 195
mineral residue 95
minerals 65-6, 82, 89, 207, 238
Minimata Bay 20, 66
mining waste 78
modified starch 70
Montreal Protocol 228-29, 311
mortality 62, 81, 324
mosquito 19, 77, 261
motor vehicle emissions 89
mould 59, 61, 71-2, 137
mushroom 152
mushroom picker's lung 152
Myalgic Encephalitis 171
mycoplasma 59
mycotoxicosis 60
mycotoxin 68

National Conservation Strategy for Australia 296
natural breakdown and decomposition 41
natural healing mechanism 62
natural wealth 298
nausea 77, 164, 166
NCI/NIOSH/NIEHS Report 98
negative ions 164
neoplasm 58
nervous complaints, degenerative 19
nervous system 154ff.
— derangement 20
nettles 121
neurotoxic effects 79
neutrophils 61
New Guinea Highlands 24
New South Wales 24
nickel 19, 31, 63, 66, 108, 146
nickel carbonyl 115
nickel poisoning 36
nicotine 4, 78
nightmares 164
nitrates 239
nitrites 72, 113
nitrogen dioxide 34, 219, 223
nitrosamines 72
noise 115, 155-56, 169, 221
non-ferrous metals 31
North Sea 240-41
nose and throat infection 165
nuclear disaster 74; *see also* Chernobyl
nuclear waste 62, 238
nutrients 83
nutrition 79, 88, 215, 254
nutritional deficiency 15

obesity 86
occupational hazards 115
oceans 46-7, 233
oil spill 238
oncogenes 93
Ordovician period 35
organic chemical 115
organic fertiliser 41
organic poison 58
organic waste 239

organochlorine 6, 3, 78, 273-74
organofluoride 6
organophosphate 73, 78, 115, 273, 274
osteoarthritis 177
overgrazing 181
oxalic acid 70
oxides 16, 29, 34, 115, 142, 167, 219,
    221, 223
oxygen 62
ozone 29, 52, 125, 167, 219, 222, 225,
    253-54, 301
— depletion potentials (ODPs) 228
— layer 14, 35

paint production 281
pancreas 16
paprika splitter's lung 152
paragonimiasis 77
parameterization 46
parasite 58
parasitic worms 68
Parathion 78
particles, alpha/beta 63
particulates 78
Pasminco-EZ zinc refinery 67
passive smoking 115, smoking 150
pathogens 77
PBBs 78
PCBs 78, 79
Pearce, David Pearce 297
peasant farmers 187
penicillin 269 272
Penicillium spp. 60
peptide 71
Permian period 35
pesticides 3, 28, 62, 70, 73, 89, 95,
    155, 210, 239, 251, 253, 268, 273,
    280
petrol 238
petroleum industry 281
pH 72
pharmaceuticals 253, 268
phenothiazine 122
phenylalanine 70
phenylethylamine 70
phenylketonuria 70
Phenylthiazine 78

phosgene 2.4.5.-T 115
phosphate 89, 244
photosensitivity 121
photosensitivity disease 132
phototoxicity 121
physical fitness 82, 86
phytates 114
pigs 282
plague 261
plankton 36
planning 299
plant hormone 73
plants and the skin 120
plastics 281
plutonium 63
pneumoconioses 142ff.
pneumonia 149
poison ivy 122
polar ice 46
polio 62
politics 292
pollen 71-2, 137, 140
polycyclic aromatic hydrocarbons
    (PAHs) 151
polynuclear aromatic compound 72
polyunsaturated vegetable oil 71
polyvinylchloride 6
poor living conditions 26, 169
population 89, 189, 201, 202, 236,
    286ff., 292
— control 189ff., 286
— density 201, 290
— dynamics 211
— growth 41, 189
— movement 236
— theory 190
porphyria 122
positive ions 164
post-viral illness 171
poverty 26, 183, 202, 290
precipitation 46-7
predation 35
pregnancy 4
premature ageing 119, 123
premature mortality 85
preventive care 265
preventive policies 254

preventive policies for Australia 316
pricing policy 187
prickly heat 14
primary pollutant 219
Pritikin, Robert 69
privacy 170
prostaglandins 72
protein 69, 104, 207-8, 212
protozoa 68, 148
psoralen 121
psoriasis 129
psychosomatic illness, 55, 161ff.
psychrophilic organisms 61
public health 235, 245, 254-55, 262-63
public water and sewer systems 257
pure air 7
pyrethrum 210

quality of life 201
Quaternary period 37

radiation 58, 62, 75, 115
— electromagnetic 17
— gamma/beta 63, 75
— infrared 28, 35
— nuclear 103
— solar 44
— ultraviolet 14, 28, 62, 103, 121,
    123-5, 218
radiation poisoning 31
radionuclides 63
radon 29, 104
Raynaud's disease 160
refrigerants 41
Renaissance period 259-60
reproductive rate 35
residues 68
resources capital 299
respiratory system 136ff.
rheumatoid arthritis 71
rhinitis 15, 137
rice 211
rice paddy 22, 41
rickettsiae 59
RNA (ribonucleic acid) 100
road traffic accidents 55
Roentgen, Wilhelm 63

Ross River fever 30, 172
rotaviral diarrhoea 77
rubber industry 74, 281
ruminants 41

safety standards 68
salicylates 70
salinization 236
salmonella 245
salmonellosis 77, 271
sanitation 4, 255, 257, 265
scabies 128
schistosomiasis 77
schizophrenia 71
Scotland 24
sea levels 38, 50
Seasonal Affective Disorder (SAD) 163
seasonal cycle 47
secondary pollutant 219
sedentary lifestyle 88
Selye's hypothesis 131
senile elastosis 123
serotonin 70
serum cholesterol 86
sewage 238, 284
sewerage 7, 20, 257
sheep 283
shellfish 284
shigella 77
sick building syndrome 165
siderosis 31
silicosis 145
skimming 246
skin 119ff; *see also* cancer of skin
— allergy 126
— and chlorine 248
— erythema 28
— papilloma 77
sleep 159
smallpox 260-61
smog 7, 16, 35, 219, 221, 222
smoke 7, 136
social class 87, 89, 97, 262
social conditions and health 9
social environment 26
social medicine 265
socio-economic groups 87, 262

sodium 89
soil erosion 289
solar emission 39
solar keratosis 123
solar wind 175
solvents 41
South Australia 24
Soviet Union *see* CIS
species loss 8
spring 162
staphylococcal diarrhoea 77
stinging emergences 121
Stockholm Conference on the Environment 293
stomach 16; *see also* cancer of stomach
stoned fruit 64
stratosphere 40, 254, 325
stress 6, 82, 89, 131, 168, 173
— psychological 56
stressors 56
stroke 55, 81
strontium-89 63
strontium-90 63
structural adjustment programmes 186
suberosis 152
sulphonamide 122, 269
sulphur dioxide 219-21, 224, 241
summer 162, 237
sun 14, 120, 122, 124, 174
sunburn 123
sunflares 46
sun-bathing 62
sun-induced dermatoses 122
sun-spots 174
sustainable development 293, 294-97, 299-300
swimming 243-44
Sydney 16

tar distillate 107
tar particle 136
Tasmania 66
tax policy 254
TCDD 275
temperature 13, 20, 47, 72, 115, 159, 164, 227, 261
Teratogen 79

tetanus 62
tetracycline 122, 270
Thalidomide 269
theobromine 70
thermophilic organisms 61
thermosphere 326
ticks 77
tin 142, 146
tinned and preserved meat 113
tobacco 27, 103, 106, 136, 150
tobacco smoke 95; *see also* cigarette smoking
tomatoes 122
Tonga 49
topography 39
Toronto Target 312
tourism 78
toxic hydrocarbons 29
toxic waste 8
toxicity symptoms 65
toxins 60, 68, 72
trace minerals 89
trachoma 77
traditional medicine 175
transport wastes 78
trauma 58, 178
Triassic period 35
triglyceride 82, 83, 110
trilobite 37
tropical cyclone 49
tropical rainforest 40, 203
Tropics of Cancer and Capricorn 43
troposphere 40, 225, 326
tuberculosis 62, 65
tulip 121
tumour 58, 60, 238, 279; *see also* brain tumour, Burkitt's tumour
turbulence 39
Tuvalu 49
typhoid fever 77, 272
tyramine 70

ugliness 167
ultrasound 65
UNCED 1992 295
undernutrition 206
unemployment 26

United States 44
— Environmental Protection Agency 18
uranium 31, 63, 238
— mining 108
urbanisation 203, 213
user pays' principle 303
Uzbekistan 280

vaccination 260
Vancouver Island 18
vectors 19
vegetables 121
vegetation 39, 46
Venus 47
vibration 115, 160
Vietnam 275
vinyl chloride 30, 74
virus 58, 68, 72
virus encephalitides 30
Visual Display Units (VDUs) 18
vitamins 14, 25, 71, 82, 122, 207
volcano 36, 38

Wales 24
warfare 184, 263, 319
wastes 7, 8, 62, 78, 173, 238-39
water 19, 61, 223-24, 233, 238
— management 306
— pollution 236
— quality 233, 241ff., 246, 305
— supply 7

water, conflicts over 235
water, politics of 234
waterways 39
wealth capital 298, 300
weather 9, 39, 72, 174
— and behaviour 177
web of causation 4
weedicide 70, 211
weight gain 162
wetlands 22
wheezing 166
Whyalla 24
wind 39, 164
winter 162, 223
witchcraft 262
Wittenoom 144
World Bank 186
World Climate Programme 308
World Conservation Strategy 296

X-ray 18, 65
xylene 115

yaws 77
yeasts 61, 173
yellow fever 30
Yorkshire 24
Yudkin, J. 86
yuppie flu 171

zinc 15, 66, 113, 115